Mammals

Rare and Endangered Biota of Florida
Ray E. Ashton, Jr., Series Editor

*Florida Committee on Rare
and Endangered Plants and Animals*

Ray E. Ashton, Jr.
FCREPA Chair (1989–91)
 and Series Editor
Water and Air Research, Inc.
6821 SW Archer Road
Gainesville, Florida 32608

Paul E. Moler
FCREPA Chair (1992–93)
Chair, Special Committee
 on Amphibians and Reptiles
Wildlife Research Laboratory
Florida Game and
 Fresh Water Fish Commission
4005 S. Main Street
Gainesville, Florida 32601

Daniel F. Austin, Co-Chair
Special Committee on Plants
Department of Biological Sciences
Florida Atlantic University
Boca Raton, FL 33431

Mark Deyrup, Co-Chair
Special Committee on Invertebrates
Archbold Biological Station
Route 2, Box 180
Lake Placid, Florida 33852

L. Richard Franz, Co-Chair
Special Committee on Invertebrates
Florida Museum of Natural History
University of Florida
Gainesville, Florida 32611

Carter R. Gilbert
Chair, Special Committee on Fishes
Florida Museum of Natural History
University of Florida
Gainesville, Florida 32611

Stephen R. Humphrey
Chair, Special Committee on Mammals
Florida Museum of Natural History
University of Florida
Gainesville, Florida 32611

Herbert W. Kale II
FCREPA Chair (1985–86)
Chair, Special Committee on Birds
Florida Audubon Society
1101 Audubon Way
Maitland, Florida 32751

Allan Stout
FCREPA Chair (1987–88)
Department of Biological Sciences
University of Central Florida
Orlando, Florida 32816

Daniel B. Ward
FCREPA Chair (1983–84)
Co-Chair, Special Committee on Plants
Department of Botany
University of Florida
Gainesville, Florida 32611

Rare and Endangered Biota of Florida

VOLUME I. MAMMALS

EDITED BY
STEPHEN R. HUMPHREY

Chair, Special Committee on Mammals
Florida Committee on Rare and Endangered
  Plants and Animals

UNIVERSITY PRESS OF FLORIDA
Gainesville, Tallahassee, Tampa, Boca Raton,
Pensacola, Orlando, Miami, Jacksonville

This volume was made possible in part by grants from Florida Power and Light Company and Save the Manatee Club.

Copyright 1992 by the Board of Regents of the State of Florida
Printed in the United States of America on acid-free paper ∞
All rights reserved

Library of Congress Cataloging-in-Publication Data

Rare and endangered biota of Florida : mammals / edited by Stephen R. Humphrey.
    p.   cm.
   Includes index.
   ISBN 0-8130-1127-2 (alk. paper). — ISBN 0-8130-1128-0 (pbk.: alk. paper)
   1. Rare mammals—Florida.   I. Humphrey, Stephen R.
QL706.83.U6R37   1992
599′.0042′09759—dc20   91-36368
                                                                             CIP

The University Press of Florida is the scholarly publishing agency of the State University System of Florida, comprised of Florida A & M University, Florida Atlantic University, Florida International University, Florida State University, University of Central Florida, University of Florida, University of North Florida, University of South Florida, and University of West Florida.

Orders for books should be addressed to
University Press of Florida
15 NW 15th St.
Gainesville, FL 32611

Contents

FOREWORD, *by Robert M. Brantly* — vii
PREFACE, *by Ray E. Ashton, Jr.* — ix
A BRIEF HISTORY OF FCREPA, *by Ray E. Ashton, Jr.* — xv
DEFINITIONS OF STATUS CATEGORIES — xvii
MAJOR TERRESTRIAL AND WETLAND HABITATS — xx
INTRODUCTION — 1
DESIGNATIONS OF STATUS — 6

Recently Extinct

Goff's pocket gopher, *Geomys pinetis goffi* — 11
Pallid beach mouse, *Peromyscus polionotus decoloratus* — 19
Chadwick Beach cotton mouse, *Peromyscus gossypinus restrictus* — 24
Florida red wolf, *Canis rufus floridanus* — 29
West Indian monk seal, *Monachus tropicalis* — 35

Recently Extirpated

Anastasia Island population of cotton mouse, *Peromyscus gossypinus gossypinus* (in part) — 41
Plains bison, *Bison bison bison* — 47

Endangered

Indiana bat, *Myotis sodalis* — 54
Gray bat, *Myotis grisescens* — 63

Lower Keys marsh rabbit, *Sylvilagus palustris hefneri*   71
Choctawhatchee beach mouse, *Peromyscus polionotus allophrys*   76
St. Andrew beach mouse, *Peromyscus polionotus peninsularis*   87
Anastasia Island beach mouse, *Peromyscus polionotus phasma*   94
Perdido Key beach mouse, *Peromyscus polionotus trissyllepsis*   102
Key Largo cotton mouse, *Peromyscus gossypinus allapaticola*   110
Key Largo woodrat, *Neotoma floridana smalli*   119
Florida saltmarsh vole, *Microtus pennsylvanicus dukecampbelli*   131
Right whale, *Eubalaena glacialis*   140
Sei whale, *Balaenoptera borealis*   148
Fin whale, *Balaenoptera physalus*   154
Humpback whale, *Megaptera novaeangliae*   160
Sperm whale, *Physeter macrocephalus*   168
Florida panther, *Felis concolor coryi*   176
Florida manatee, *Trichechus manatus latirostris*   190
Key deer, *Odocoileus virginianus clavium*   201

Threatened

Florida mastiff bat, *Eumops glaucinus floridanus*   216
Big Cypress fox squirrel, *Sciurus niger avicennia*   224
Sherman's fox squirrel, *Sciurus niger shermani*   234
Southeastern beach mouse, *Peromyscus polionotus niveiventris*   242
Florida mouse, *Podomys floridanus*   250
Florida black bear, *Ursus americanus floridanus*   265

Species of Special Concern

Round-tailed muskrat, *Neofiber alleni*   276

Rare

Southeastern big-eared bat, *Plecotus rafinesquii macrotis* — 287
Eastern chipmunk, *Tamias striatus striatus* — 294
Lower Keys population of rice rat, *Oryzomys palustris natator* (in part) — 300
Southeastern weasel, *Mustela frenata olivacea* — 310
Florida weasel, *Mustela frenata peninsulae* — 315
Southern Florida population of mink, *Mustela vison mink* (in part) — 319

Status Undetermined

Sherman's short-tailed shrew, *Blarina carolinensis shermani* — 328
Southeastern brown bat, *Myotis austroriparius* — 335
Big brown bat, *Eptesicus fuscus fuscus* — 343
Northern yellow bat, *Lasiurus intermedius floridanus* — 349
Brazilian free-tailed bat, *Tadarida brasiliensis cynocephala* — 357

CONTRIBUTORS — 369

INDEX — 372

Foreword

In 1978, not long after I became executive director of the Florida Game and Fresh Water Fish Commission, I was privileged to be associated with the production of the initial six-volume Rare and Endangered Biota of Florida series. That series has enjoyed enormous popularity (each volume was reprinted at least once, most two or three times). It has served as the definitive reference compendium on endangered and threatened species in Florida and is widely recognized as among the most authoritative and comprehensive such works in the nation. I am proud the commission was integrally involved in that initial work, and likewise proud that we were involved in producing this revised series.

In the forewords to the initial volumes, I and my predecessor, Dr. O. E. Frye, Jr., acknowledged the momentum of endangered species conservation to that point, and how the series was a significant contribution in that regard, but admonished that we must not rest on our laurels—much remained to be done. Although much has indeed been done in the interim, I am disappointed that we have not approached the level of progress I had hoped we would attain by now. As the species accounts herein clearly demonstrate, many Florida species are perilously near extinction, and many of the factors leading to that dire circumstance are still with us. The composition of the current official state lists—42 endangered, 28 threatened, and 47 special concern animals, along with 199 endangered and 283 threatened plants—is compelling evidence in and of itself that our progress has been relatively minor (by comparison, there were 31 endangered and 54 threatened species in 1978). There are several reasons for this much-less-than-hoped-for progression, but primarily it has been related to insufficient funding at both the state and federal levels. And without proper funding, the necessary manpower and other resources cannot be emplaced to address many critical needs. So we face the dilemma of either addressing the needs of only a few species so as to maximize effect, or spreading our resources thinly among many species, minimizing the effects on an individual basis.

This is not to say, however, that we have not made some substantial strides forward in the last decade or so. Through an innovative translocation strategy, we have reestablished in Florida the previously extirpated

Perdido Key beach mouse and significantly expanded the range of the Choctawhatchee beach mouse; because of stringent protection and rigorous application of "Habitat Management Guidelines for the Bald Eagle in the Southeast Region," Florida's bald eagle nesting population has grown to more than 550 pairs (as of the 1990–91 nesting season); the brown pelican and Pine Barrens treefrog have been delisted because of increasing populations and/or because our research efforts have provided new insight into those species' true status; nearly 50 manatee sanctuaries have been established in which boat speeds are restricted during the winter congregation period; our research since 1978 has resulted in more knowledge about endangered species biology, habitat needs, and the like, than during all previous time cumulatively; considerable endangered species habitat has been secured through CARL (Conservation and Recreation Lands), Save Our Coasts, Save Our Rivers, Save Our Everglades, and other land acquisition programs; and various information/education programs have resulted in a significant increase in public awareness and support for endangered species conservation. These few examples demonstrate what can be done with adequate resources and commitment, but in fact represent only the proverbial drop in the bucket in light of the total needs.

I hope this revised series reinvigorates our resolve and commitment to endangered and threatened species conservation and we will be able to cite a multitude of such examples by the time a third revision is necessary. These volumes provide an authoritative and comprehensive database from which to embark on such a course, and I congratulate and personally thank each researcher, writer, editor, and individual whose committed efforts have culminated in this exemplary work.

Colonel Robert M. Brantly, Executive Director
State of Florida Game and Fresh Water Fish Commission

Preface

"Thirty years ago Florida was one of the most extraordinary states in the Union, but being flat and quite park-like in character (a large part of the country consisted of open pinelands) it was an easy state for man to ruin, and he has ruined it with ruthless efficiency." This quote from Thomas Barbour's *That Vanishing Eden, A Naturalist's Florida*, written in 1944, is ever more appropriate today. He continues his lament—"A large part of Florida is now so devastated that many of her friends are disinclined to believe that she ever could have been the Paradise which I know once existed." Barbour was talking about the loss of natural habitat in Florida from 1915 to the early 1940s. Imagine what he would think today!

Within the FCREPA volumes, the emphasis is on specific plants and animals that the committee considers to be endangered, threatened to become endangered, or species of special concern (those species apparently in danger but about which we need more information). However, as one reads through the species accounts, there is a continuing theme of habitat loss or alteration by man. Since Barbour's days of study in Florida, the loss and degradation of natural habitats have accelerated beyond human comprehension. We are faced with the possible reality that the only thing which will cause a decline in the loss is that there will soon be no land left to develop.

We are also faced with the fact that we actually know very little about the fauna and flora of this state. When challenged to protect a species or develop regulations to prevent extinction, we are inevitably confronted with the fact that we know little about their life histories, let alone what is needed to preserve a population through biological time. We are also faced with the dilemma that there probably is not enough time or money to allow us to study these organisms, let alone experiment with management techniques. Our ecological knowledge of interspecific interactions and biological communities is even less. Yet we do know that once certain biological needs are not met, we lose another species and another community. The biological communities of this state are being compromised time and again by all levels of government, simply to serve the hunger for growth and development.

We are the first generation to realize that not only do we have local or regional environmental concerns but we now have to be aware of serious global degradations of our air and water. Global warming, acid rain, increased ultraviolet radiation, and degradation of our oceans are making us realize for the first time that our species may well be jeopardizing itself as well as the lowly gopher tortoise and tree snail. We are realizing that the world's biodiversity and the biological engine that drives many of the necessities of all life are being used up or changed by our overpopulated species. If we know so little about individual species and communities, how can we be prepared to understand the complexities of the biosphere, let alone the cause and effect of our actions.

Alarms have sounded in the minds of many people around the world, including Florida. The first step toward a solution to all of this is acknowledging that we are causing problems. Loss of uplands not only means loss of wildlife but also that we affect our water supplies, river systems, and ultimately the health of our coastal systems. Our state agencies are in the fledgling stages of creating regulations on development and the organized effort of protecting biological diversity and natural communities. Hopefully these agencies and the people of Florida will begin to recognize that we must increase our efforts to protect our environment and the creatures who inhabit it, not just to use for recreation but for the sake of preserving the machinery that makes our lives as living things possible.

It is these concerns that have been the driving force behind the volunteer-biologists who have unselfishly spent so many long hours putting together the information in these volumes. We hope that through the information provided here more biologists will turn their thoughts from the test tube to the laboratory in the field, funding agencies will realize the need for this basic knowledge, and government agencies will begin to think more on the biological community level and not the species or individual organism level. Most important, we hope that these volumes serve to educate the citizens of Florida so that we may all recognize the need to learn more and work together to make prudent decisions about our "Vanishing Eden."

Ray E. Ashton, Jr.
FCREPA Chair and Series Editor

A Brief History of FCREPA

The Florida Committee on Rare and Endangered Plants and Animals, FCREPA, was founded in 1973. The original group of 100 scientists, conservationists, and concerned citizens was organized by James Layne, Peter Pritchard, and Roy McDiarmid. The chairs of the Special Committees on Terrestrial Invertebrates, Marine Invertebrates, Plants, Reptiles and Amphibians, Fish, Birds, Mammals, and Liaison made up the first Endangered Species Advisory Board to the Florida Game and Fresh Water Fish Commission. These special committees were made up of concerned biologists who were living and/or working in the state of Florida. The first FCREPA meeting was called for biologists to discuss and evaluate the status of Florida's wildlife and to determine which species should be considered for special classification and concern. From this conference, five volumes—The Rare and Endangered Biota of Florida series—were produced. These were edited by Peter Pritchard of the Florida Audubon Society and Don Wood of the Florida Game and Fresh Water Fish Commission. Section editors for the first series included Roy McDiarmid, reptiles and amphibians; Herb Kale, birds; James Layne, mammals; Carter Gilbert, fish; Howard Weems, terrestrial invertebrates; Joe Simon, marine invertebrates; and Dan Ward, plants. Before its completion, the invertebrate volumes were combined under the editorship of Richard Franz.

Following the production of the FCREPA volumes by the University Presses of Florida in 1976, FCREPA continued to meet and support a special section of papers at the annual meeting of the Florida Academy of Sciences. The affiliation of FCREPA was organized under the guidance of Dan Ward, director of the herbarium at the University of Florida.

In the fall of 1986, it became obvious that the original publications were becoming dated and the demand for the publication was great (the volumes had been reprinted repeatedly). Then chair, Herbert Kale, vice president of the Florida Audubon Society, convened the second FCREPA conference at the youth camp in the Ocala National Forest. The committees on each group met and deliberated on the status of the species in their charge. It was decided at that meeting to rewrite the FCREPA series since considerable changes in our knowledge and in the state of the natural environment in Florida made much of the information produced more than 13 years before out of date. Editors for each of the volumes called together those knowledgeable individuals and potential contributors to the future volumes to discuss the status of the taxa covered in their volume. Their recommendations on the status (and the criteria used) of various species were discussed by everyone present at the 1986 meeting.

Under the direction of Jack Stout, University of Central Florida, and each of the section chairs and editors, the arduous task of preparing the new manuscripts was undertaken. Each section chair served as compiler and editor for each volume. Individual species accounts were prepared by biologists who were among the most qualified to write about the status of that species.

Ray Ashton, vertebrate zoologist, Water and Air Research, Inc. was appointed by the section chairs as managing editor of the series in 1988. Paul Moler, research biologist, Florida Game and Fresh Water Fish Commission, was voted as chair-elect (1992–1993). Four years of preparation and coordination, fund raising, and gentle prodding of the seven volunteer editors and the many contributors have produced the second FCREPA series.

Without the thousands of volunteer hours given by many outstanding Florida biologists, and the support from the Florida Game and Fresh Water Fish Commission, Save the Manatee Club, and Florida Power and Light Company, this effort would not have been possible. Royalties from the sales of these volumes and donations to the FCREPA effort are used to keep all the volumes in print and to fund future work.

Ray E. Ashton, Jr.
FCREPA Chair and Series Editor

Definitions of Status Categories

Categories used to designate the status of the organisms included in the Florida list of rare and endangered species are defined below. In the case of species or subspecies whose ranges extend outside the state, the category to which the form is assigned is based on the status of its population in Florida. Thus, a plant or animal whose range barely reaches the state ("peripheral species") may be classified as endangered, threatened, or rare as a member of the Florida biota, although it may be generally common elsewhere in its range.

In the following definitions, "species" is used in a general sense to include (1) full taxonomic species, (2) subspecies (animals) or varieties (plants), and (3) particular populations of a species or subspecies that do not have formal taxonomic status. This use of the term agrees with that of the Endangered Species Act of 1973.

Endangered.—Species in danger of extinction or extirpation if the deleterious factors affecting their populations continue to operate. These are forms whose numbers have already declined to such a critically low level or whose habitats have been so seriously reduced or degraded that without active assistance their survival in Florida is questionable.

Threatened.—Species that are likely to become endangered in the state within the foreseeable future if current trends continue. This category includes: (1) species in which most or all populations are decreasing because of overexploitation, habitat loss, or other factors; (2) species whose populations have already been heavily depleted by deleterious conditions and, while not actually endangered, are nevertheless in a critical state; and (3) species that may still be relatively abundant but are being subjected to serious adverse pressures throughout their range.

Rare.—Species that, although not presently endangered or threatened as defined above, are potentially at risk because they are found only within a restricted geographic area or habitat in the state or are sparsely distributed over a more extensive range.

Species of special concern.—Species that do not clearly fit into one of the preceding categories yet warrant special attention. Included in this category are: (1) species that, although they are perhaps presently relatively abundant and widespread in the state, are especially vulnerable to certain types of exploitation or environmental changes and have experienced long-term population declines; and (2) species whose status in Florida has a potential impact on endangered or threatened populations of the same or other species outside the state.

Status undetermined.—Species suspected of falling into one of the above categories for which available data are insufficient to provide an adequate basis for their assignment to a specific category.

Recently extirpated. Species that have disappeared from Florida since 1600 but still exist elsewhere.

Recently extinct.—Species that have disappeared from the state since 1600 through extinction.

Major Terrestrial And Wetland Habitats

In discussing habitat needs of various organisms, it is often convenient to refer to classifications of plant communities as a starting reference point. A number of classifications that address Florida plant communities have been developed, usually with different purposes or different constraints imposed on the classifier (i.e., "differences must be discernible on aerial photography"). Because plants' and animals' habitat needs rarely fit these plant community classification schemes very closely, considerable additional description is usually necessary to adequately define an organism's habitat.

Commonly used plant classification schemes include the following:

Davis, John H. 1967. General map of natural vegetation of Florida. Agr. Exp. Sta., University of Florida, Gainesville.

Soil Conservation Service. 1980. General map ecological communities, state of Florida.

Florida Natural Areas Inventory. Undated. List of plant communities used by FNAI in collecting plant community data.

Florida Department of Transportation. 1985. Florida land use, cover and forms classification system.

Plant community types used in the narrative loosely follow the general map of natural vegetation of Florida by John H. Davis. Descriptions of each habitat are based on this map plus published sources and personal experience. Several of Davis' vegetation categories are grouped together for convenience and brevity. Marl prairies and Everglades sawgrass are combined with wet prairies and marshes while southern slash pine forests are included with flatwoods. Other categories are added, further subdivided or omitted in order to provide a brief overview of major plant communities in Florida. The narratives on Florida terrestrial and wetland plant communities are grouped as follows:

Upland plant communities
 1. Coastal strand
 2. Dry prairies

3. Pine flatwoods
4. Sand pine scrub
5. Longleaf pine–xerophytic oak woodlands (sandhill communities)
6. Mixed hardwood–pine
7. Hardwood hammocks
8. Tropical hammocks

Wetland plant communities
9. Coastal marshes
10. Freshwater marshes and wet prairies
11. Scrub cypress
12. Cypress swamps
13. Hardwood swamps
14. Mangrove swamps

1. Coastal Strand

The coastal strand includes beaches and the vegetation zones of beaches and adjacent dunes or rock. This vegetation type is most commonly associated with shorelines subjected to surf and high winds but may sometimes be found bordering calmer bays and sounds.

The vegetation of the beaches and foredunes is characterized by pioneer plants able to establish themselves in the shifting sand. Typical species include railroad vines, beach cordgrass, and sea oats. Inland from the foredune, saw palmetto and dwarf scrubby oaks are found and, in southern Florida, sea grape and other tropical vegetation as well. The vegetation tends to change from grassy to woody from the foredune inland to the more protected back dunes, and the composition of the vegetation of these back dunes is often similar to that of sand pine scrub habitat found inland on old dunes.

Strand communities are adapted to the severe stresses of shifting sands, a highly saline environment, and high winds. In some instances, salt spray plays a role similar to fire in other ecosystems by retarding succession indefinitely at a grass or shrubby stage.

Historically, impacts to coastal strand plant communities (sometimes the total loss of the community) have resulted from beachfront residential development; invasion by exotic vegetation, primarily Australian pine; and accelerated erosion of beaches due to maintenance of inlets or nearby residential and tourist development.

2. Dry Prairies

Dry prairies are vast, treeless plains, often intermediate between wet grassy areas and forested uplands. Scattered bayheads, cypress ponds, or cabbage palm hammocks often occur in prairie areas. The largest areas of dry prairies occur north and west of Lake Okeechobee.

This community is dominated by many species of grasses such as wiregrass and broomsedge. Palmettos are the most common shrubby plant over large areas, with fetterbush, staggerbush, and dwarf blueberry common in places. A number of sedges and herbs are also found on the dry prairies.

Relatively little has been published on the ecology of dry prairies. They have often been compared to flatwoods minus the overstory trees, and the similar vegetative groundcover would seem to justify this idea.

Fire is important in determining the nature of the vegetation and its suitability for different species of wildlife. Winter burning associated with cattle operations may have shifted this community from grasses and forbs to saw palmetto. Absence of fire may result in shrubby communities, while frequent growing season fires yield a more herbaceous environment.

Large areas of native dry prairies have been converted to improved pasture, and this trend is continuing. Eucalyptus plantations have also been established on some former dry prairie sites although this does not appear to be a continuing trend. Expansion of citrus production southward is probably responsible for most dry prairie losses at this time.

3. Pine Flatwoods

Pine flatwoods are characterized by one or more species of pine as the dominant tree species and occur on level areas. The soils of flatwoods are sandy with a moderate amount of organic matter in the top few centimeters and an acid, organic hardpan 0.3–1.0 m (1–3 ft) beneath the surface.

Three major types of flatwoods occur in Florida: **longleaf pine flatwoods** found on well-drained sites and characterized by longleaf pine as the dominant overstory tree; **slash pine flatwoods** with slash pine as the dominant overstory species and usually in areas of intermediate wetness; and **pond pine flatwoods** with the pond pine as the dominant tree species and typically occurring in poorly drained areas. South Florida slash pine tends to replace both slash pine and longleaf pine in central to southern peninsular Florida.

Southern slash pine forest is found on the sand flatlands and rocklands of extreme southern Florida and is characterized by an overstory of the south Florida variety of the slash pine. This association often has tropical components in its understory.

Considerable overlap in understory plants exists among the three major types of flatwoods, with many species found in all three communities. Generally, however, gallberry and saw palmetto dominate the understory in slash pine flatwoods; wiregrasses, blueberries, and runner oaks are especially prevalent in longleaf pine flatwoods; and several of the bay trees are characteristic of pond pine areas. Flatwoods also often include intermingled cypress domes, bayheads, and small titi swamps.

Pine flatwoods are the most widespread of major plant communities, occurring throughout the relatively level Pleistocene marine terraces in both peninsular and panhandle Florida. Their suitability for growing pine trees has resulted in vast areas being incorporated into industrial forests. Changes in both fire and moisture regimes have resulted in changes in plant species composition and wildlife values. In south Florida, residential development is rapidly eliminating this plant community (and all other upland communities).

4. Sand Pine Scrub

Sand pine scrub is a plant community found almost exclusively in Florida on relict dunes or other marine features created along present and former shorelines or other marine features. Sand pine scrub communities occur along the coasts on old dunes, in the Ocala National Forest, and along the Lake Wales Ridge extending through Polk and Highlands counties. The soil is composed of well-washed, sterile sands.

This community is typically two-layered, with sand pine occupying the top layer and various scrubby oaks and other shrub species making up a thick, often clumped, understory. Little herbaceous groundcover exists, and large areas of bare sand occur frequently. Groundcover plants, when present, frequently include gopher apple and Florida bluestem grass. Deermosses are often common. Typical understory plants include myrtle oak, inopina oak, sand live oak, Chapman's oak, rosemary, scrub holly, and silkbay.

Where sand pines are absent, this community is often referred to as evergreen oak scrub. **Scrubby flatwoods** is a scrub-like association often occurring on drier ridges in typical flatwoods or near coasts. The understory species of this vegetation type are similar to those of sand pine scrub, but the sand pine is replaced by slash pine or longleaf pine.

The sand pine scrub is essentially a fire-based community. Ground vegetation is extremely sparse and leaf fall is minimal, thus reducing the chance of frequent ground fires so important in the sandhill community. As the sand pines and scrub oaks mature, however, they retain most of their branches and build up large fuel supplies in the crowns. When a fire does occur, this fuel supply, in combination with the resinous needles and high stand density, ensures a hot, fast-burning fire. Such fires allow for regeneration of the sand pine community and associated oak scrub, which would otherwise pass into a xeric hardwood community.

Sand pine scrub and its ecologically important variations are seriously threatened. Residential development and especially citrus production have eliminated much of this plant community. In addition, isolation from fire has resulted in succession to xeric hardwood hammock with a relatively closed canopy, thereby reducing its value to most endemic plants and animals.

5. *Longleaf Pine–Xerophytic Oak Woodlands (Sandhill Communities)*

Sandhill communities (the **longleaf pine–turkey oak** association being one major subtype of this community) occur on well-drained, white to yellowish sands.

Longleaf pines form a scattered overstory in mature natural stands. In many areas, xeric oaks such as turkey oak, bluejack oak, southern red oak, and sand post oak, which were originally scattered or small understory trees, now form the overstory as the result of cutting of the pines and prevention of fire. In some areas of southern peninsular Florida, south Florida slash pine replaces longleaf pine in the overstory. Although tree species diversity in sandhills is low, there is a wide variety of herbaceous plants such as wiregrass, piney woods dropseed, golden aster, partridge pea, gopher apple, bracken fern, and paw paw, which provide fairly complete groundcover.

Sandhills were second in area only to flatwoods in Florida's predevelopment landscape, occurring widely throughout the panhandle and the northern half of the peninsula.

Fire is a dominant factor in the ecology of this community. The interrelationships of the sandhill vegetation types, particularly the longleaf pine–wiregrass relationship, are dependent on frequent ground fires. The longleaf pine is sensitive to hardwood competition, and wiregrass plays a major role in preventing the germination of hardwood seeds while ensuring that there is sufficient fuel buildup on the floor of the community to carry a fire over large areas.

Very little longleaf pine sandhill remains. Commercial foresters have attempted to convert large areas to slash pine with poor success but have had better success converting sandhills to a closed canopy monoculture of sand pine. In many cases the wiregrass groundcover has been destroyed, making restoration of this community type problematic. Large areas have been converted to improved pasture or citrus. The well-drained soils make attractive development sites and the majority of sandhill community in peninsular Florida in private ownership has either been developed, is being developed, or is platted and subdivided for future development.

6. Mixed Hardwood–Pine

The mixed hardwood–pine community of Davis is the southernmost extension of the piedmont southern mixed hardwoods and occurs on the clay soils of the northern panhandle.

Younger growth may be primarily pine, with shortleaf and loblolly pine predominant. As succession proceeds the various hardwoods become dominant and constitute the natural climax vegetation of much of the area, especially wetter, yet well-drained sites. The overstory is characterized by a high species diversity and includes American beech, southern magnolia, white oak, sweetgum, mockernut hickory, pignut hickory, basswood, yellow poplar, white ash, and spruce pine. The understory includes many young overstory species plus dogwood, red mulberry, hop hornbeam, blue beech, and sweetleaf.

Historically, fire played a role in the function of this community by limiting its expansion into higher, better-drained sites. Later, agriculture served a similar function and limited this community to slopes and creek bottoms. The best examples, with the most diversity of tree species, tend, therefore, to occur in creek bottoms or on moist but well-drained slopes.

Residential subdivisions and other aspects of urbanization and conversion to loblolly pine plantations and agriculture are resulting in continued losses of this plant community. Locally significant losses result from stream or river impoundments, clay mining, and highway construction.

7. Hardwood Hammocks

The hardwood hammock community constitutes the climax vegetation of many areas of northern and central Florida. Hardwoods occur on fairly rich, sandy soils and are best developed in areas where limestone or phosphate outcrops occur. Hardwood forests are similar to the mixed hardwood and pine of the panhandle but generally lack the shortleaf pine, American beech, and other more northern species, have a lower overstory species diversity, and tend to have a higher proportion of evergreen species. Southern magnolia, sugarberry, live oak, laurel oak, American holly, blue beech, and hop hornbeam are characteristic species of this association. Variations in the species composition of hardwood hammocks are partially due to differences in soil moisture.

Major variations of this vegetative association include coastal hammocks, live oak–cabbage palm hammocks, and maritime hammocks. **Coastal hammocks** are relatively wet hardwood communities that occur in narrow bands along parts of the Gulf and Atlantic coasts and often extend to the edge of coastal marshes. **Live oak–cabbage palm hammocks** often border larger lakes and rivers and are scattered throughout the prairie region of central Florida. Either the oak or palm may almost completely dominate in any one area. **Maritime hammocks** occur behind sheltering beachfront dunes and are often dominated by live oak.

Notable examples of hardwood hammocks are in public ownership but residential development is widespread in better-drained hammocks within the ever-increasing range of urban centers. Historically, large areas of coastal hardwood hammocks have been site-prepared and planted to pine for pulpwood production. The current rate of loss due to this land use has apparently declined. Hammocks continue to be lost to agricultural conversion although the near-surface limestone of many hammocks sometimes makes them unattractive for agriculture.

8. *Tropical Hammocks*

Tropical hammocks are found on many of the tree islands in the Everglades and on many of the Florida Keys. Remnants of these habitats occur north to Palm Beach on the east coast and Sarasota on the west coast.

Tropical hammocks typically have very high plant diversity, containing over 35 species of trees and almost 65 species of shrubs and small trees. Typical tropical trees are the strangler fig, gumbo limbo, mastic, bustic, lancewood, the ironwoods, poisonwood, pigeon plum, and Jamaica dogwood. Vines, air plants, and ferns are often abundant. Tropical hammocks of the Florida Keys contain a number of plants that are extremely rare in the United States, including mahogany, lignum vitae, and thatch palms.

The tropical hardwood forest is the successional climax for much of the uplands of extreme south Florida. Because of susceptibility to frequent fires, this association is largely confined to islands or slightly wetter areas but may invade drier areas if fire is removed for any length of time.

Tropical hammocks have been largely lost to residential development in most areas of southern Florida. Relatively large areas remain on north Key Largo, where intensive efforts to buy or regulate this community have occurred.

9. Coastal Marshes

Coastal marshes occur on low wave-energy shorelines north of the range of the mangroves and are also interspersed with mangroves in many areas. Salt marshes may also extend into tidal rivers and occur as a narrow zone between the mangroves and freshwater marshes in the southern areas of the state.

Many areas within salt marshes are dominated by one plant such as saltgrass, smooth cordgrass, or blackrush. The species existing in any one area depends largely on the degree of inundation by tides.

Smooth cordgrass typically occupies the lower areas and often borders tidal creeks and pools. Blackrush occurs over vast areas, particularly along the Gulf coast, and is inundated less frequently, while the highest areas of the marsh are vegetated by saltgrass or such succulents as saltwort, glasswort, and sea ox-eye daisy.

The functioning of salt marshes centers primarily on tides and salinity. The harsh conditions associated with daily inundation, desiccation, and high salinities contribute to a low plant and animal species diversity. Those organisms that have adapted to this environment can be very productive, however. Tides also provide a close ecological relationship with adjacent estuaries.

Coastal marshes have been affected primarily by waterfront residential development. Current wetland regulatory programs appear to be successful in preventing major losses of this community for the time being, although scattered losses continue.

10. Freshwater Marshes and Wet Prairies

Freshwater marshes are herbaceous plant communities occurring on sites where the soil is usually saturated or covered with surface water for one or more months during the growing season.

Wet prairies are characterized by shallower water and more abundant grasses, and usually fewer of the tall emergents, such as bulrushes, than marshes. This category also includes the wet to dry marshes and prairies found on marl areas in south Florida.

Upwards of 15 separate types of marshes or wet prairies have been described in Florida. Major ones include sawgrass marshes; flag marshes dominated by pickerel weed, arrowhead, fire flags, and other nongrass herbs; cattail marshes; spike rush marshes; bulrush marshes; maidencane prairies; grass, rush, and sedge prairies; and switchgrass prairies dominated by taller grasses. Any single marsh may have different sections composed of these major types and there is also almost complete intergradation among the types.

Fire and water fluctuations, the two major ecosystem managers of Florida, are important in the maintenance of marshes and wet prairies. Fire, especially when combined with seasonal flooding, serves to stress plants not adapted to these conditions and reduces competition from more upland species.

Historic major marsh systems include the Everglades, Upper St. Johns River, Kissimmee River floodplain, and Lake Apopka/Oklawaha marshes. Drainage for agriculture has been the dominant factor in marsh losses. Existing wetland regulatory programs and a relatively small amount of agriculturally suitable major marsh systems remaining in private ownership have reduced past rates of loss of these large systems. Major wetland acquisition and restoration projects are underway in the examples cited. Ephemeral, isolated, smaller marshes are more vulnerable to both agricultural and urban development and drainage or use as stormwater holding basins.

11. Scrub Cypress

Scrub cypress areas are found on frequently flooded rock and marl soils in south Florida. The largest areas occur in the Big Cypress region of eastern Collier County and northern Monroe County.

Scrub cypress forests are primarily marshes with scattered, dwarfed pond cypress. Much of the vegetation is similar to other relatively sterile marshes with scattered sawgrass, beakrushes, St. John's-wort and wax myrtle occurring commonly. Bromeliads, as well as orchids and other epiphytes, are often abundant on the cypress trees.

Most scrub cypress in the Big Cypress is in public ownership and does not appear threatened.

12. Cypress Swamps

Cypress swamps are usually located along river or lake margins or interspersed through other habitats such as flatwoods or dry prairies. In addition, they also occur as strands along shallow, usually linear drainage systems. These swamps have water at or above ground level for a considerable portion of the year.

Bald cypress is the dominant tree along lake and stream margins and may be the only tree that occurs in significant numbers in these locations. Other trees that are found within bald cypress swamps include water tupelo, ogeechee tupelo, and Carolina ash. Pond cypress occurs in cypress heads or domes that are typically found in flatwoods or dry prairies. Associated trees and shrubs include slash pine, blackgum, red maple, wax myrtle, sweetbay, and buttonbush. Other plants include various ferns, epiphytes, poison ivy, greenbrier, and lizard's tail, with arrowhead, pickerel weed, sawgrass, and other marsh plants often found in the open water within cypress domes or strands.

Cypress swamps occur in submerged or saturated soils. Fire is an additional factor in drier cypress heads or domes. These factors are important in reducing competition and preventing the community from advancing to one dominated by evergreen hardwood trees (the bayhead community). There has apparently been a shift from cypress to hardwood swamps in areas where heavy harvesting of cypress has occurred in the past and the surviving hardwoods subsequently prevented cypress regeneration.

Bald cypress swamps are reasonably well protected by wetland regulations and the high cost of converting them to other land uses. Pond cypress swamps, while extremely widespread, have less protection because of their smaller size and more isolated nature. Cypress heads and ponds are susceptible to draining associated with industrial pine management, dredging for open water sites in residential development, and increased flooding when used to store stormwater runoff.

13. *Hardwood Swamps*

Deciduous hardwood swamps are found bordering rivers and lake basins where the forest floor is saturated or submerged during part of the year. Other names for this community include floodplain forest, bottomland hardwoods, and river swamp.

The wettest portions of these forests usually overlap with bald cypress swamps and consist largely of water tupelo, Carolina ash, and ogeechee tupelo. In slightly higher areas this community is characterized by such hardwoods as pop ash, pumpkin ash, red maple, overcup oak, sweetgum, and water hickory. On terraces or other higher portions of the floodplain, the overstory includes a variety of more mesic species such as spruce pine, swamp chestnut oak, and diamond-leaf oak. Understory trees and shrubs include dahoon holly, buttonbush, blue beech, and hop hornbeam. Groundcover is sparse in most of these swamps.

Two distinctive additions to this major category are **bay swamps** (bayheads or baygalls) and **titi swamps**. The former are broadleaf evergreen swamps occurring in shallow drainage ways and depressions, particularly in pine flatwoods. Loblolly bay, red bay, and sweet bay are the major tree species. Water levels are relatively stable, and the soil is usually an acidic peat. Titi swamps are dominated by one or more of three titi species and occur as strands or depressions in flatwoods or along the borders of some alluvial swamps in north Florida.

The periodic flooding of the river swamps is a dominant factor in the functioning of the system, and different communities will become established if these fluctuations are eliminated. All species within this community must be able to withstand or avoid the periodic stresses imposed by high water.

Hardwood swamps share common threats with cypress swamps. The wetter and the more contiguous with open waters, the stronger the regulatory protection. Bay swamps are occasionally mined for peat or lost in phosphate-mining operations, and receive comparatively less wetland regulation protection than other wetlands. The rate of loss is unknown.

14. Mangrove Swamps

Mangroves occur along low wave-energy shorelines on both coasts from Cedar Keys on the Gulf to St. Augustine on the Atlantic. Some of the best examples of mangrove forests are located in the Ten Thousand Islands area of southwest Florida.

Three species of mangroves dominate the composition of mangrove swamps. The red mangrove, with its stilt root system, is typically located on the outermost fringe with the most exposure to salt water. Further inland, but usually covered by water at high tides, are the black mangroves, with white mangroves yet farther inland. Buttonwood trees are often found above the reach of salt water. Other plants commonly found among the mangroves include saltwort, glasswort, and a variety of other saltmarsh species.

The mangrove community contributes to the productivity of bordering estuaries. Leaf fall from the mangroves provides food or substrate for countless organisms, ranging from bacteria to large fish such as the striped mullet. Detritus-feeding organisms support much of the estuarine trophic structure in mangrove areas including such gamefish as snook, tarpon, and spotted sea trout.

Mangrove swamps are largely in public ownership and the remainder are reasonably well protected by wetland regulations although losses due to marina and residential developments occur on a relatively small scale.

Prepared by: Brad Hartman, Florida Game and Fresh Water Fish Commission, Tallahassee, FL 32399-1600.

Introduction

The 1978 publication of the Rare and Endangered Biota of Florida volume on mammals, stimulated and edited by James N. Layne of Archbold Biological Station, was a formative event in the conservation of mammals in Florida. It brought together for the first time the state of knowledge on the biology, survival status, and management of Florida mammals thought to be in danger of extinction. That publication focused the attention of our society on these mammals, ultimately providing more protection than would have occurred otherwise. Public and private agencies have mounted extensive conservation programs for certain species, preserves have been acquired or enlarged, and some real estate developments have been modified or even prohibited. Meanwhile, the community of researchers and managers has made remarkable progress in expanding our knowledge of what the problems are and what can be done about them.

This second edition reprises the enhanced state of knowledge. We hope this edition stimulates further improvement of the knowledge base and responsive management actions. The past decade's surge of conservation work recorded herein also provides a fuller understanding of the consequences of human action for wild Florida. A critical reading between the lines clearly shows our successes and failures, and it indicates opportunities still available or forever foreclosed.

The information contained on the following pages should be valuable to all interested in the status, biology, and management needs of the listed species—students, naturalists, tourists, researchers, natural resource managers, land-use planners, and interested citizens. The species accounts are organized to be useful to both those who need technical information on the characteristics of the species at hand and those who need to know what has been and could be done to protect the resource. Using an organizational approach designed by Don A. Wood, endangered species coordinator for the Florida Game and Fresh Water Fish Commission, each account is divided into two groups of information. The first contains identity, survival status, and place in the environment, with these categories of data: taxonomy, description, population size and trend, distribution and history of distribution, geographic status (peripheral, endemic, migratory, vagrant, etc.), habitat requirements and habitat trend, and vulnerability of species and habitat. The second reviews aspects of the biology of the species relevant to the present cultural context: causes of threat, responses to habitat modification, demographic characteristics, key behaviors, conservation measures taken, and conservation measures proposed.

What sorts of mammals have become extinct or extirpated from Florida, and from what causes? Were humans always involved? Why were development and conservation incompatible, and what alternative pathways are now available? One species, five subspecies, and one undifferentiated population formerly thought to be a subspecies of mammal have been recorded as extinct or extirpated from Florida. Three (the Florida red wolf, *Canis rufus floridanus*; the West Indian monk seal, *Monachus tropicalis*; and the Plains bison, *Bison bison bison*), all overhunted or intentionally eradicated, were recognized as lost in the 1978 edition. These losses occurred at a time when rapacious use of natural resources was our cultural norm, and market hunting and extermination of "vermin" were accepted practices.

Remarkably, loss of the other four (Goff's pocket gopher, *Geomys pinetis goffi*; the pallid beach mouse, *Peromyscus polionotus decoloratus*; the Chadwick Beach cotton mouse, *Peromyscus gossypinus restrictus*; and the Anastasia Island population of the cotton mouse, *Peromyscus gossypinus gossypinus*) was recognized only after publication of the 1978 edition, even though all apparently were lost before then. All four had restricted distributions, and conversion of these local areas to human uses is implicated in all four cases (although the exact roles of the cascading effects of habitat conversion is a matter of debate in three of the four cases). Only in the case of Goff's pocket gopher was all the occupied habitat obliterated.

The uncertainty over what mechanisms caused extinction or extirpation in the three cases where patches of suitable habitat remained is informative: although destruction of wild habitat is the most obvious problem, detrimental influences clearly extend well beyond the boundary of lands directly used for human dwelling, industry, and agriculture. Clearly we had no idea of the indirect effects of human progress. We still are unable to specify exactly what happened, so our ability to predict future losses is limited.

Furthermore, as a society we are only dimly aware of the natural entities around us. Had it not been for a flurry of collecting and study by museum scientists from 1890 to 1945, the subspecific status of these four populations would have been unknown and their loss would have been unnoticed or unpublicized. It is difficult to play the conservation game if the scorecard is wrong.

Finally, several of these four were not listed as endangered or threatened until after their loss had been declared. Our ability to monitor changes in the biota is very poor.

The list of mammals considered endangered contains few surprises in the sense that it conforms to the list of entities legally protected as endangered by the U.S. Fish and Wildlife Service and the Florida Game and Fresh Water Fish Commission. This edition's list, however, indicates an alarming trend: two subspecies have been removed from the list in the 1978 edition by extinction, and two subspecies have been added. Again the difficulty of monitoring and responding to survival status is implicated—it takes several years to a decade of concerted effort by a variety of people to obtain resources and complete field surveys, publish the results, and complete a public process leading to formal government protection. Most of the endangered listings are of coastal or island mammals, reflecting the fact that 80 percent of Floridians live within 10 miles of the coast. For example, six of the seven coastal subspecies of beach mice (*Peromyscus polionotus*) in Florida are now considered to be extinct, endangered, or threatened. Alarming harbingers of worsening conditions are contained in the details of the new status reports. For example, the population of the Key deer (*Odocoileus virginianus clavium*) has declined since the 1978 edition. And recently the experts concluded that Florida's state mammal, the panther (*Felis concolor coryi*), will become extinct in 30–35 years without the radical intervention of captive rearing and reestablishment of several new populations in the animal's former range.

The list of mammals considered threatened has been greatly refined. Some mammals listed as threatened in the 1978 edition now are considered endangered; some are judged to be in more favorable condition than previously thought; and a few are no longer considered to be distinct subspecies. The most remarkable feature of the list of threatened species is that half the list (Sherman's fox squirrel, *Sciurus niger shermani*; the Florida mouse, *Podomys floridanus*; and the Florida black bear, *Ursus americanus floridanus*) consists of species or subspecies that once were widespread and common but whose habitat has been so severely reduced that the remnants must be considered threatened. Between 1936 and 1986, the area of longleaf pine savanna (the primary habitat for Sherman's fox squirrel) declined by more than 90 percent, to 951,000 ac. The area of scrub vegetation (the primary habitat for the Florida mouse) in central Florida's Lake Wales Ridge has declined about 70 percent; it is the heart of the range of Florida's only endemic genus of mammal.

The list of species of special concern is the only one that has not changed since the 1978 edition. Only the round-tailed muskrat (*Neofiber alleni*) is included. This mammal is special because it is the only species in its genus, its range is limited to Florida and southernmost Georgia, and it lives only in marsh habitat. Fortunately marshes in Florida are well protected by existing regulations, so the survival of this mammal should be secure as long as these regulations persist.

The list of mammals considered rare is shorter than in 1978, because some now are considered to be more common than previously thought. The list of mammals of undetermined status has changed almost completely, with the notable exception of Sherman's short-tailed shrew (*Blarina carolinensis shermani*). Virtually nothing has been learned about this mammal in a decade despite at least two surveys. The current list contains four bats that actually may be common but whose status may have changed for the worse. These deserve closer examination.

The disposition of those in the 1978 edition but omitted from this one shows how much the state of knowledge has improved. Inclusion of the eastern gray wolf (*Canis lupus lycaon*) was a case of mistaken identity—reports of the gray wolf now are thought to have been misidentifications of the Florida red wolf. The Key Vaca raccoon (*Procyon lotor auspicatus*) is no longer considered to be a distinct subspecies but rather simply a population of a much more widely distributed, common subspecies. Surveys indicate that the Lower Keys and insular cotton rats (*Sigmodon hispidus exsputus* and *S. h. insulicola*) are more common and less vulnerable than previously thought. The southeastern and Homosassa shrews (*Sorex longirostris longirostris* and *S. l. eionis*) have been found to be more abundant and less habitat-specific than previously thought, and the range of the latter has been found to be much larger. The hoary bat (*Lasiurus cinereus*) has been found to be only a migrant in Florida, traveling from abundant summer populations in the eastern United States and Canada to points unknown in winter. Keen's bat appears to be so rare in Florida that it is functionally not a Florida species at all—students of birds would consider this species to occur in the state only "accidentally." The Florida mink (*Mustela vison lutensis*), divided since the 1978 edition into an Atlantic coast subspecies and a Gulf coast subspecies (*M. v. halilimnetes*), now is considered to be common and not threatened, even though the two have relatively small ranges and occur only in saltmarsh habitat. And the Pine Island rice rat (*Oryzomys palustris planirostris*) is now considered to be an undifferentiated population of a common, widespread, mainland subspecies. Although of course some will debate these judgments and new information will dictate further refinements, the improved state of knowledge has empowered us to focus on the highest priorities of the conservation agenda and avoid diverting resources to lesser issues.

This volume reports which Florida species are on the brink of extinction and why. It also reports steps taken by citizens of the nation and state to slow or stop the course of destruction. By proposing new conservation measures, it raises the question of whether ways can be found for both the people and the rest of the fauna to prosper. Considering that Florida's human population is expected to grow to 17.5 million in 2010, the answer to this question may be negative. Given Florida's recent, strong protection of wetland habitats, most of this human population growth will affect upland habitats. Because U.S. law grants ownership of plant life to property owners but reserves ownership of wildlife to the state, protection by endangered-species legislation necessarily is more effective for animals than for their habitat. In Florida, large-scale subdivision and building projects are regulated, and planning of all construction and infrastructural developments is becoming a standard practice. The great exception is conversion of native habitats to agriculture, ranching, citrus, and forestry, which does not require permits unless water consumption and discharge of pollutants are major issues. This regulatory system means that the pathway to urban development is through agriculture, prompting the ecologist's joke that natural succession in modern Florida has been transformed from the oldfield–savanna–forest sequence to oldfield–orange groves–cities. "Scrub rubbing" became a common practice in the 1980s—bulldozing the vegetation of ancient scrub in preparation for citrus development, so no endangered species would be found in an environmental-impact survey if a housing development were proposed instead.

A casual look at the trajectory of land use in Florida shows that the future of endangered wildlife is to occur on islands of wild land in a cultural sea. Few kinds of endangered species will live in the urban matrix, and most will not even cross it. For those interested in devising strategies by which endangered species will survive indefinitely, the immediate future should be devoted to establishing preserves in the most important wildlife-population areas and to optimizing the configurations of existing preserves. These steps should be taken before all the surrounding land is converted to human uses and before land prices rise out of reach.

I thank each of the authors of the species accounts for contributing their professional judgment and technical work. They and I also are grateful to those whose previous work we have cited. Without this accumulated knowledge, and all the work done to record it, this volume would not exist. All of us thank John Wollinka for preparing the maps to illustrate the geographic distribution of the animals.

Designations of status

The following are the Florida mammal species treated in this volume according to the 1989 list (FCREPA) and subsequent actions of the U.S. Fish and Wildlife Service (Federal) and the 1990 list of the Florida Game and Fresh Water Fish Commission (State). Also included are those that appeared on the previous FCREPA list (1978) but have no current status. UR2 indicates a species is formally under review for listing, but substantial evidence of biological vulnerability of threat is lacking. UR3 indicates a species is still formally under review for listing, but no longer being considered for listing due to evidence of extinction.

Current FCREPA status Species	Federal	State	1978 FCREPA
Recently extinct			
Goff's pocket gopher (*Geomys pinetis goffi*)	UR3	End.	End.
Pallid beach mouse (*Peromyscus polionotus decoloratus*)	UR3	End.	End.
Chadwick Beach cotton mouse (*Peromyscus gossypinus restrictus*)	UR3	End.	None
Florida red wolf (*Canis rufus floridanus*)	End.	None	Ext.
West Indian monk seal (*Monachus tropicalis*)	None	None	Ext.
Recently extirpated			
Cotton mouse, Anastasia Island pop. (*Peromyscus gossypinus gossypinus*)	UR3	None	SU
Plains bison (*Bison bison*)	None	None	Extp.

Ext.=Extinct, Extp.=Extirpated, End.=Endangered, Thr.=Threatened, SSC=Species of special concern, SU=Status undetermined, UR2=Under review 2, UR3=Under review 3, None=no status

Current FCREPA status

Species	Federal	State	1978 FCREPA
Endangered			
Indiana bat (*Myotis sodalis*)	End.	End.	End.
Gray bat (*Myotis grisescens*)	End.	End.	End.
Lower Keys marsh rabbit (*Sylvilagus palustris hefneri*)	End.	End.	None
Choctawhatchee beach mouse (*Peromyscus polionotus allophrys*)	End.	End.	Thr.
St. Andrew beach mouse (*Peromyscus polionotus peninsularis*)	End.	End.	None
Anastasia Island beach mouse (*Peromyscus polionotus phasma*)	End.	End.	None
Perdido Key beach mouse (*Peromyscus polionotus trissyllepsis*)	End.	End.	Thr.
Key Largo cotton mouse (*Peromyscus gossypinus allapaticola*)	End.	End.	End.
Key Largo woodrat (*Neotoma floridana smalli*)	End.	End.	End.
Florida saltmarsh vole (*Microtus pennsylvanicus dukecampbelli*)	End.	End.	None
Right whale (*Eubalaena glacialis*)	End.	End.	End.
Sei whale (*Balaenoptera borealis*)	End.	End.	End.
Fin whale (*Balaenoptera physalus*)	End.	End.	End.
Humpback whale (*Megaptera novaeangliae*)	End.	End.	End.
Sperm whale (*Physeter macrocephalus*)	End.	End.	End.
Florida panther (*Felis concolor coryi*)	End.	End.	End.
Florida manatee (*Trichechus manatus latirostris*)	End.	End.	End.
Key deer (*Odocoileus virginianus clavium*)	End.	End.	End.

Ext.=Extinct, Extp.=Extirpated, End.=Endangered, Thr.=Threatened, SSC=Species of special concern, SU=Status undetermined, UR2=Under review 2, UR3=Under review 3, None=no status

Current FCREPA status Species	Federal	State	1978 FCREPA
Threatened			
Florida mastiff bat (*Eumops glaucinus floridanus*)	UR2	None	None
Big Cypress fox squirrel (*Sciurus niger avicennia*)	UR2	Thr.	End.
Sherman's fox squirrel (*Sciurus niger shermani*)	UR2	SSC	Thr.
Southeastern beach mouse (*Peromyscus polionotus niveiventris*)	Thr.	Thr.	None
Florida mouse (*Podomys floridanus*)	UR2	SSC	Thr.
Florida black bear (*Ursus americanus floridanus*)	UR2	Thr.*	Thr.
Species of special concern			
Round-tailed muskrat (*Neofiber alleni*)	UR2	None	SSC
Rare			
Southeastern big-eared bat (*Plecotus rafinesquii macrotis*)	UR2	None	Rare
Eastern chipmunk (*Tamias striatus striatus*)	None	None	Rare
Rice rat, Lower Keys pop. (*Oryzomys palustris natator*)	End.	End.	End.
Southeastern weasel (*Mustela frenata olivacea*)	None	None	Rare
Florida weasel (*Mustela frenata peninsulae*)	UR2	None	Rare
Mink, Southern Florida pop. (*Mustela vison mink*)	UR2	Thr.	Rare
Status undetermined			
Sherman's short-tailed shrew (*Blarina carolinensis shermani*)	UR2	SSC	SU
Southeastern brown bat (*Myotis austroriparius*)	UR2	None	None

*Not applicable in Baker and Columbia counties and the Apalachicola National Forest.

Ext.=Extinct, Extp.=Extirpated, End.=Endangered, Thr.=Threatened, SSC=Species of special concern, SU=Status undetermined, UR2=Under review 2, UR3=Under review 3, None=no status

Current FCREPA status

Species	Federal	State	1978 FCREPA
Big brown bat (*Eptesicus fuscus fuscus*)	None	None	Rare
Northern yellow bat (*Lasiurus intermedius floridanus*)	None	None	None
Brazilian free-tailed bat (*Tadarida brasiliensis cynocephala*)	None	None	None

No current status

Species	Federal	State	1978 FCREPA
Homosassa shrew (*Sorex longirostris eionis*)	UR2	SSC	Rare
Southeastern shrew (*Sorex longirostris longirostris*)	None	None	Rare
Keen's bat (*Myotis septentrionalis*)	None	None	Rare
Hoary bat (*Lasiurus cinereus*)	None	None	Rare
Pine Island rice rat (*Oryzomys palustris planirostris*)	UR2	None	SU
Lower Keys cotton rat (*Sigmodon hispidus exsputus*)	UR3	None	Thr.
Insular cotton rat (*Sigmodon hispidus insulicola*)	UR2	None	SU
Eastern gray wolf (*Canis lupus lycaon*)	End.	None	Extp.
Key Vaca raccoon (*Procyon lotor auspicatus*)	UR2	None	Thr.
Florida mink (*Mustela vison lutensis*)	UR2	None	Rare
Southern mink (*Mustela vison mink*)	None	None	Rare

Ext.=Extinct, Extp.=Extirpated, End.=Endangered, Thr.=Threatened, SSC=Species of special concern, SU=Status undetermined, UR2=Under review 2, UR3=Under review 3, None=no status

Recently Extinct

Goff's Pocket Gopher
Geomys pinetis goffi
FAMILY GEOMYIDAE
Order Rodentia

TAXONOMY: *G. p. goffi* is known from 25 specimens (Wilkins 1985) taken between 1897 and 1955. It was described by Sherman (1944) on the basis of a longer body than that of other Florida pocket gophers. This difference he attributed to the occurrence of an extra body segment, reflected by 13 pairs of ribs instead of the usual 12. Compared with 127 specimens of *G. pinetis* from other parts of Florida and Georgia, the ratio of skull length to head and body length is 2 percent lower in *goffi* (Sherman 1944). Based only on features of the skull, more recent studies (Williams and Genoways 1980; Wilkins 1985) were unable to differentiate this population from other populations of *G. pinetis*. Under this interpretation, *goffi* would be a synonym of *G. p. pinetis*. Apparently these more recent workers have not reexamined the characteristics of body length and an extra pair of ribs (and, presumably, an extra vertebra). The latter trait is what separates other domestic horses from the Arabian breed. Probably this gopher taxon should be retained until such a major diagnostic feature has been reviewed.

DESCRIPTION: Pocket gophers have a cylindrical body plan analogous to that of a subway train. They have very large claws on the forefeet, large incisors protruding from the mouth (which is normally closed behind them), and external fur-lined cheek pouches, in which they carry foodstuffs. This subspecies cannot be distinguished from other *G. pinetis* by color. Its fur is sepia, shaded on the sides of the shoulders and flanks with orange-cinnamon. White hairs are present on the throat and forearms, a white patch extends from the forehead nearly to the nostrils, and about half the specimens have a white spot on the upper side of the head. The underparts are grayish. Males are larger than females. Average mea-

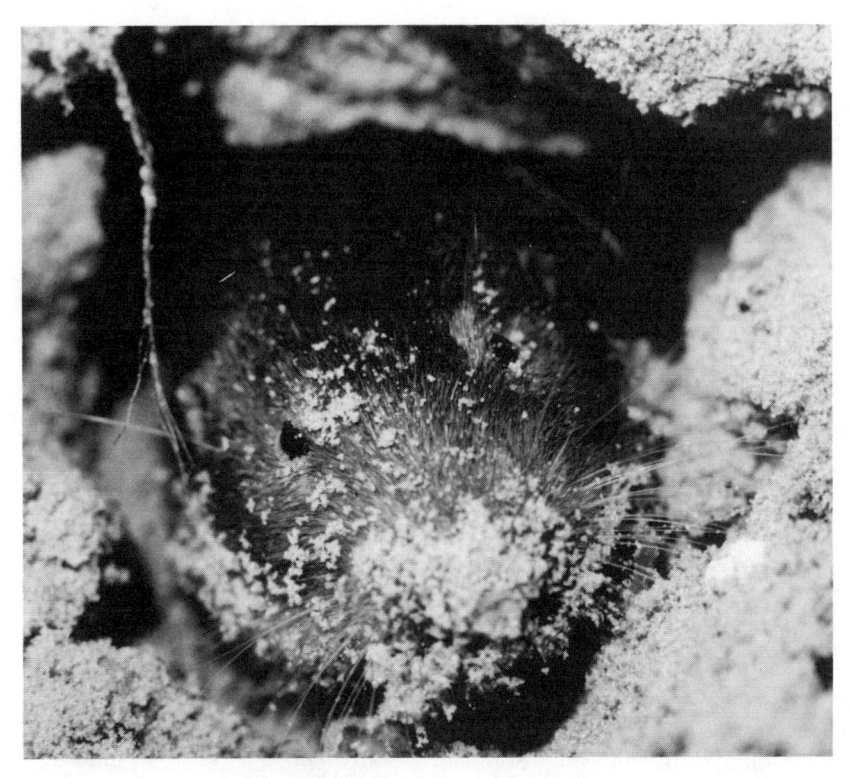

Southeastern pocket gopher, *Geomys pinetis*. (Photo by Stephen R. Humphrey)

surements of 8 males and 7 females are, respectively, 290 and 261 mm in total length, 92 and 86 mm in tail length, 36 and 34 mm in length of hind foot, 50.7 and 44.9 mm in condylobasal length of skull, and 1.3 and 3.3 mm in least distance between temporal ridges (Sherman 1944).

POPULATION SIZE AND TREND: No population counts have been made, but Bangs (1898) considered the population to be abundant. This observation is consistent with the number of specimens in museums. The last observation of Goff's pocket gopher in the field was by Albert Schwartz (personal communication), who found only one small colony (a few mounds) and collected two specimens in a day of searching on 31 December 1955. This colony was less than 1 km from the old downtown area of Eau Gallie, in an uncleared lot within a new housing development on the outskirts of the town. No evidence of the population has been recorded since 1955. Ehrhart (1978) searched unsuccessfully for these

Geomys pinetis goffi

animals in February 1974. Other unsuccessful searches have included an exhaustive survey from October 1978 to March 1979 (Humphrey 1981) and a cursory one in July 1986 (Humphrey et al. 1987). Like any other null hypothesis, extinction can be disproven but never proven. Nonetheless, these three surveys constitute substantial evidence that this subspecies has become extinct.

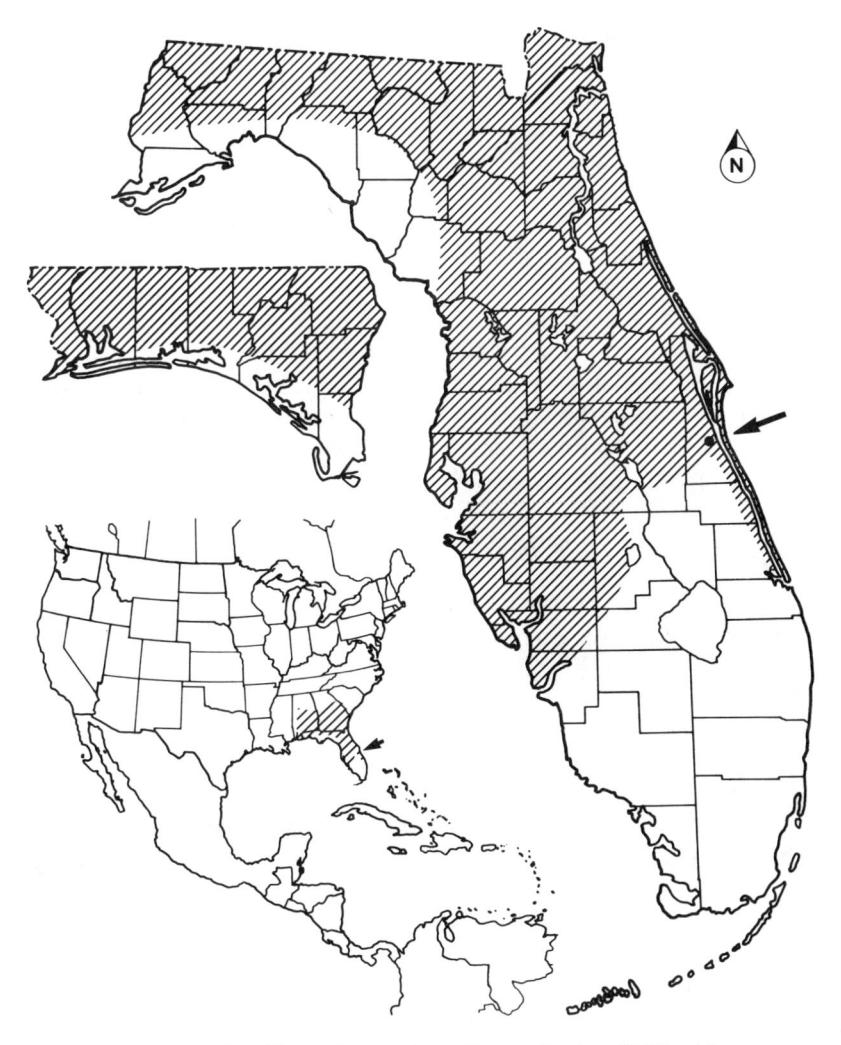

Distribution map of Goff's pocket gopher, *Geomys pinetis goffi*. Hatching represents the species; the dot represents the subspecies.

DISTRIBUTION AND HISTORY OF DISTRIBUTION: All records of *G. p. goffi* are from the type locality of Eau Gallie, described by Sherman (1944: 38) as "part of a strip of Norfolk and St. Lucie fine sands which border the Indian River. Evidence of *Geomys* has been found for a distance of about two miles both north and south of Eau Gallie and for a distance of about two miles inland from the Indian River." This soil is distributed as a long, narrow landform, the Pineda Ridge, which extends from north of Bonaventure through the city of Melbourne and south to Malabar. The former town of Eau Gallie is now part of the city of Melbourne, and most of the area within a 3.2-km radius of the old town is urban or suburban habitat.

GEOGRAPHICAL STATUS: This subspecies was endemic to the Pineda Ridge, and in modern times the population was isolated from other pocket gophers by unsuitable soils along the St. Johns River. Wilkins (1985) pointed out that, during the Sangamonian interglacial period about 122,000 yBP, when sea level was about 8 m above present levels, the Pineda Ridge was isolated as one of several large, offshore islands in the Pamlico Sea. Wilkins hypothesized that during the subsequent Wisconsinan glaciation about 19,000 yBP, when sea level dropped to about 90–130 m below the present level, vast expanses of land became suitable habitat for pocket gophers, and all the populations of peninsular Florida achieved genetic interchange. For a more detailed treatment of this oscillation in sea level, see Kennett (1982:268–273).

HABITAT REQUIREMENTS AND HABITAT TREND: A moderate amount of suitable habitat still occurs on roadsides, railroad right-of-way, a few pastures, fields along airport runways, parks, golf courses, baseball fields, and cemeteries. In the immediate area of Eau Gallie, the largest tract of potential habitat occurs on the fenced grounds of Melbourne Regional Airport, but this is mostly flatwoods and hence floods occasionally; the rest is in golf courses, cemeteries, and other areas where gophers are likely to have been actively exterminated. Some native habitat also persists in the extremities of the Pineda Ridge, but it has been so long protected from fire that the vegetation forms an extremely dense understory and (in the case of sand pine scrub) a closed tree canopy.

VULNERABILITY OF SPECIES AND HABITAT: Because the distribution of this subspecies was so small, it was vulnerable to habitat conversion. It happened to occur where a major city developed.

CAUSES OF THREAT: Conversion of its habitat to human uses appears to be the cause of its extinction. Probably the immediate causes were a combination of habitat loss, predation by domestic dogs, and poisoning.

RESPONSES TO HABITAT MODIFICATION: The details are unknown, but pocket gophers clearly do not survive under urban and suburban conditions.

DEMOGRAPHIC CHARACTERISTICS: Unknown, but presumably like other subspecies of the southeastern pocket gopher. In northern Florida, *G. pinetis* has litters of one to three young, averaging 1.5, the smallest litter size known in pocket gophers (Barrington 1940; Wing 1960). At least some *G. pinetis* have two litters per year. Although breeding occurs year-round, peaks in pregnancy and lactation occur in March and July-August. All adult males are fertile from January through June, and some males are sexually active during the rest of the year. Neither sex may become sexually mature until at least 6 months old (Wing 1960).

KEY BEHAVIORS: Because of their subterranean habits, the behavior of pocket gophers is poorly known. Individuals are solitary most of the time (Hubbell and Goff 1939), except when seeking mates or rearing young. They are thought to feed on plant roots, tubers, bulbs, and stems, which they transport in their cheek pouches to underground storage chambers. Each animal digs an extensive system of tunnels, with portions constantly being added and abandoned in the search for food. Gopher tunnels are always deep enough to be undetectable at the surface, but the excavated soil is pushed to the surface through lateral, ascending branches off the main tunnel and piled in large mounds. Usually the approximate course of the tunnel can be identified by the position of these mounds.

Hubbell and Goff (1939) described a large tunnel system occupied by an adult female southeastern pocket gopher. About 167 m of open tunnel occupied a rectangular area of about 0.2 ha. Sixty-seven mounds were scattered above the tunnel, mostly along terminal branches rather than the central portion. Depth of the tunnel ranged from 0.1 to 1 m. It contained no nest chamber or food stores, but about 1 l of feces was deposited in a side chamber. They also described a smaller tunnel system that contained a nest made of grasses. Brown and Hickman (1973) described 40 tunnels of *G. pinetis*, varying from 20 to 525 ft in length and averaging 145. Each tunnel consisted of a main linear passage with short lateral tunnels branching off at infrequent intervals, each ending in a nest chamber, food cache, or a dead-end. Most of each system was 6–18 in below the surface, but some intervals were as shallow as 2 in or as deep as 5 ft. Most tunnels contained one or two tightly spiraling sections of three to four loops connecting deep passages to shallow ones. No more than one nest occurred per tunnel. Nests, made of shredded grasses and stems, usually were in the deepest part of the tunnel, near the spiral section. Food caches contained almost entirely tubers of bahia grass (*Paspalum notatum*). Fresh droppings were found throughout the tunnels, and large dung masses commonly occurred 2–5 ft from nest chambers. Special chambers packed with fecal pellets, dirt, and dried grass were uncommon; these appeared to be old chambers that became repositories for debris during tunnel renovations and housekeeping activities.

Pocket gophers play several important roles in ecosystem functioning. First, pocket gophers retard eluviation by returning leached nutrients to the surface of the soil, casting up to 81,600 kg/ha of burrow soil per year (Kalisz and Stone 1984). Second, soil mounds made by mammals promote a rich herbaceous flora by creating numerous microsites for colonization and secondary succession (Platt 1975) within grasslands, sandhills, and scrub. Third, the southeastern pocket gopher supports a top carnivore by serving as the primary prey of the pine snake (*Pituophis melanoleucas mugitus*) where the ranges of the two coincide. Here pine snakes spend about half of their lives in tunnels of pocket gophers (Franz 1986). Two torpid pine snakes have been dug from pocket gopher tunnels (I.J. Stout, personal communication). The ascending tunnels to gopher mounds are plugged with soil except when in active use for mound casting, but pine snakes are adept at finding and penetrating the plugged burrows by probing with their conical heads. Long-tailed weasels (*Mustela frenata*) also have been recorded in tunnels of *G. pinetis* (Sherman 1929; Barrington 1940), but their importance as predators is unknown. Fourth, the tunnels of the southeastern pocket gopher create habitat for a sizable subset of subterranean, commensal invertebrates, and the animal itself supports parasites that occur nowhere else (Hubbell and Goff 1939; Woodruff 1973). Hubbell and Goff found many kinds of invertebrates in pocket gopher burrows in Florida, of which 14 are obligate tunnel-dwellers associated with this rodent. By far the most abundant is a blind cricket (*Typhloceuthophilus floridanus*). The tunnel system mentioned above contained at least 92 of these blind crickets when it was excavated. Gopher tunnels contain abundant food for invertebrates, including rootlets, humus, feces, food stores, nest material, and fungi. Fifth, 20 species of amphibians and reptiles use the surface mounds of pocket gophers as habitat (Funderburg and Lee 1968). Of these, the crowned snake (*Tantilla coro-*

nata) and mole skink (*Eumeces egregius*) are permanent residents of gopher mounds and are seldom encountered elsewhere.

CONSERVATION MEASURES TAKEN: This animal is listed as endangered by the Florida Game and Fresh Water Fish Commission (1990). It once was being considered for listing by the U.S. Fish and Wildlife Service, but it no longer is being considered because of the evidence of extinction.

CONSERVATION MEASURES PROPOSED: No action is required. Additional surveys offer little hope that the population will be found to survive.

ACKNOWLEDGMENTS: Llewellyn M. Ehrhart, who prepared the account on Goff's pocket gopher for the first edition of this volume, focused attention on the loss of this animal and on some interesting features of its life. I am grateful to Albert Schwartz for sharing his observations, the last to be recorded for this animal. Thanks are due to James N. Layne and I. Jack Stout for helpful reviews of the manuscript.

Literature Cited

Bangs, O. 1898. The land mammals of peninsular Florida and the coastal region of Georgia. Boston Soc. Nat. Hist. Proc. 28:157–235.

Barrington, B. A. 1940. The natural history of pocket gophers. Unpubl. M.S. thesis, University of Florida, Gainesville. 49 pp.

Brown, L. N., and G. C. Hickman. 1973. Tunnel system structure of the southeastern pocket gopher. Florida Sci. 36:97–103.

Ehrhart, L. M. 1978. Goff's pocket gopher. Pp. 6–7 *in* J. N. Layne, ed., Rare and endangered biota of Florida. Vol. 1. Mammals. University Presses of Florida, Gainesville. xx + 52 pp.

Florida Game and Fresh Water Fish Commission. 1990. Official lists of endangered and potentially endangered fauna and flora in Florida, Tallahassee, Florida. 23 pp.

Franz, R. 1986. Florida pine snakes and gopher frogs as commensals of gopher tortoise burrows. Pp. 16–20 *in* D. R. Jackson and R. J. Bryant, eds., The gopher tortoise and its community. Proc. 5th Ann. Mtg., Gopher Tortoise Council. 93 pp.

Funderburg, J. B., and D. S. Lee. 1968. The amphibian and reptile fauna of pocket gopher (*Geomys*) mounds in central Florida. J. Herpetology 1:99–100.

Hubbell, T. H., and C. C. Goff. 1939. Florida pocket-gopher burrows and their arthropod inhabitants. Proc. Florida Acad. Sci. 4:127–166.

Humphrey, S. R. 1981. Goff's pocket gopher (*Geomys pinetis goffi*) is extinct. Florida Sci. 44:250-252.

Humphrey, S. R., W. H. Kern, Jr., and M. E. Ludlow. 1987. Status survey of seven Florida mammals. Unpubl. final report to U.S. Fish and Wildlife Service, 3100 University Boulevard, South Suite 120, Jacksonville. 39 pp.

Kalisz, P. J., and E. L. Stone. 1984. Soil mixing by scarab beetles and pocket gophers in north-central Florida. Soil Sci. Soc. Amer. J. 48:169-172.

Kennett, J. P. 1982. Marine geology. Prentice-Hall, Inc., Englewood Cliffs, New Jersey. 813 pp.

Platt, W. J. 1975. The colonization and formation of equilibrium plant species associations on badger disturbances in a tall-grass prairie. Ecol. Monogr. 45: 285-305.

Sherman, H. B. 1929. Notes on Florida mammals. J. Mammal. 10:258-259.

Sherman, H. B. 1944. A new subspecies of *Geomys* from Florida. Proc. New England Zool. Club 23:37-40.

Wilkins, K. T. 1985. Variation in the southeastern pocket gopher, Geomys pinetis, along the St. Johns River in Florida. Amer. Midland Nat. 114:125-134.

Williams, S. L., and H. H. Genoways. 1980. Morphological variation in the southeastern pocket gopher, *Geomys pinetis* (Mammalia: Rodentia). Ann. Carnegie Mus. 47:541-570.

Wing, E. S. 1960. Reproduction in the pocket gopher in north-central Florida. J. Mammal. 41:35-43.

Woodruff, R. E. 1973. The scarab beetles of Florida (Coleoptera: Scarabaeidae), part 1. The Laparosticti (Subfamilies: Scarabaeinae, Aphodiinae, Hybosorinae, Ochodaeinae, Geotrupinae, Acanthocerinae). Florida Dept. Agric., Div. of Plant Industry. Arthropods of Florida and Neighboring Land Areas 8:1-220.

Prepared by: Stephen R. Humphrey, Florida Museum of Natural History, University of Florida, Gainesville, FL 32611.

Recently Extinct

Pallid Beach Mouse
Peromyscus polionotus decoloratus
FAMILY CRICETIDAE
Order Rodentia

TAXONOMY: The pallid beach mouse was described on the basis of 13 specimens taken in 1925 at Ponce Park, just north of Mosquito Inlet (now Ponce de Leon Inlet), Volusia County (Howell 1939). Although it is not clear exactly what comparisons Howell made, he cited similarities and differences of adjoining subspecies, and the differences appear substantial. Populations along the Atlantic coast were not included in Bowen's (1968) taxonomic revision of oldfield and beach mice.

DESCRIPTION: This is a small, whitish mouse whose color closely matches the color of the beach sand on which it lives. Reduction of pigment in the fur of this species has been studied in detail by F. B. Sumner, W. W. Bowen, and M. H. Smith, among others, and it is thought to result from selective pressure from intense predation, by owls in particular. The pallid subspecies is paler than the form from Anastasia Island and much paler than the subspecies from the southeastern coast of Florida. The head and dorsum are pale pinkish buff, slightly darkened with wood brown on the back. The underparts and sides of the head are pure white. Average measurements are: total length 135.5 mm, tail length 52.4 mm, hind foot length 18.4 mm, greatest length of skull 23.3 mm, zygomatic breadth 12 mm, interorbital breadth 3.9 mm, length of nasals 9.3 mm, and length of maxillary tooth row 3.4 mm (Howell 1939). The skull is smaller than that of the southeastern subspecies and has a rostrum slightly broader at the tip than the Anastasia Island form.

POPULATION SIZE AND TREND: The pallid beach mouse was abundant in the 1920s on the sand dunes near the ocean at Ponce Park (Howell 1939). Ehrhart (1978) reported a dozen or more unsuccessful searches for a surviving population from 1959 to 1974. At least the last of these efforts was quite thorough. Humphrey and Barbour (1981) made an exhaustive, unsuccessful search in 1979. These surveys constitute substantial evidence that the taxon is extinct.

DISTRIBUTION AND HISTORY OF DISTRIBUTION: The pallid beach mouse was known from two sites: the type locality just north of Mosquito Inlet, at what is now the town of Ponce Inlet, and at Bulow, presumably at or near Bulow Plantation Historical Site, Flagler County. Howell (1939) considered the probable range to be from Mosquito Inlet to Matanzas Inlet. The distribution has diminished to nothing.

GEOGRAPHICAL STATUS: This subspecies was endemic to the barrier beach between Matanzas and present-day Ponce de Leon inlets.

HABITAT REQUIREMENTS AND HABITAT TREND: Howell (1939:364) reported occupied habitat as "the sand dunes near the ocean." This is a grassland vegetated mainly by sea oats (*Uniola paniculata*) and dune panic grass (*Panicum amarulum*), and adjoined by a dense woodland "scrub," characterized by sand pine (*Pinus clausa*), oaks (*Quercus* sp.), and palmetto (*Serenoa repens*). Very large amounts of this habitat have been transformed for the development of highways, cities, vacation homes, and condominiums around Daytona Beach. Nonetheless, both Ehrhart (1978) and Humphrey and Barbour (1981) noted that much suitable habitat remained in the historic range of the pallid beach mouse. Apparently conversion of habitat to other uses was one but not the only cause of extinction.

Habitat loss at the type locality has been episodic, although no relation to the subspecies' extinction is postulated. In the 1920s the foundation and walls of a hotel were built literally in the type locality, but construction was terminated because financial backing was withdrawn in response to overspeculation in real estate development. After the inlet near the type locality was deepened by dredging and stabilized with jetties, changes in the alongshore current and trapping of surges during winter storms from the northeast caused erosion of some dune habitat at Ponce Park.

VULNERABILITY OF SPECIES AND HABITAT: Extinction makes this issue moot. Evidently the species was more vulnerable than its habitat.

Peromyscus polionotus decoloratus

CAUSES OF THREAT: The cause of extinction of the pallid beach mouse is undocumented, but Humphrey and Barbour (1981) speculated that competitive exclusion by house mice (*Mus musculus*) was responsible. They showed that these exotic mice have colonized the habitat of beach

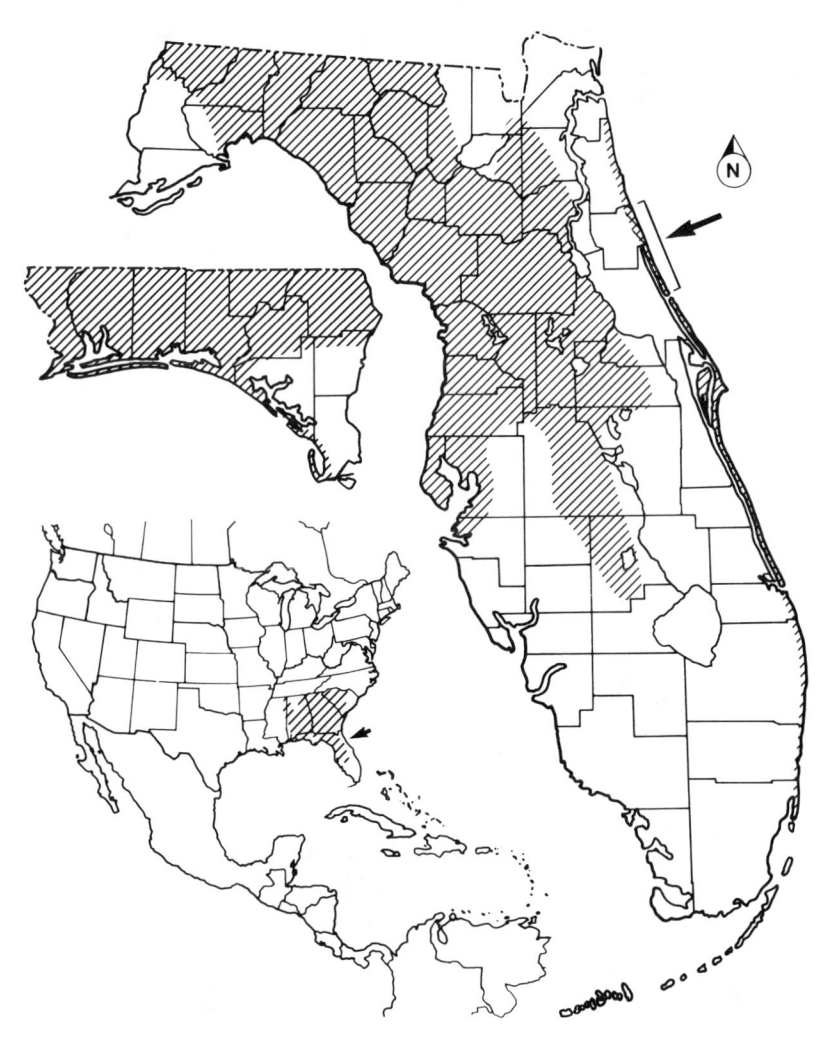

Distribution map of the pallid beach mouse, *Peromyscus polionotus decoloratus*. Hatching represents the species; crosshatching and bracket represent the subspecies.

mice in many parts of Florida, and they found a statistically significant negative correlation of local distributions of the two species. They inferred that house mice use human dwellings and trash in and next to dune habitat as points of introduction and refuge from which to colonize the dune grassland. A related issue is that free-ranging house cats (*Felis catus*) may be important predators on beach mice, although the evidence (Bowen 1968) is even more scanty than for the role of house mice. The effect of these two exotic species on survival of beach mouse populations may be quite important. Either a competitor or a predator alone can eliminate another species, and the effects of a competitor and a predator together could be additive. Both processes reduce the population of the species of interest, competition by supressing its reproduction, and predation by removing individuals while stimulating reproduction of the survivors. If these biological interactions are the cause of extinction of beach mice, managers should recognize that proximity of human activity and structures is nonetheless the driving force behind the commensal agents. Beach mice may require essentially wilderness conditions on a local scale to persist.

RESPONSES TO HABITAT MODIFICATION: Because the influence of commensal house mice and house cats extends considerably beyond the limits of human land use, beach mice are extremely sensitive to habitat modification. The survival of beach mouse populations appears to depend on the wilderness quality of their habitat, and preserves designed for them may need to be several square kilometers in size.

DEMOGRAPHIC CHARACTERISTICS: See the species account for the Choctawhatchee beach mouse, *Peromyscus polionotus allophrys*.

KEY BEHAVIORS: See the species account for the Choctawhatchee beach mouse, *Peromyscus polionotus allophrys*.

CONSERVATION MEASURES TAKEN: This animal is listed as endangered by the Florida Game and Fresh Water Fish Commission (1990). It once was being considered for listing by the U.S. Fish and Wildlife Service, but it no longer is being considered because of the evidence of extinction.

CONSERVATION MEASURES PROPOSED: None.

ACKNOWLEDGMENTS: I thank I. Jack Stout for suggesting improvements to the manuscript.

Literature Cited

Bowen, W.W. 1968. Variation and evolution of Gulf coast populations of beach mice, *Peromyscus polionotus*. Bull. Florida State Museum, Biol. Sciences 12:1–91.

Ehrhart, L.M. 1978. Pallid beach mouse. Pp. 8–9 *in* J.N. Layne, ed., Rare and endangered biota of Florida. Vol. 1. Mammals. University Presses of Florida, Gainesville. xx + 52 pp.

Florida Game and Fresh Water Fish Commission. 1990. Official lists of endangered and potentially endangered fauna and flora in Florida. Tallahassee, Florida. 23 pp.

Howell, A.H. 1939. Descriptions of five new mammals from Florida. J. Mammal. 20:363–365.

Humphrey, S.R., and D.B. Barbour. 1981. Status and habitat of three subspecies of *Peromyscus polionotus* in Florida. J. Mammal. 62:840–844.

Prepared by: Stephen R. Humphrey, Florida Museum of Natural History, University of Florida, Gainesville, FL 32611.

Recently Extinct

Chadwick Beach Cotton Mouse
Peromyscus gossypinus restrictus
FAMILY CRICETIDAE
Order Rodentia

TAXONOMY: The Chadwick Beach cotton mouse was described on the basis of 15 specimens taken at Chadwick Beach, near Englewood, Sarasota County (Howell 1939). They occurred in the sea oats (*Uniola paniculata*) along the beach and in the adjacent cabbage palm (*Sabal palmetto*) forest. No specimens of the Chadwick Beach cotton mouse appear to have been collected since the type series was taken (Layne et al. 1977). A taxonomic review of this population might be worthwhile, because the comparisons made in the type description were qualitative. The published descriptions, however, appear to support the validity of this taxon. With so few specimens of this subspecies available, a new evaluation of geographic variation would require a major statistical comparison with other subspecies in the region. In view of the apparent extinction of the taxon, such an exercise would be strictly academic.

DESCRIPTION: This is a large mouse with a dark brown back. The Chadwick Beach subspecies is smaller and paler than that of the nearby mainland (*P. g. palmarius*), and its mid-dorsal stripe is narrower; the skulls of these two forms are similar. The upper parts are pinkish-cinnamon, mixed with fuscous especially in the middle of the back; the underparts are white with a wash of pale pinkish buff on the chest; and the tail is brown above and buffy below. Average measurements of adults are: total length 172 mm, tail length 72.5 mm, length of hind foot 22.3 mm, length of the ear from notch 22.3 mm, greatest length of skull 27.6 mm, zygomatic breadth 13.9 mm, interorbital breadth 4.4 mm, length of nasals 10.9 mm, and length of the maxillary tooth row 3.9 mm (Howell 1939).

POPULATION SIZE AND TREND: No data are available on population size or trend. A thorough survey in 1985 of the Chadwick Beach area, however, produced substantial evidence that the former population no longer exists (Repenning and Humphrey 1986). It is conceivable but unlikely that a remnant population of Chadwick cotton mice persists somewhere on Manasota Key.

DISTRIBUTION AND HISTORY OF DISTRIBUTION: The place-name "Chadwick Beach" does not occur on modern maps. As determined through conversation with realtors and long-term residents, its location was near the southern end of Manasota Key in present-day Englewood Beach (Repenning and Humphrey 1986), Charlotte County. Examination of a 1965 street map of Englewood, found in the University of Florida Map Library, showed that present-day Highway 776, which leads to Englewood Beach, was called only "Beach Road" at that time. This map also shows a street named "Chadwick Place" in Englewood Beach, between Stanford and Meredith drives. The name "Chadwick Beach" does not occur on old topographic, nautical, or county road maps, but the latter two show that prior to the use of the name "Englewood Beach," the settlement around the landing of Beach Road on Manasota Key was named "Punta Gorda Beach," and the key itself was named "Peninsula Key." Nothing more is known about the original distribution of this subspecies. Based on a survey by Repenning and Humphrey (1986), the distribution appears to have been reduced to nothing.

GEOGRAPHICAL STATUS: *P. g. restrictus* is an endemic subspecies. Manasota Key is a peninsula, not an island. The primary habitat of this subspecies is essentially insular, however, because the coastal forest is very narrow at the neck of the peninsula.

HABITAT REQUIREMENTS AND HABITAT TREND: Cotton mice prefer closed-canopy forest, but as noted in the type description of *P. g. restrictus*, cotton mice also occupy grassland such as sea oats adjacent to coastal forest. The maritime forest of Manasota Key is a very narrow strip at the northern end, becoming wider southward. The forest edge is affected by the harsh coastal environment and is composed only of cabbage palms, which are relatively tolerant of salt and wind. Where the forest is narrow, only the palm edge occurs. Where the forest widens farther south, its interior reflects a successional history and is dominated by live oaks (*Quercus virginiana*) and southern red cedar (*Juniperus silvicola*). The greatest tree-species and structural diversity now occurs at the southern end of Sarasota County. The sea oats strip along the coast of Manasota Key has been reduced to a few remnants, mostly in publicly owned parks dedicated to beach recreation. Landward, numerous remnants of the forest remain, divided into many small, privately owned parcels developed for residences and vacation homes. Some of the larger lots in Sarasota County were developed by clearing vegetation for a house site but leaving forest to buffer neighboring properties. The southernmost portion of the forest,

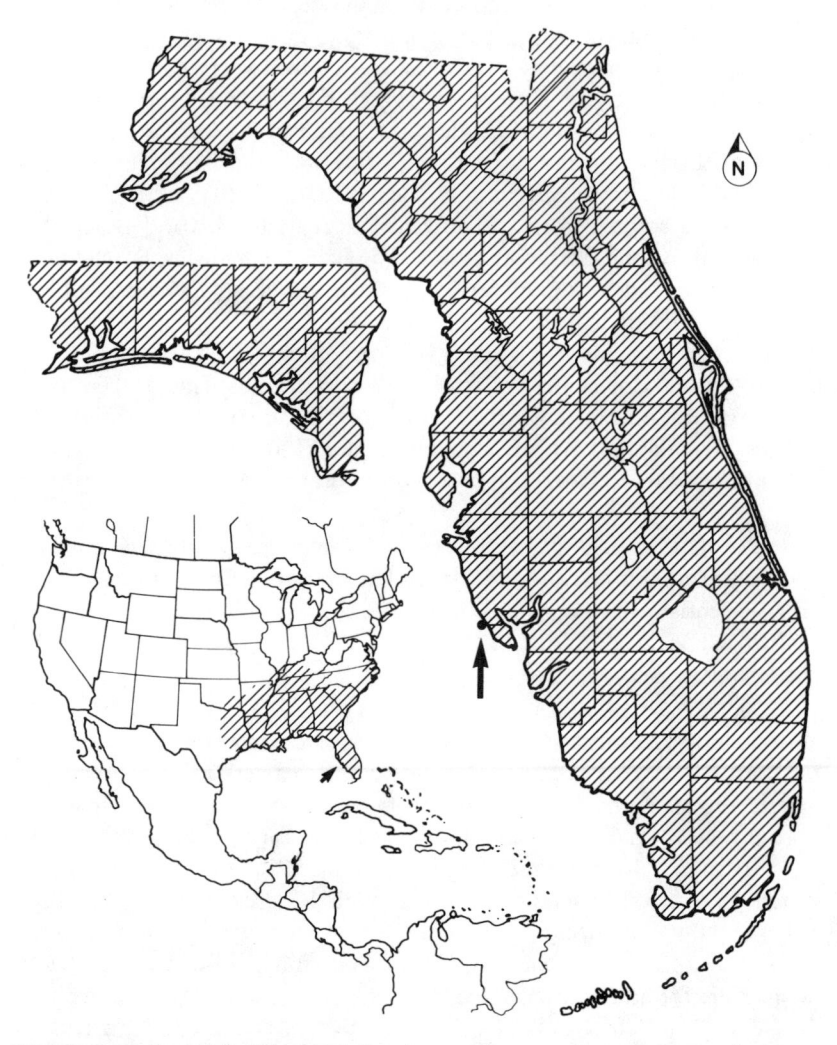

Distribution map of the Chadwick Beach cotton mouse, *Peromyscus gossypinus restrictus*. Hatching represents the species; the dot represents the subspecies.

in Charlotte County, has been completely destroyed by conversion to human uses.

VULNERABILITY OF SPECIES AND HABITAT: In view of the amount of forest in Sarasota County, it is surprising that no cotton mice appear to remain on Manasota Key. A mechanism that would have made the remaining woodlots uninhabitable for cotton mice has not been demonstrated. Predation by the large number of house cats (*Felis catus*) associated with the continuously distributed human residences may have had an important effect on cotton mice. Black rats (*Rattus rattus*) were found at only one site at which cotton mice were expected, so this potential competitor does not seem to be a ubiquitous element in the forest of Manasota Key.

CAUSES OF THREAT: Unknown, but the point is moot if the subspecies is extinct.

RESPONSES TO HABITAT MODIFICATION: Unknown.

DEMOGRAPHIC CHARACTERISTICS: Unknown, but presumably similar to other subspecies of cotton mice. See the species account on the Key Largo cotton mouse, *Peromyscus gossypinus allapaticola*.

KEY BEHAVIORS: The behavior of this subspecies has not been studied, but cotton mice elsewhere are both subterranean and arboreal, but not terrestrial. They spend very little time on the surface of the ground, perhaps because that is where they would be most vulnerable to predators. Instead, they nest underground in cavities made by roots of trees, and they forage in the forest canopy. Individuals may have large home ranges, which they traverse by traveling from tree to tree. Cotton mice are strictly nocturnal, so they are seldom seen by man unless specifically sought.

CONSERVATION MEASURES TAKEN: At the time the field survey was undertaken, the Chadwick Beach cotton mouse was listed as a species of special concern by the state of Florida (Florida Game and Fresh Water Fish Commission 1985). It was subsequently listed as endangered (Florida Game and Fresh Water Fish Commission 1987). It once was being considered for listing by the U.S. Fish and Wildlife Service, but it no longer is being considered because of the evidence of extinction.

CONSERVATION MEASURES PROPOSED: Because the Chadwick cotton

mouse appears to be extinct, the only appropriate action is to continue to search for a remnant population.

ACKNOWLEDGMENTS: James N. Layne pointed to the need to investigate the status of this subspecies, David J. Wesley and Michael Bentzien made a search possible, and Robert W. Repenning and Mark E. Ludlow did the field work. Mark E. Ludlow also suggested improvements to the manuscript.

Literature Cited

Florida Game and Fresh Water Fish Commission. 1985. Official lists of endangered and potentially endangered fauna and flora in Florida. Tallahassee, Florida. 24 pp.

Florida Game and Fresh Water Fish Commission. 1987. Official lists of endangered and potentially endangered fauna and flora in Florida. Tallahassee, Florida. 19 pp.

Howell, A. H. 1939. Descriptions of five new mammals from Florida. J. Mammal. 20:363–365.

Layne, J. N., J. A. Stallcup, G. E. Woolfenden, M. N. McCauley, and D. J. Worley. 1977. Fish and wildlife inventory of the seven-county region included in the central Florida phosphate industry areawide environmental impact study. Archbold Biological Station, Lake Placid, Florida. 1279 pp.

Repenning, R. W., and S. R. Humphrey. 1986. The Chadwick Beach cotton mouse (Rodentia: *Peromyscus gossypinus restrictus*) may be extinct. Florida Sci. 49: 259–262.

Prepared by: Stephen R. Humphrey, Florida Museum of Natural History, University of Florida, Gainesville, FL 32611.

Recently Extinct

Florida Red Wolf
Canis rufus floridanus
FAMILY CANIDAE
Order Carnivora

TAXONOMY: Other names: black wolf. Synonyms: *Canis niger niger* and *Canis floridanus*. The Florida red wolf was one of three subspecies of the red wolf. The type is from Horse Landing, St. John's River, about 19 km south of Palatka, Putnam County (Miller 1912).

The taxonomy of the red wolf has been reviewed frequently because of its potential for hybridization with other species of *Canis* and its high variability in physical appearance. Young and Goldman (1944) regarded the red wolf as a full species. Lawrence and Bossert (1967) thought that the Florida red wolf (which they identified as *C. niger*) was a local form of the gray wolf, *C. lupus*. Paradiso (1968) suggested that the gray wolf and the coyote, *C. latrans*, may be conspecific. Nowak (1970) regarded the red wolf to be a full species. Hybridization can and has occurred between coyotes and red wolves (Paradiso and Nowak 1972, Warren Parker, personal communication).

DESCRIPTION: The red wolf is a small, relatively primitive species that arose in the early Pleistocene (Paradiso and Nowak 1982). Positive identification of the species is not easy because all canids vary geographically, as well as by age and sex. In many of the cranial and dental measurements, red wolves and coyotes are similar (Paradiso and Nowak 1972). Red wolves are generally larger than coyotes when specimens of the same sex are compared, and smaller than the gray wolf (Paradiso and Nowak 1972; Paradiso and Nowak 1982).

Total length of 67 specimens of red wolves collected in the field varied from 1355 to 1650 mm. Weights varied from 21 to 41 kg for males, and 16 to 29 kg for females (Paradiso and Nowak 1972). The red wolf has a narrower physique and shorter fur than the gray wolf. Young (1946) commented on the long-legged appearance of red wolves, likening them to greyhounds. Field biologists frequently commented on the long legs of red wolves in Texas (Paradiso and Nowak 1972).

Pelage is variable. The red wolf generally resembles the gray wolf in color, but the most common color phase in a large series of skins collected in the southeastern United States prior to 1930 was more reddish and sparsely haired than either coyotes or gray wolves (Paradiso and Nowak 1972, 1982). The pelage of Florida red wolves has been described as black, black and white, mixed colors, and brown shading to gray (Bartram 1791; Cory 1896).

POPULATION SIZE AND TREND: Historically unknown, but now extinct.

DISTRIBUTION AND HISTORY OF DISTRIBUTION: The Florida red wolf once occurred in Florida, Alabama, and Georgia (Layne 1978) and is believed to have occurred in the Carolinas and Virginia prior to A.D. 1600 (Paradiso and Nowak 1972). It was found throughout Florida in a variety of habitat types. Townshend (1875) reported wolves howling along the lower 10 km of the Manatee River in 1874, and he also reported seeing one in the vicinity of the Myakka River. The last known wolf killed in the Gulf Hammock area (Levy County) was in 1875. The last known wolf killed in Lee County was in 1890. Cory (1896) reported a female and two pups taken from the Big Cypress in 1894 and one animal that was seen southwest of Lake Worth in 1895. He considered them to be common in these areas. Howell (1940) reported records of wolves "on the prairie next to the Everglades" in 1903, in the Kissimmee Valley and around Lake Okeechobee, on Indian Prairie and Parker Island southeast of Lake Istokpoga in 1917, and around Fort Christmas in 1922.

The Florida red wolf is extinct throughout its range. By 1972 the species was known to occur in genetically pure form only in the coastal prairies and marshes of southeastern Texas and adjacent Louisiana (Paradiso and Nowak 1972). During the 1970s all known red wolves were trapped and removed from the area for captive breeding and reintroduction. A reintroduced population now exists on Alligator River National Wildlife Refuge (ARNWR) in North Carolina (Venters 1989).

GEOGRAPHICAL STATUS: This was a Florida race of a southeastern species.

HABITAT REQUIREMENTS AND HABITAT TREND: The range of the red wolf was mainly restricted to the humid division of the Lower Austral

Canis rufus floridanus 31

Life Zone, and it preferred warm, moist, densely vegetated habitat. It occurred in both upland and wetland areas, occupying pine forests, bottomland hardwoods, and coastal marshes. The Florida red wolf likewise occupied a variety of habitat types (Paradiso and Nowak 1972; Layne 1978).

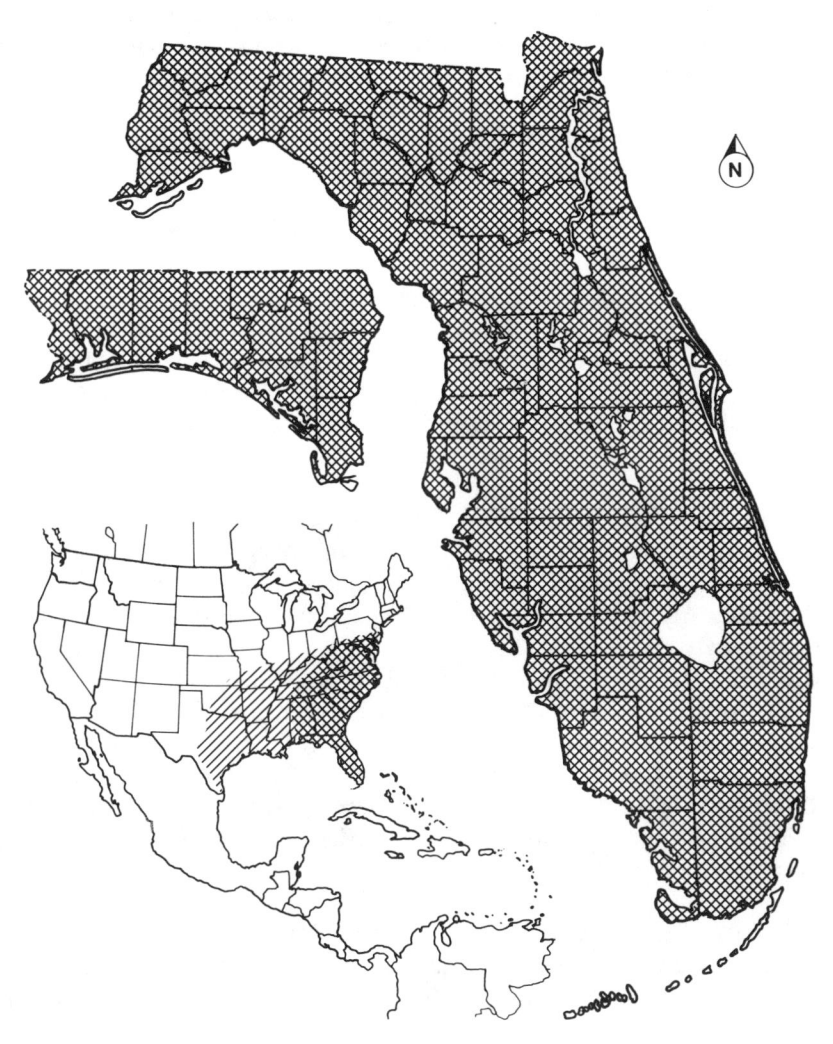

Distribution map of the Florida red wolf, *Canis rufus floridanus*. Hatching represents the species; crosshatching represents the subspecies.

VULNERABILITY OF SPECIES AND HABITAT: The red wolf was eliminated from most of its range by the combined effects of heavy hunting, trapping, poisoning, and widespread environmental change. The species reportedly was secretive but much easier to trap than coyotes. Land clearing associated with agriculture and settlement eliminated essential habitat while enhancing conditions for the immigration of coyotes and subsequent hybridization between red wolves and coyotes (Paradiso and Nowak 1972; Venters 1989; Warren Parker, personal communication). Captive and reintroduced wolves are subject to severe heartworm and intestinal parasite infections, reducing their ability to withstand stress. Poor pup survival has been attributed to high parasite loads (Warren Parker, personal communication).

CAUSES OF THREAT: Same as above.

RESPONSES TO HABITAT MODIFICATION: Unknown.

DEMOGRAPHIC CHARACTERISTICS: Red wolves become sexually mature when 2–3 years old. Breeding occurs from December through February, and young are born in April, May, and early June. The average litter size is seven (Paradiso and Nowak 1972).

KEY BEHAVIORS: On the Texas Gulf coast, red wolves were most active at night and bedded down in weedy fields or brushy pastures during the day. Mated pairs and an occasional extra male usually traveled together. Pairs regularly moved around in their established territories, which were marked by scent and scratch marks. Dens were in hollow logs, stumps, road culverts, and embankments (Paradiso and Nowak 1972).

Red wolves apparently were not a major predator of large mammals. Various authors reported evidence of red wolf predation on rabbits, rodents, small domestic livestock, deer, and hogs (Paradiso and Nowak 1972). Analysis of scats of red wolves reintroduced at ARNWR indicates a diet of raccoons, rodents, frogs, and turtles. Marsh rabbits are a favorite prey item (Venters 1989).

CONSERVATION MEASURES TAKEN: By 1970, red wolves were restricted to a small area of coastal Texas and Louisiana. Recognizing the genetic erosion and imminent extinction of the species, the U.S. Fish and Wildlife Service (USFWS) removed about 20 wolves and placed them in breeding facilities in Tacoma, Washington (Paradiso and Nowak 1982; Venters 1989). In the late 1970s mated pairs were experimentally released on Bull's Island near Charleston, South Carolina (Anonymous 1986). The success of this experiment led to the release of four pairs of wolves on ARNWR in September 1987 (Parker 1987). By 1989 there were 14 free-ranging wolves at Alligator River and several more in acclimation pens. Additionally, USFWS has established three "island nurseries" for the production of wild-born young for permanent reintroduction elsewhere. These are on Bull's Island, Horn Island in Mississippi, and St. Vincent Island in Florida. No permanent reintroduction sites in addition to ARNWR have been announced to date, although potential sites are being investigated.

CONSERVATION MEASURES PROPOSED: The state of Florida, in concert with the USFWS and other interested research and conservation organizations, should evaluate the biological, social, and political feasibility of a permanent reintroduction of red wolves in Florida. Geographically Florida represents a significant part of the historic range of the species, and several large, publicly owned tracts of land contain suitable habitat. Information obtained from the program at ARNWR should be evaluated to help determine those sites best meeting the biological requirements for reintroduction. In addition to biological requirements, potential social and political problems must be evaluated. Experiences at ARNWR show, however, that with an understanding of public perceptions and concerns, these problems can be addressed to ensure the support of affected citizens for a reintroduction program. Although the Florida race is extinct, wolves available from Texas and Louisiana are similar in behavior and appearance to the Florida race and could serve as a suitable source for reintroductions.

Literature Cited

Anonymous. 1986. Red wolf proposed for reintroduction. Endangered Species Tech. Bull. 11:8–9.

Bartram, W. 1791. Travels through North and South Carolina, Georgia, East and West Florida, the Cherokee country, the extensive territories of the Muscogulges or Creek Confederacy, and the country of the Choctaws; containing an account of the soil and natural productions of those regions, together with observations in the manner of the Indians. Philadelphia, Pennsylvania. 24 + 522 pp.

Cory, C. B. 1896. Hunting and fishing in Florida. Estes and Lauriat, Boston, Massachusetts. 304 pp.

Howell, A. E. 1940. Unpubl. manuscript on Florida mammals. In files of the U.S. Fish and Wildlife Service, Washington, D.C. 114 pp.

Lawrence, B., and W. H. Bossert. 1967. Multiple character analysis of *Canis lupus, latrans,* and *familiaris*, with a discussion of the relationship of *Canis niger*. Amer. Zool. 7:223-232.

Layne, J. N. 1978. Florida red wolf. Pp. 48-49 *in* J. N. Layne, ed., Rare and endangered biota of Florida. Vol. 1. Mammals. University Presses of Florida, Gainesville. xx + 52 pp.

Miller, G. S. 1912. The names of two North American wolves. Proc. Biol. Soc. Wash. 25:95.

Nowak, R. M. 1970. Report on the red wolf. Defenders of Wildl. News 45:82-94.

Paradiso, J. L. 1968. Canids recently collected in east Texas with comments on the taxonomy of the red wolf. Amer. Midland Nat. 80:529-534.

Paradiso, J. L., and R. M. Nowak. 1972. *Canis rufus*. Mammalian Species 22:1-4.

Paradiso, J. L., and R. M. Nowak. 1982. Wolves: *Canis lupus* and allies. Pp. 460-474 *in* J. A. Chapman and G. A. Feldhamer, eds., Wild mammals of North America. Johns Hopkins University Press, Baltimore. 1147 pp.

Parker, W. 1987. Red wolves return to the wild. Endangered Species Tech. Bull. 12:11-12.

Townshend, F. T. 1875. Wild life in Florida, with a visit to Cuba. Hurst and Blackett, Publishers, London. 319 pp.

Venters, V. 1989. Return of the red wolf. Wildl. in North Carolina 53:19-23.

Young, S. P. 1946. The wolf in North American history. Caxton Printers, Ltd., Caldwell, Idaho. 149 pp.

Young, S. P., and E. A. Goldman. 1944. The wolves of North America. Amer. Wildl. Inst., Washington, D.C. 20 + 636 pp.

Prepared by: Mark S. Robson, Florida Game and Fresh Water Fish Commission, 551 North Military Trail, West Palm Beach, FL 33415.

Recently Extinct

West Indian Monk Seal
Monachus tropicalis
FAMILY PHOCIDAE
Order Pinnipedia

TAXONOMY: The species was originally assigned to *Phoca tropicalis* by Gray (1850). The type locality is Pedro Cays, 80 km south of Jamaica.

The West Indian monk seal belongs in the tribe Monachini along with the Mediterranean and Hawaiian monk seals (*Monachus monachus* and *M. schauinslandi*) and in the subfamily Monachinae with southern phocids, excluding the southern elephant seal *Mirounga*. The three species of *Monachus* are very similar but are regarded as distinct species because of their complete geographic isolation (Scheffer 1958). The West Indian and the Hawaiian monk seals are the most alike anatomically.

Other common names are Caribbean monk seal and West Indian seal.

DESCRIPTION: The adult pelage is described by Ward (1887:260) as "grayish brown or grisled on the back, a result of the Vandyke-brown hairs being tipped with light horn-color, the lower surface ochreous-yellow to yellowish white." Very young animals were black. Characteristics of the skull are that the nasal process of the premaxilla is broadly in contact with the nasal bone and the frontomaxillary suture is shorter than the nasomaxillary suture. The incisors are constricted at the junction between the root and the crown. The molars have two roots, a low central cusp, a single small anterior cusp, and two small posterior cusps (Scheffer 1958).

POPULATION SIZE AND TREND: None were seen or reported during a survey in 1951 by Gilmore (1959), during an extensive aerial search made by Kenyon (1977) in 1973, or during a cruise made by LeBoeuf, Kenyon, and Villa-Ramirez (1986). These surveys were along the Mexican, Middle American, and southern Jamaican coasts, across the Gulf of

Mexico with a visit to islands where they used to be plentiful. A survey, based on interviews with fishermen, was conducted along the north coast of Haiti in 1986 by Woods (1987). One informant reported seeing a seal in 1981.

DISTRIBUTION AND HISTORY OF DISTRIBUTION: The last report of a sighting believed to be monk seal was at the mouth of the Baie de l'Acul on the north coast of Haiti in 1981 (Woods and Hermanson 1987). In 1974 a sighting was made by fishermen of at least two animals near Cay Verde and Cay Burro, southeastern Bahamas. This stimulated the undertaking of a two-week expedition in 1980 to the southeastern Bahamas but no monk seals were seen. Prior to that a sighting was made by C. B. Lewis on the Serranillo Bank south of Jamaica in 1952 (Rice 1973). Previous reported sightings include two animals near Jamaica in 1949. A seal was killed in 1939 on Pedro Bank south of Jamaica (Caldwell and Caldwell 1978). Other reported records in this century include sightings in the Lower Laguna Madre of Texas in 1932 (Gunter 1947) and one animal killed in Key West in 1922 (Moore 1953; Townsend 1923). Some recent alleged reports may (Rice 1973) be sightings of feral California sea lions, *Zalophus californicus*. Extralimital sightings of juveniles of the hooded seal, *Cystophora cristata*, which look like monk seals, have been reported from the Atlantic coast of southern Florida: Cape Canaveral (Miller 1917) and Fort Lauderdale in 1984 (D. K. Odell, unpubl. data).

During the mid-1800s, the population was apparently viable but heavily hunted on its breeding grounds in the Arrecifes Triangulos in the Bay of Campeche (Ward 1887).

The range of the monk seal, while contemporaneous with man, extended along the coast of Middle America from southern Texas to Honduras, east throughout the Greater and northern Lesser Antilles and Bahamas, and along the coast of Florida probably as far north as Tampa Bay but certainly as far north as Charlotte Harbor on the west coast and Cumberland Island in southeastern Georgia on the east coast.

Scattered reports of the monk seal come from early historic documents and prehistoric contexts. During early Spanish exploration in the West Indies eight monk seals were killed south of Haiti by sailors accompanying Columbus on his second voyage in 1494 (Ward 1887), and 14 were killed during the travels of Juan Ponce de León on Las Tortugas in 1513 (Herrera 1935). Remains of monk seal were recovered from two Spanish sites, a ranch and a mission, in southeastern Texas (Raun 1964). Archaeological finds documenting the prehistoric exploitation of the monk seal are reported from South Indian Field, Melbourne, Brevard County,

Monachus tropicalis 37

Florida (Cumbaa 1980), an Indian site near Lake Hellen Blazes, Brevard County (Ray 1961), possible archaeological association of remains dredged at St. Petersburg, Pinellas County (Ray 1961), Wightman site on Sanibel Island, Lee County (A. Fradkin, personal communication), a

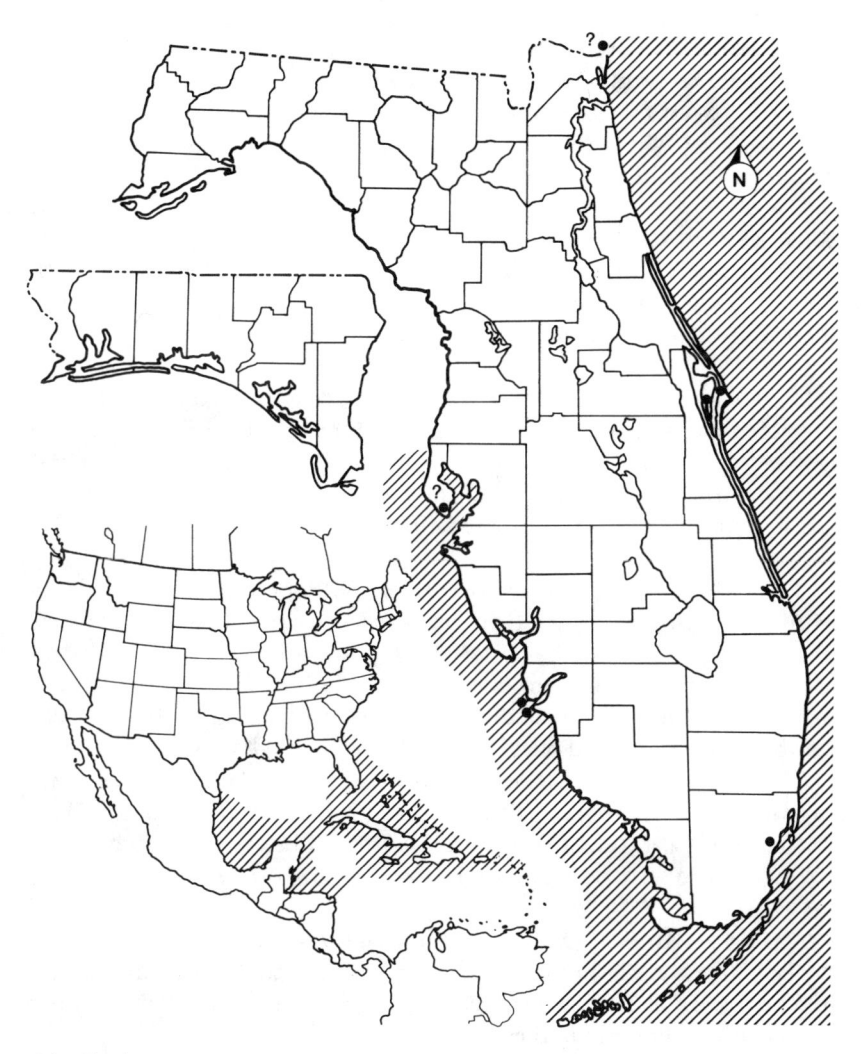

Distribution map of the West Indian monk seal, *Monachus tropicalis*. Hatching represents the historic range of the species; dots represent archaeological sites containing bones of the species.

site on Key Marco, Lee County (Cumbaa 1980), and the Grenada site in present-day Miami, Dade County (Wing and Loucks 1984). A tentative identification of monk seal was made of remains excavated from a site on Cumberland Island in southeasternmost Georgia (Milanich 1971). Monk seal remains have been found in archaeological sites on Puerto Rico, St. Eustatius (H. van der Klift, personal communication), and Nevis, West Indies. None have been encountered in Middle American sites. Although they appear to be abundant in the archaeological record, they occur in relatively few of the archaeological sites from which faunal remains have been studied. Where they are found they are rare, represented by only one or two specimens of bone or tooth.

GEOGRAPHICAL STATUS: Once widespread in the Gulf of Mexico, the Caribbean Sea, and extending up the Atlantic coast of the southeastern United States. Other species of the genus *Monachus* are represented in the Mediterranean Sea and around the Hawaiian Islands.

HABITAT REQUIREMENTS AND HABITAT TREND: Unknown, but presumably they ate fish and shellfish and required predator-free beaches to rest and to bear and nurse young. Animals used beaches to haul out and for rookeries. With increased human use of beaches, the seals were subject to human disturbance and predation.

VULNERABILITY OF SPECIES AND HABITAT: The Caribbean species is extinct. Unrestrained killing of this species at their breeding colonies during the past centuries depressed the population levels (Westermann 1953). The remaining animals succumbed during this century with the increase in human activity and the thorough penetration of the Gulf and Caribbean by motorized boats.

The Hawaiian species, *Monachus schauinslandi*, is the only species that survives in any numbers to this day. The great vulnerability of the West Indian monk seal to human disturbance should be taken into account in attempts to preserve its Hawaiian relative.

CAUSES OF THREAT: Only six possible sightings have been reported in the last 30 years. All monk seals are very sensitive to a disturbance. Their numbers were already greatly reduced by the end of the 19th century, particularly by hunting in their rookeries.

RESPONSES TO HABITAT MODIFICATION: Unknown.

DEMOGRAPHIC CHARACTERISTICS: Unknown.

KEY BEHAVIORS: Apparently, monk seals could be easily approached and killed. The report by Ward (1887) of their behavior on Arrecifes Triangulos may not be entirely typical, as they were about to give birth and also had heavy parasite infestations. He describes "the whole character of this seal is that of tropical inactivity" (Ward 1887:261).

CONSERVATION MEASURES TAKEN: Attempts have been made to locate vestiges of the seal population (Gilmore 1959; Kenyon 1977; LeBoeuf et al. 1986; Woods and Hermanson 1987).

CONSERVATION MEASURES PROPOSED: The measures proposed by the IUCN *Mammal Red Data Book* (Thornback and Jenkins 1982:412) are appropriate: "If a viable colony of seal were to be located, the following urgent steps are advocated: a) complete legal protection of the species by the government concerned, hauling-out grounds and adjacent waters being declared an inviolate refuge; b) posting of a full-time warden to ensure that the seals are not molested; and c) initiation by an experienced pinniped biologist of a detailed observational study of the colony."

Literature Cited

Caldwell, D. K., and M. C. Caldwell. 1978. West Indian monk seal. P. 49 *in* J. N. Layne, ed., Rare and endangered biota of Florida. Vol. 1. Mammals. University Presses of Florida, Gainesville. xx + 52 pp.

Cumbaa, S. L. 1980. Aboriginal use of marine mammals in the southeastern United States. Proceedings of the 30th Southeastern Archaeological Conference, Bulletin 17:6–10.

Gilmore, R. M. 1959. Is the West Indian monk seal extinct? Sea Frontiers 5:225–236.

Gray, J. E. 1850. Seals (part 2). *In* Catalogue of the specimens of Mammalia in the collection of the British Museum. London, Brit. Mus., 3 parts.

Gunter, G. 1947. Sight records of the West Indian seal, *Monachus tropicalis* (Gray), from the Texas coast. J. Mammal. 28:289–290.

Herrera, A. de. 1934–57. Historia general de los hechos de los Castellanos en las islas y tierrafirma del mar oceano. [A reprint in 17 volumes of the 1720 edition by La Real Academia de la Historia.] Tipografia de Archivos, Madrid. 1935, 3:324.

Kenyon, K. W. 1977. Caribbean monk seal extinct. J. Mammal. 58:97–98.

LeBoeuf, B. J., K. W. Kenyon, and B. Villa-Ramirez. 1986. The Caribbean monk seal is extinct. Marine Mammal Science 2:70–72.

Milanich, J. T. 1971. The Deptford Phase: An archaeological reconstruction. Unpubl. Ph.D. diss., University of Florida, Gainesville. 237 pp.

Miller, G. S., Jr. 1917. A hooded seal in Florida. Proc. Biol. Soc. Wash. 30:121.

Moore, J. C. 1953. Distribution of marine mammals to Florida waters. Amer. Midland Nat. 49:117–158.

Raun, G. G. 1964. West Indian seal remains from two historic sites in coastal south Texas. Bull. of the Texas Arch. Soc. 35:189–192.

Ray, C. E. 1961. The monk seal in Florida. J. Mammal. 42:113.

Rice, D. W. 1973. Caribbean monk seal (*Monachus tropicalis*). Pp. 98–112 *in* Seals. International Union for Conservation of Nature and Natural Resources, publ. New Series, Suppl. Papers 39:1–176.

Scheffer, V. B. 1958. Seals, sea lions, and walruses: A review of the Pinnipedia. Stanford University Press, Stanford, California 179 pp.

Thornback, J., and M. Jenkins, compilers. 1982. Caribbean monk seal. Pp. 411–412 *in* The IUCN mammal red data book, part 1. International Union for Conservation of Nature and Natural Resources, Gland, Switzerland.

Townsend, C. H. 1923. The West Indian seal. J. Mammal. 4:55.

Ward, H. L. 1887. Notes on the life history of *Monachus tropicalis*, the West Indian seal. Amer. Nat. 21:257–264.

Westermann, J. H. 1953. Nature preservation in the Caribbean. Found. for Scientific Research in Surinam and the Netherlands Antilles, Utrecht. 106 pp.

Wing, E. S., and L. J. Loucks. 1984. Grenada site faunal analysis. Pp. 259–345 *in* J. W. Griffin, ed., Excavations at the Grenada site. Florida Division of Archives, History and Records Management, Tallahassee, Florida. 412 pp.

Woods, C. A., and J. Hermanson. 1987. An investigation of possible sightings of Caribbean monk seals (*Monachus tropicalis*), along the north coast of Haiti. Report to U.S. Marine Mammal Commission, National Technical Information Service PB 87-164307, Washington, D.C. 10 pp.

Prepared by: Elizabeth S. Wing, Florida Museum of Natural History, University of Florida, Gainesville, FL 32611.

Recently Extirpated

Anastasia Island Population of Cotton Mouse
Peromyscus gossypinus gossypinus (in part)
FAMILY CRICETIDAE
Order Rodentia

TAXONOMY: The Anastasia Island population of the cotton mouse was named as *Peromyscus anastasae* by Bangs (1898) and assigned to *Peromyscus gossypinus anastasae* by Osgood (1909). Osgood also included in *P. g. anastasae* the cotton mouse populations from Cumberland Island, Camden County, Georgia, which had been named as *P. insulanus* by Bangs (1898). Occurrence of the same mouse on these two distant islands could be explained only as a relict distribution from a more extensive range in the past. Alternatively, such an arrangement may represent the lumping of relatively unrelated populations that are convergent in color, adapted to the light color of coastal soils.

Osgood referred only 18 specimens from Anastasia Island and 36 from Cumberland Island to *anastasae*. However, he examined *P. gossypinus* from other nearby areas, including St. Marys, Georgia; Amelia Island; Burnside Beach (on the coast of Duval County north of Anastasia Island); Carterville (on the mainland west of Anastasia Island); and Summer Haven (formerly El Peñon Island, now the point south of Matanzas Inlet). Osgood's remarks about *anastasae* (1909:141) offer perspective on the validity of the taxon: "Although the pale forms from Anastasia and Cumberland islands, respectively, are entirely isolated from each other and from the mainland forms, they seem to be absolutely alike and also are not different from certain aberrant (intermediate?) specimens from the mainland. Moreover, the mainland specimens most similar to them are not from localities immediately adjacent to the islands in question, specimens from St. Marys, Burnside Beach, etc., being typical *gossypinus*." Further doubt about this arrangement is cast by Osgood's remarks under *P. g. palmarius* (1909:140), whose range includes most of the Atlantic coast of Florida: "The type of *palmarius* and a very small percentage of the large series of topotypes are unusually pale and scarcely distinguishable from comparable specimens of *anastasae* . . . the great preponderance of dark specimens from the [*palmarius*] type locality [Oak Lodge, near Floridana Beach, Brevard County] tends to indicate that the type is probably an aberrant specimen rather than the representative of a well-defined form. The case might be construed also to the effect that pale coast forms are undergoing parallel differentiation at several points and that the same character (paleness) has been established independently on Anastasia and Cumberland islands and is only in its incipiency on the peninsula opposite Micco."

Recently, J. L. Boone, J. Laerm, and M. H. Smith (personal communication) evaluated genetic and cranial characters of cotton mice and concluded that neither the Anastasia nor Cumberland Island populations warrant recognition as separate subspecies, even though each is somewhat different from all others genetically. On this basis, the Anastasia Island population is treated here as a disjunct, undifferentiated population of the mainland subspecies, *P. g. gossypinus*.

DESCRIPTION: This is a large, dark brown, white-footed mouse. The Anastasia Island population is smaller and paler (more buffy in color) than typical mainland animals. The upper parts are brown on the mid-dorsal stripe and buff on cheeks, lower sides, inside of flanks, and base of tail; the underparts are grayish white; and the tail is dusky above and white below. Average measurements of seven adults are: total length 167.1 mm, tail length 69.5 mm, and length of hind foot 21.3 mm (Bangs 1898). To this description, Osgood (1909) added that the combined populations of Anastasia and Cumberland islands have upper parts "pale ochraceous buff rather lightly mixed with dusky, which is slightly or not at all concentrated in the mid-dorsal region." Osgood gave average measurements of three adults from Cumberland Island as total length 171.7 mm, tail length 68 mm, and length of hind foot 21.6 mm.

POPULATION SIZE AND TREND: Cotton mice on Anastasia Island apparently were not common when first encountered—Bangs (1898) took only 9 individuals. Elliot (1901, cited in Pournelle and Barrington 1953) reported 10 specimens taken at "Espanita," the home of a Mr. Middleton, located 2–3 mi (presumably north) from the type locality. Since 1901, at least five unsuccessful efforts have been made to find this population. Pournelle and Barrington (1953) failed to catch any in 306 trap-nights in 1948, even though the purpose of their trip was to obtain topotypes. Michael H. Smith (personal communication) failed to catch

Peromyscus gossypinus gossypinus (in part) 43

any in 3553 trap-nights in October 1978; this effort was directed at cotton mice and cotton rats, so trapping was in habitats other than dunes. Joshua Laerm (personal communication) failed to find this animal in approximately 5000 trap-nights on Anastasia Island in the early 1980s.

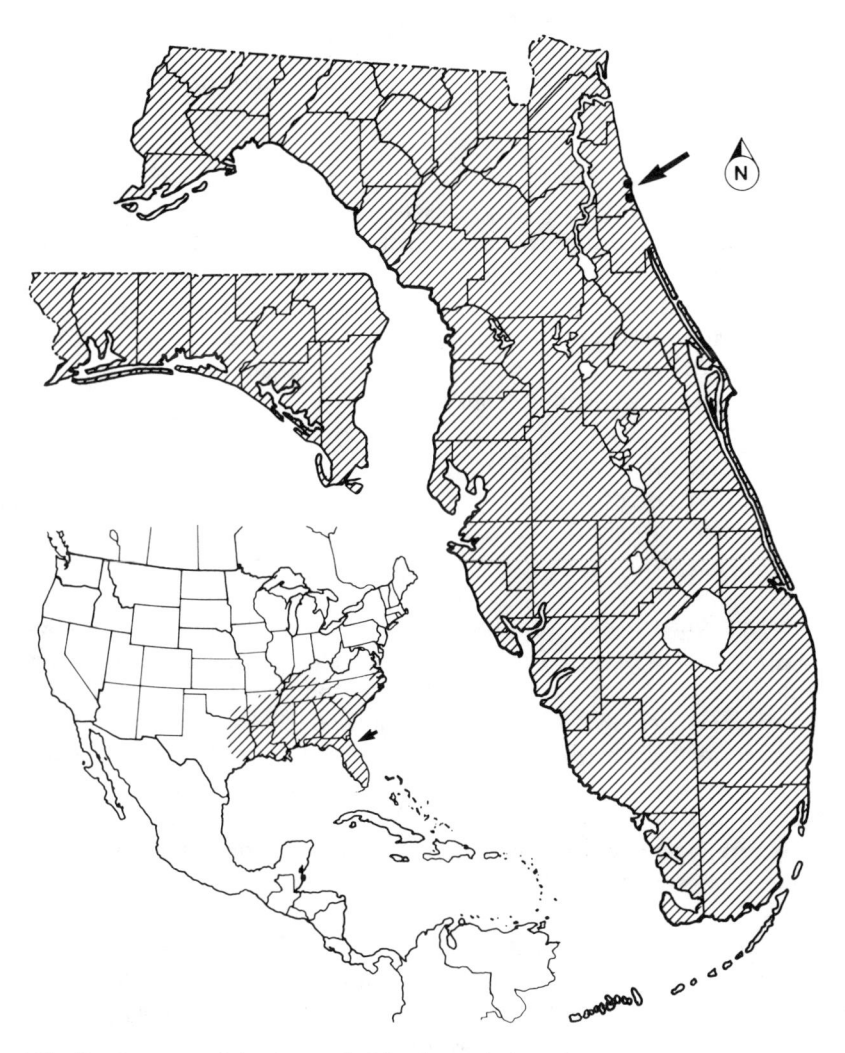

Distribution map of the Anastasia Island population of cotton mouse, *Peromyscus gossypinus gossypinus* (in part). Hatching represents the species; dots represent the subspecies.

Humphrey et al. (1988) failed to catch any in 880.5 trap-nights in suitable habitat. Phillip A. Frank obtained a single *P. gossypinus* from Anastasia Island in >10,000 trap-nights during 1989. This animal was not distinguishable from mainland individuals; perhaps it was brought from the mainland by human transportation. This evidence indicates that the original population on Anastasia Island has been extirpated for decades. According to the taxonomic arrangement established by Osgood, however, the subspecies survives, because cotton mice are extant and abundant on Cumberland Island.

DISTRIBUTION AND HISTORY OF DISTRIBUTION: The original distribution of *P. g. anastasae* was Anastasia Island. As revised by Osgood (1909), its distribution expanded northward as far as Cumberland Island, Camden County, Georgia. If one accepts this taxonomic arrangement, *P. g. anastasae* is apparently extirpated from Anastasia Island but remains widespread and common elsewhere in its range. If one rejects the current taxonomic arrangement, the Anastasia Island population of the cotton mouse appears to be extinct.

GEOGRAPHICAL STATUS: The geographic status of this population is uncertain because of the taxonomic problem presented above. It appears to be endemic to either one island or two, but other arrangements are possible.

HABITAT REQUIREMENTS AND HABITAT TREND: Cotton mice on Anastasia Island occurred mostly in thickets of wax myrtle (*Myrica cerifera*) and Spanish bayonet (*Yucca aloifolia*) and occasionally in the dune grassland (Bangs 1898). Based on habitat used by cotton mice elsewhere, the maritime live oak forest probably also was occupied. Suitable habitat remains on Anastasia Island, but in small, fragmented patches. Considering the range of cotton mice north of Anastasia Island, the available habitat is very extensive and includes coastal scrub, xeric oak forest, mesic forest, flatwoods, and hydric hammock.

VULNERABILITY OF SPECIES AND HABITAT: Although much of the habitat of cotton mice on Anastasia Island has been converted to human uses, the apparent extirpation of the population points to vulnerability of the species to other threats, the identity of which is uncertain.

CAUSES OF THREAT: Pournelle and Barrington (1953) cited development by man as the probable cause of the scarcity or absence of cotton mice at Anastasia Island. Perhaps more important is the fact that the remaining patches of suitable habitat now are occupied by black rats (*Rattus rattus*; Humphrey et al. 1988). Although we have no direct evidence of competition between these two species, a causal relation between colonization of *Rattus* and extinction of *Peromyscus* is possible.

RESPONSES TO HABITAT MODIFICATION: Unknown.

DEMOGRAPHIC CHARACTERISTICS: See the species account for the Key Largo cotton mouse, *Peromyscus gossypinus allapaticola*.

KEY BEHAVIORS: See the species account for the Key Largo cotton mouse, *Peromyscus gossypinus allapaticola*.

CONSERVATION MEASURES TAKEN: It once was being considered for listing by the U.S. Fish and Wildlife Service, but it no longer is being considered because of the evidence of extirpation from Anastasia Island and abundance on Cumberland Island.

CONSERVATION MEASURES PROPOSED: If the subspecies is extinct, management questions are moot. If a study of geographic variation in cotton mice confirms the current taxonomic arrangement, reintroduction of the subspecies from Cumberland Island to Anastasia Island could be considered. Research is needed on whether and under what conditions a strong competitive interaction occurs between black rats and cotton mice in coastal forests.

ACKNOWLEDGMENTS: I thank Philip A. Frank and Mark E. Ludlow for reviewing the manuscript. The author of the species account in the 1978 edition was Hans N. Neuhauser.

Literature Cited

Bangs, O. 1898. The land mammals of peninsular Florida and the coast region of Georgia. Boston Soc. Nat. Hist. Proc. 28:157–235.
Elliot, D. G. 1901. A list of mammals obtained by Thaddeus Surber in North and South Carolina, Georgia, and Florida. Field Columbian Mus., Zool. Ser. 3:39–57.
Humphrey, S. R., W. H. Kern, Jr., and M. E. Ludlow. 1988. The Anastasia Island cotton mouse (Rodentia: *Peromyscus gossypinus anastasae*) may be extinct. Florida Sci. 51:150–155.

Osgood, W. H. 1909. Revision of the mice of the American genus *Peromyscus*. North Amer. Fauna 28:1–285.

Pournelle, G. H., and B. A. Barrington. 1953. Notes on mammals of Anastasia Island, St. Johns County, Florida. J. Mammal. 34:133–135.

Prepared by: Stephen R. Humphrey, Florida Museum of Natural History, University of Florida, Gainesville, FL 32611.

Recently Extirpated

Plains Bison
Bison bison bison
FAMILY BOVIDAE
Order Artiodactyla

TAXONOMY: The systematics of bison have been controversial for at least a century. At present, two subspecies are recognized, the wood or mountain bison and the plains bison. Other names include American bison and buffalo.

DESCRIPTION: The bison is the largest terrestrial mammal in North America. Adults of the plains subspecies weigh 360–1000 kg and stand 150–190 cm at the shoulder (Reynolds et al. 1982). Males are larger and have more massive skulls than females. The bison is distinguished from true buffalo in having a high hump at the shoulders, while the hindquarters appear disproportionately small. It has a massive head, short neck, and short, up-curved horns. The forehead is broad and short, the muzzle is narrow, and the nasal bones are pointed and do not contact the premaxillae. The hair is brown to light brown, but that on the head, neck, and shoulders is a darker brownish-black. Long hair forms a beard on the chin, and the short tail also is tufted with long hair.

POPULATION SIZE AND TREND: The original population of bison in North America was estimated at 30–75 million animals (Seton 1900; McHugh 1972). The scattered literature on bison in Florida has been ably summarized by Rostlund (1960). He concluded that bison were rare or absent from Florida and the extreme Southeast during the 16th century but were common about a century later. Despite detailed accounts of the landscape, plants, and animals of the region produced from army expeditions of Hernando de Soto in the 1540s, Tristan de Luna in the 1550s, and Juan Pardo in the 1560s, not a single observation of a bison is recorded. (At best, bison were very rare in the region at the time: the

Plains bison, *Bison bison bison*. (Photo by Stephen R. Humphrey)

first historic records by Nicolas Le Challeux in 1565 near Fort Caroline, in the vicinity of Mayport, Duval County, and by d'Escalante Fontaneda in 1575 for the Tallahassee region, are of uncertain validity, although plausible in view of later reports.) After a century in which no eye-witness reports appeared, bison were reported commonly beginning in the 1670s. The conclusion that bison were absent from much of the Southeast during the early explorations by Europeans was also reached by Gallatin (1836) and Allen (1876).

The population of bison in the Southeast probably peaked in numbers about A.D. 1700. Reports of their abundance became commonplace—herds of 100 or more, and hides became an export commodity (Rostlund 1960).

By the latter part of the 18th century, bison again were becoming rare in the Southeast. For example, in 1773 William Bartram said that "buffiloe . . . is become scar[c]e in East Florida, yet there remain a few in the Point [the peninsula]." In contrast to many reports of abundant bison in Georgia in the early 1700s, William Bartram said of the Georgia piedmont in 1774, "The buffalo, once so very numerous, is not at this day to be seen in this part of the country" (citations in Rostlund 1960:399, 400). The last bison in Louisiana was shot in 1803, and the last one known on the Atlantic coast was shot near Brunswick, Georgia, in the early 19th century (citations in Rostlund 1960:399, 406).

Bison bison bison 49

DISTRIBUTION AND HISTORY OF DISTRIBUTION: The oldest clear records in Florida are by Rodrigo de la Barreda in 1674 and Gabriel Diaz Vara Calderon in 1675. Thereafter, numerous reports occurred. Most were from the panhandle. The farthest east was at Newnan's Lake, Alachua County (two killed in 1716, reported by Diego Pena; Boyd 1949). The

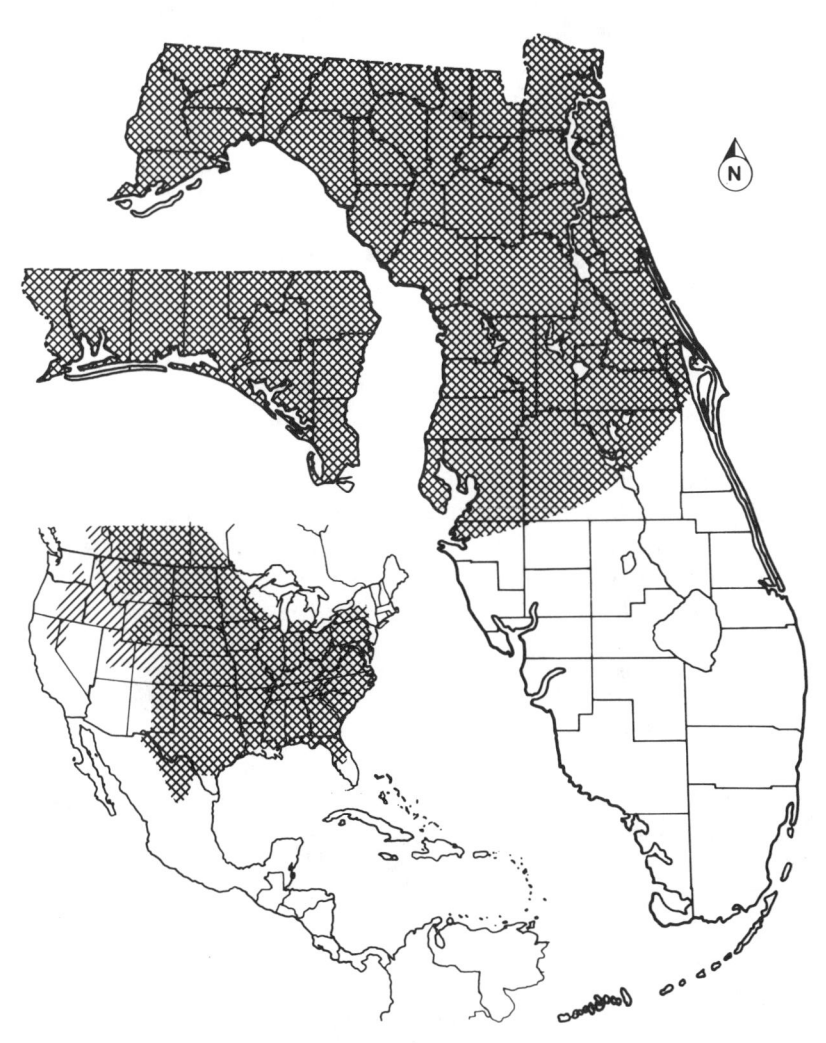

Distribution map of the plains bison, *Bison bison bison*. Hatching represents the species; crosshatching represents the subspecies.

farthest south was footprints of a small herd northeast of Tampa Bay, Hillsborough County, in 1772 (Romans 1775). The distributions of bison in Florida given in the monographic work of Allen (1876) and by Hall (1981) and Reynolds et al. (1982) are incorrect (Boyd 1936, 1937; Swanton 1938, 1941; Rostlund 1960). Some modern commentators reject the available evidence regarding bison distribution in peninsular Florida as unverifiable, but it seems unreasonable to ignore a substantial narrative documentation of such a distinctive animal, all by people familiar with domestic cattle, made centuries to decades before Audubon's and Bachman's introduction of natural science methodology to the region.

Former place-names embodying the term "buffalo" usually originated with the presence of the animal nearby (Allen 1876). In Florida, these sites include Sivelo or Buffalo Point in Tampa Bay (Gadsden Point), Buffalo Ford in Polk County (19 km east of Lake Hancock), Buffalo Bluff in Putnam County (7 km south of Palatka), and Buffalo Mill Creek in Santa Rosa County (18 km WNW of Milton). None of these toponyms are particularly old, providing further evidence of the late arrival of the species in the region (Rostlund 1960).

Despite an abundant archaeological record for the Southeast and Florida, remains of bison are few, and none are from early sites or horizons (Rostlund 1960). A gorget or pendant made from a bison femur was excavated near the surface of the Pierce Mound west of Apalachicola, Franklin County (Moore 1902). Part of a bison humerus and four carpals were found in an excavation of Fort Pupo, which existed from 1716 to 1740, 5 km south of Green Cove Springs, Clay County (Goggin 1951; Sherman 1954). Considering the impossibility of differentiating *Bison* from *Bos* (cattle) bones without a skull with good horn cores, however, it probably is just as well that no more archaeological evidence exists over which to argue.

B. bison is considered to be a modern species. All fossil records of bison (Webb and Wilkins 1984), beginning early in the Rancholabrean land mammal age, about 122,000 yBP, are of a different species, *B. antiquus*. A particularly interesting record is a piece of a paleoindian projectile point found in a *B. antiquus* skull dated at 11,000 yBP, in a late Pleistocene fossil site (Webb et al. 1984).

GEOGRAPHICAL STATUS: Rostlund (1960) concluded that the bison was a recent immigrant to Florida, arriving in synchrony with the depopulation of the aboriginal humans (between A.D. 1540 and 1700) and with the deforestation of the region.

HABITAT REQUIREMENTS AND HABITAT TREND: It is not entirely clear which habitats bison used in Florida. Presumably they could feed in all the habitats that will support livestock on open range.

VULNERABILITY OF SPECIES AND HABITAT: Many historical narratives report that bison in the Southeast and Florida were heavily hunted, including reports of extirpation and killing of the last animal in a region or state. The entire Southern Plains population (a deme) was extirpated in 1875.

CAUSES OF THREAT: None.

RESPONSES TO HABITAT MODIFICATION: Unknown in the region.

DEMOGRAPHIC CHARACTERISTICS: Most Plains bison reach sexual maturity when 2–3 years old, but some breed as yearlings. Gestation takes 270–300 days, and normally single calves are produced 2 of every 3 years. Typically mating occurs between July and October, and calves are born between April and June. Full body size is not reached until 6 years for males and 4 years for females. Longevity in the wild is poorly documented, but 15 years is considered quite old (Reynolds et al. 1982).

KEY BEHAVIORS: Rostlund (1960) calculated that the margin of the range of bison shifted about 400 mi in 100 years, from Tennessee to northern Florida. Such a rate of immigration is typical for mammals.

Bison are diurnal grazers, mainly of grasses and sedges, although in some places they eat some forbs and woody browse. They feed by biting off forage with forward-projecting incisor teeth that are present only in the lower jaw. The grazing of bison helps maintain shortgrass prairie and its associated fauna (Weaver and Clements 1938; Larson 1940; Krueger 1986). The bison's thick fur is ideally suited for dispersal of barbed, awned, or sticky seeds. Bison wallows create water catchments in flat terrain, which can still be seen in the southern Great Plains a century after the animals' extirpation there.

Bison are highly gregarious, forming matriarchal groups of 11–20 animals, bull groups ranging from solitary males to a few bulls, and larger breeding groups of about 10–500 animals. Before they were confined to nature preserves, bison typically undertook annual migrations of up to several hundred km between winter and summer ranges.

CONSERVATION MEASURES TAKEN: Substantial conservation measures have been taken in other portions of the species' range (Reynolds et al. 1982). Given that many authorities do not even realize that bison occurred in Florida, it is not surprising that little has been done there. A small herd has been established at Payne's Prairie State Preserve, Alachua County, from the population in the Wichita Mountains National Wildlife Refuge, Oklahoma. This was primarily a test to see if bison could maintain themselves on open range under minimal management. Because of a mineral deficiency in the soil, it became necessary to supplement the diet with minerals. From an original 10 animals, the herd grew to about 35, when they contracted brucellosis from cattle on a neighboring property. The herd was culled of infected animals and numbered six animals in 1989 (J. Gilliam, personal communication). Once bison become dense relative to range conditions, the impossibility of keeping them fenced in makes a serious reintroduction impractical in present-day Florida.

CONSERVATION MEASURES PROPOSED: None.

ACKNOWLEDGMENTS: I am grateful to James Weimer, Daniel K. Odell, and Davis S. Maehr for reviewing the manuscript. The author of the species account in the 1978 edition was Llewellyn M. Ehrhart.

Literature Cited

Allen, J. A. 1876. The American bisons, living and extinct. Memoirs Mus. Comparative Zool., Harvard Univ. 4(10):1–246.
Boyd, M. F. 1936. The occurrence of the American bison in Alabama and Florida. Science 84:203.
Boyd, M. F. 1937. The expedition of Marcos Delgado from Apalache to the Upper Creek country in 1686. Florida Historical Quart. 16:2–32.
Boyd, M. F. 1949. Diego Pena's expedition. Florida Historical Quart. 28:1–27.
Gallatin, A. 1836. A synopsis of the Indian tribes within the United States east of the Rocky Mountains, in the British and Russian possessions in North America. Archaeologia Americana 2:1–422.
Goggin, J. M. 1951. Fort Pupo: A Spanish frontier outpost. Florida Historical Quart. 30:139–192.
Hall, E. R. 1981. The mammals of North America. 2d ed. John Wiley and Sons, New York. 2:601-1181 + 90 pp.
Krueger, K. 1986. Feeding relationships among bison, pronghorn, and prairie dogs: An experimental analysis. Ecology 67:760–770.

Larson, F. 1940. The role of the bison in maintaining the shortgrass plains. Ecology 21:113–121.

McHugh, T. 1972. The time of the buffalo. Alfred A. Knopf, New York. 339 pp.

Moore, C. B. 1902. Certain aboriginal remains of the northwest Florida coast, part 2. J. Acad. Natural Sci. Philadelphia, second series, 12:127–355.

Reynolds, H. W., R. D. Glaholt, and A. W. L. Hawley. 1982. Bison. Pp. 972–1007 *in* J. A. Chapman and G. A. Feldhamer, eds., Wild mammals of North America. Johns Hopkins University Press, Baltimore. 1147 pp.

Romans, B. 1775. A concise natural history of east and west Florida. New York. 1–89 pp., 1–342 pp.

Rostlund, E. 1960. The geographic range of the historic bison in the Southeast. Annals Assoc. Amer. Geographers 50:395–407.

Seton, E. T. 1900. Lives of game animals. 8 vols. Reissued 1953, by Charles T. Branford Co., Boston.

Sherman, H. B. 1954. The occurrence of bison in Florida. Quart. J. Florida Acad. Sci. 17:228–232.

Swanton, J. R. 1938. Notes on the occurrence of bison near the Gulf of Mexico. J. Mammal. 19:379–380.

Swanton, J. R. 1941. Occurrence of bison in Florida. J. Mammal. 22:322.

Weaver, J. W., and F. E. Clements. 1938. Plant ecology. McGraw-Hill Book Co., New York. 413 pp.

Webb, S. D., and K. T. Wilkins. 1984. Historical biogeography of Florida Pleistocene mammals. Pp. 370–383 *in* H. H. Genoways and M. R. Dawson, eds., Contributions in Quaternary vertebrate paleontology: A volume in memorial to John E. Guilday. Spec. Publ. Carnegie Mus. 8:1–538.

Webb, S. D., J. T. Milanich, R. Alexon, and J. S. Dunbar. 1984. A *Bison antiquus* kill site, Wacissa River, Taylor County, Florida. Amer. Antiquity 49:384–392.

Prepared by: Stephen R. Humphrey, Florida Museum of Natural History, University of Florida, Gainesville, FL 32611.

Endangered

Indiana Bat
Myotis sodalis
FAMILY VESPERTILIONIDAE
Order Chiroptera

TAXONOMY: The Indiana bat was described by Miller and Allen (1928) and contains no subspecies (Hall 1962; LaVal 1970). The species also is called the social bat because it hibernates in dense clusters, although many other species also form clusters.

DESCRIPTION: This species is difficult for the casual observer to differentiate from many other species of its genus, but it has many distinctive features. The fur is a dull chestnut-gray color dorsally (not glossy), and the fur around the shoulder is a darker umber. The ventral fur is buffy- to grayish-white. The fur is darker at the base than at the tips. The skin of the rostral area is pinkish-brown (useful for inspecting clustered animals during hibernation). The foot is short (9 mm), and the hairs between the toes are short and sparse, never extending as far as the tips of the claws. The calcar has a well-developed keel. The ear measures 15 mm, and the tragus is not particularly long, measuring slightly less than half the length of the ear. The forearm is 39 (range 35–41) mm long. Other useful measurements are: breadth of braincase 6.4–7.2 mm, interorbital breadth 3.3–4.3 mm, and mass 6 g (up to 9 g if pregnant or laden with fat for hibernation; Hall 1962; Barbour and Davis 1969; Thomson 1982).

POPULATION SIZE AND TREND: The Indiana bat is abundant. The populations known in 1980–81 were estimated to total about 550,000 individuals (Richter et al. 1978; U.S. Fish and Wildlife Service 1983).

The earliest population estimate of 640,000 animals in 1954–66 (Humphrey 1978) probably was too low, because it assumed that populations discovered thereafter had been stable, whereas some probably had declined. The true population trend is unknown, because at least until the

Myotis sodalis

Indiana bat, *Myotis sodalis*. (Photo by Stephen R. Humphrey)

early 1980s new winter populations were being discovered. However, many known populations have decreased in size (Humphrey 1978; Brady 1980; U.S. Fish and Wildlife Service 1983). It seems reasonable to conclude that the trend is substantially downward.

DISTRIBUTION AND HISTORY OF DISTRIBUTION: The Indiana bat is widespread in several midwestern states, especially Missouri, Illinois, Indiana, Ohio, and Kentucky, and it extends as far west as eastern Oklahoma and Kansas, as far north as southern Wisconsin and Michigan, southeasternmost Ontario, and southern New England, and as far south as central Arkansas and the Florida panhandle. It appears to be uncommon in the eastern and southern portions of its range.

In Florida, the Indiana bat is known from only one specimen taken in October 1955 in Old Indian Cave, Florida Caverns State Park, Jackson County (Jennings and Layne 1957). Whether the species occurs in the state regularly or only during the breeding season is unknown.

GEOGRAPHICAL STATUS: The range of this species is peripheral to Florida. The distributional status in Florida is unknown; possibilities range from occasional visits to permanent residence.

HABITAT REQUIREMENTS AND HABITAT TREND: Winter habitat consists of caves and mines that have cool and stable temperatures. To achieve a metabolic rate low enough to ensure that finite fat reserves outlast the

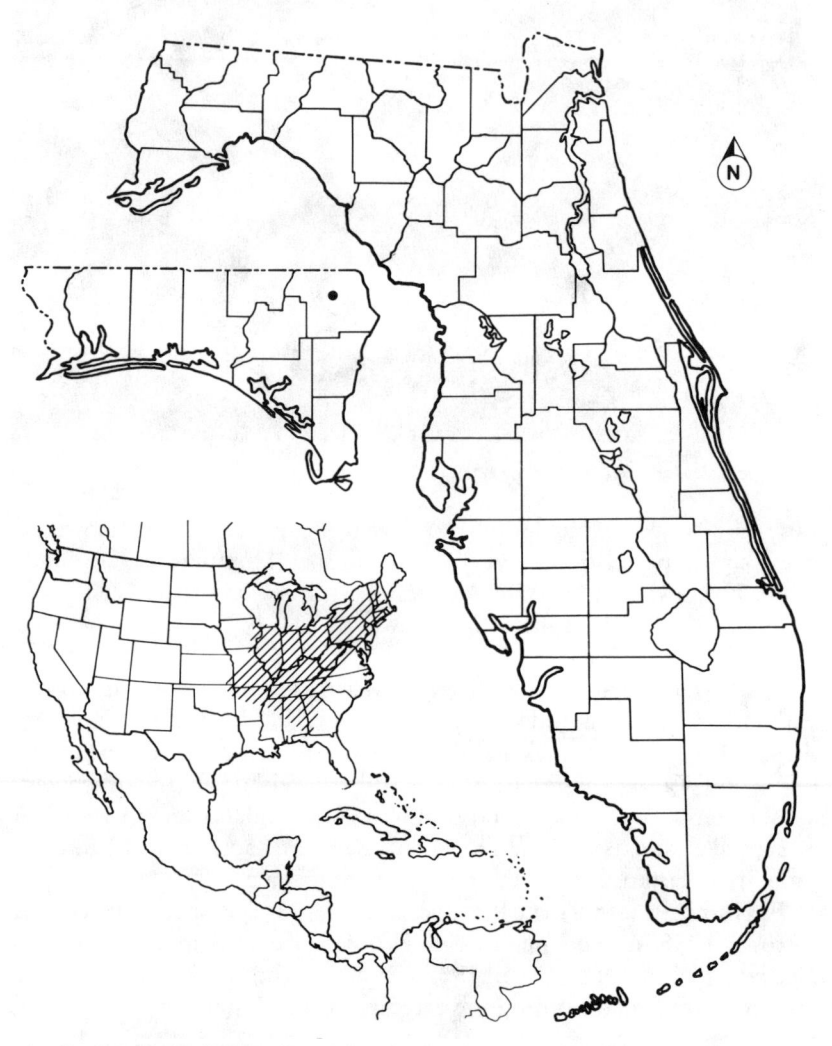

Distribution map of the Indiana bat, *Myotis sodalis*. Hatching represents the species; the dot represents its only Florida record.

foodless winter period, the species selects sites that are as cool as possible yet warm enough to preclude freezing. Freezing would cause either death or excessive energy use for arousal and relocation. Hibernation sites proven by the test of survival are used every year, often throughout an individual's lifetime. Body temperature is controlled mainly by radiative heat exchange with the rock on which the bats hang (McNab 1974). The species has been found at sites with rock temperature as high as 17°C, but the preferred range is 4–8°C (Hall 1962; Myers 1964; Humphrey 1978). Winter cave temperatures fall below regional mean-annual temperature only where cold air is trapped in cave passages that lie below entrances, so only certain caves have configurations providing suitable habitat (Humphrey 1978). Florida's Old Indian Cave is suitable because its three entrances allow free flow of cold winter air, and temperature at the bat roost reaches 7.8°C, instead of the 20.0–22.8°C typical of Florida caves. Probably very few Florida caves can be used by this species for hibernation.

Small nursery populations roost under loose bark on dead trees during summer. Tree roosts are occupied each year until they decay too much to be reused. Optimum foraging habitat appears to be the foliage and air space in and under crowns of riparian and floodplain trees (Humphrey et al. 1977). Colonies and foraging bats usually are found near small to medium-sized streams and drainage ditches; streams without riparian trees appear to be unsuitable (Cope et al. 1978). The best sites have mature trees on both banks, overhanging the stream by more than 3 m. The bats forage at a height of 2–30 m. The diet is mostly moths and aquatic insects (Belwood 1979; LaVal and LaVal 1980; Brack and LaVal 1985). The population studied by Humphrey et al. (1977) occupied a linear strip of creek 0.82 km long, or about 60 adults and young per km. The foraging area was 1.5–4.5 ha, and foraging density was 11–17 individuals per ha. A few Indiana bats have been found foraging in upland habitats (Easterla and Watkins 1969; LaVal et al. 1977; LaVal and LaVal 1980; Bowles 1980).

VULNERABILITY OF SPECIES AND HABITAT: The species is extraordinarily vulnerable to catastrophes from natural or human causes, because individuals are highly concentrated in a few caves during hibernation (Humphrey 1978). Nearly half the total population winters in just two caves, and 85 percent occur in seven (U.S. Fish and Wildlife Service 1983). Just as this winter habitat is unusually vulnerable, however, it is unusually accessible to conservation action. Summer habitat of the species also is vulnerable to loss or degradation by human uses.

CAUSES OF THREAT: Indiana bats face natural hazards, notably drowning when caves flood during hibernation (Hall 1962; DeBlase et al. 1965), freezing of hibernation sites (Humphrey 1978), and collapse of mine ceilings (Brady 1980). Such events can be catastrophic, as when an estimated 300,000 Indiana bats died in a flood of Bat Cave, Mammoth Cave National Park, Kentucky. Nonetheless, the species probably can withstand these incidental hazards indefinitely.

Instead, the major cause of threat is a variety of human activities. The most serious are losses of winter habitat caused by construction of doors, walls, and even buildings at cave entrances to control human access. These structures prevent access by bats, and they block or impede air circulation, changing roost temperatures. Known cases were estimated to have caused about half the population decline reported by Humphrey (1978).

Outright killing of large numbers of hibernating bats by people has been reported (Humphrey 1978). Probably this practice has become rare, as the public has become better informed and bat caves have been targeted for wildlife management.

There is some evidence that banding by biologists and passive observation by cave enthusiasts affect bats adversely (Humphrey 1978). Human visits under low clusters of Indiana bats cause loss of irreplaceable fat reserves, when the bats produce metabolic heat, arouse, fly, and recluster. Careful cavers who leave the hibernation area after a brief visit do not observe this effect and wrongly assume that they have had no effect on the bats. Handling by biologists presumably causes greater levels of fat loss.

The other major effect of humans on Indiana bats is destruction of summer habitat by deforestation and stream channelization (Humphrey et al. 1977; U.S. Fish and Wildlife Service 1983). The extent of this damage has not been estimated, but it is considerable. For example, 30 percent of the 18,737 km of interior streams of Illinois were channelized by 1976 and an additional 15.5 percent were scheduled to be (Conlin 1976).

RESPONSES TO HABITAT MODIFICATION: Negative responses to habitat modification have been documented fairly well, whereas positive responses to management practices are largely unknown. Loss of populations to degraded winter microclimate include about 95,000 Indiana bats at Coach Cave, Kentucky, about 19,000 at Colossal Cave, Kentucky, and an uncertain number at Wyandotte Cave, Indiana (Humphrey 1978). During the 95,000-bat decline at Coach Cave, populations of Indiana bats at nearby caves in Mammoth Cave National Park increased by about 13,000; unless most individuals moved to sites unknown to humans, they failed to survive (Humphrey 1978). Populations in several hibernacula in Missouri declined despite installation of gates designed to allow free air circulation and access by bats, perhaps because of too-frequent visits to census the bats (U.S. Fish and Wildlife Service 1983).

DEMOGRAPHIC CHARACTERISTICS: Nursery populations consist of adult females and their young. The three populations that have been found ranged from 50 to 100 individuals, including young (Humphrey et al. 1977; Cope et al. 1978). No adult males occur in these roosts, and few are netted in nearby foraging areas. Each female bears a single young. In one study, the young incurred 8 percent mortality from birth to weaning (Humphrey et al. 1977). In ordinary circumstances, adults have high survival rates and long lives. Females appear to live longer than males; maximum recorded life span is 14.8 years for a female and 13.5 years for a male (Humphrey and Cope 1977). Humphrey et al. (1977) reported that a cold summer delayed the recruitment of flying young by 2.5 weeks and the completion of migration by 3 weeks, exposing bats to freezing weather at the nursery and possibly affecting mortality, autumn mating, or fat storage for winter.

KEY BEHAVIORS: The Indiana bat is a migratory species, traveling as far as 483 km between winter and summer habitats. Therefore two types of habitat are required to complete an annual cycle.

Great numbers of Indiana bats swarm about cave entrances nightly during late August, September, and early October (Cope and Humphrey 1977). Despite much activity, few bats roost in the caves early in this period. Most breeding occurs at this time (Hall 1962), although some additional breeding takes place during winter and as the bats leave caves in spring. Individuals that are widely dispersed in small summer populations, by aggregating to mate at a few sites, readily find mates and avoid inbreeding.

Females enter hibernation soon after arrival at the caves, fat storage, and mating. Occupation of the caves is uninterrupted from November through March. All the females depart for summer habitat in April (Cope and Humphrey 1977). Many males remain until May, and a few occupy these caves through the summer (Hall 1962).

Riparian habitat is occupied from mid-May to mid-September. One intensively studied roost was a dead tree exposed to the sun and hence warm during sunny days. On some occasions, the bats moved to a less exposed tree nearby, which stayed warmer during cold nights. Parental care included a circling behavior around the roost tree that appeared to involve communication with young, carrying young from one place to another in the tree, and tandem flights to foraging areas of mothers with newly mobile young (Humphrey et al. 1977).

Although divided into numerous winter and summer populations, the species is organized in a very few demes (discrete interbreeding populations), of which three have been delimited. Indiana bats wintering in Missouri spend the summer in Missouri and southern Iowa (Myers 1964). Those wintering in north-central Kentucky reside in central Kentucky, Indiana, and southwestern Michigan in summer; Indiana bats wintering in northeastern Kentucky summer in northeastern Kentucky, western Ohio, and southern Michigan (Barbour and Davis 1969). These are the entities around which management programs should be designed.

CONSERVATION MEASURES TAKEN: This species is listed as endangered by the Florida Game and Fresh Water Fish Commission (1990) and the U.S. Fish and Wildlife Service (1989). A recovery plan for the species has been completed and approved by the U.S. Fish and Wildlife Service (1983). Winter habitat thermally altered by structures at cave entrances can be restored simply by removing the structures, and many have been. Old Indian Cave, the only known locality for this species in Florida, is protected by state park ownership and monitored as a hibernaculum of the endangered gray bat, *Myotis grisescens*.

CONSERVATION MEASURES PROPOSED: The primary need is to gain control of important hibernacula and protect them from disturbance by humans. A secondary need is to protect and restore foraging habitat, especially riparian forest. A routine program to monitor population numbers is needed to measure the effectiveness of protective efforts (U.S. Fish and Wildlife Service 1983).

The Indiana Bat Recovery Plan establishes criteria for reclassifying the species, as goals for recovery. Reclassification from endangered to threatened would be permitted if seven of the eight top-priority caves are protected and if populations are shown to be stable or increasing in three consecutive census periods. The species could be delisted altogether if the above criteria are met and if 50 percent of the second-priority caves are protected and if their populations are shown to be stable or increasing in three consecutive census periods. Censuses are to be conducted every two years, between mid-January and mid-February (U.S. Fish and Wildlife Service 1983). Humphrey (1978) recommended that a management program should consider each regionally distinct breeding population separately, but the recovery plan does not do so. Needs for the major regional populations are well defined by the available information, but needs for other regional populations should be defined by banding studies.

The Florida portion of the total population, as presently known, is so small that no conservation measures are proposed.

ACKNOWLEDGMENTS: I thank Jacqueline J. Belwood and Jeffrey A. Gore for suggesting improvements to the manuscript. The authors of the species account in the 1978 edition were Stephen R. Humphrey and Sylvia J. Scudder.

Literature Cited

Barbour, R. W., and W. H. Davis. 1969. Bats of America. University Press of Kentucky, Lexington. 286 pp.

Belwood, J. J. 1979. Feeding ecology of an Indiana bat community with emphasis on the endangered Indiana bat, *Myotis sodalis*. Unpubl. M.S. thesis, University of Florida, Gainesville. 103 pp.

Bowles, J. B. 1980. Preliminary report, ecological studies on the Indiana bat in south-central Iowa in summer 1980. Unpubl. report to Iowa Conservation Commission, Des Moines.

Brack, V., Jr., and R. K. LaVal. 1985. Food habits of the Indiana bat in Missouri. J. Mammal. 66:308–315.

Brady, J. T. 1980. Status and management of the Indiana bat (*Myotis sodalis*). Proceedings of the Fifth National Cave Management Symposium. Mammoth Cave National Park, Park City, Kentucky. 7 pp.

Conlin, M. 1976. Stream channelization in Illinois—1976 update. Illinois Dept. of Conserv., Springfield.

Cope, J. B., and S. R. Humphrey. 1977. Spring and autumn swarming behavior in the Indiana bat, *Myotis sodalis*. J. Mammal. 58:93–95.

Cope, J. B., A. R. Richter, and D. A. Seerley. 1978. A survey of bats in Big Blue Lake project area in Indiana. Unpubl. report to U.S. Army Corps of Engineers, Louisville, Kentucky. 51 pp.

DeBlase, A. F., S. R. Humphrey, and K. S. Drury. 1965. Cave flooding and mortality in bats in Wind Cave, Kentucky. J. Mammal. 46:96.

Easterla, D. A., and L. C. Watkins. 1969. Pregnant *Myotis sodalis* in north-western Missouri. J. Mammal. 50:372–373.

Florida Game and Fresh Water Fish Commission. 1990. Official lists of endangered and potentially endangered fauna and flora in Florida. Tallahassee, Florida. 23 pp.

Hall, J. S. 1962. A life history and taxonomic study of the Indiana bat, *Myotis sodalis*. Sci. Publ., Reading Public Mus. and Art Gallery 12:1–68.

Humphrey, S. R. 1978. Status, winter habitat, and management of the endangered Indiana bat, *Myotis sodalis*. Florida Sci. 41:65–76.

Humphrey, S. R., and J. B. Cope. 1977. Survival rates of the endangered Indiana bat, *Myotis sodalis*. J. Mammal. 58:32–36.

Humphrey, S. R., A. R. Richter, and J. B. Cope. 1977. Summer habitat and ecology of the endangered Indiana bat, *Myotis sodalis*. J. Mammal. 58:334–346.

Jennings, W. L., and J. N. Layne. 1957. *Myotis sodalis* in Florida. J. Mammal. 38:259.

LaVal, R. K. 1970. Infraspecific relationships of bats of the genus *Myotis austroriparius*. J. Mammal. 51:542–552.

LaVal, R. K., R. L. Clawson, M. L. LaVal, and W. Caire. 1977. Foraging behavior and nocturnal activity patterns of Missouri bats, with emphasis on the endangered species *Myotis grisescens* and *Myotis sodalis*. J. Mammal. 58:592–599.

LaVal, R. K., and M. L. LaVal. 1980. Ecological studies and management of Missouri bats, with emphasis on cave-dwelling species. Missouri Dept. Conserv., Terrestrial Series 8:1–53.

McNab, B. K. 1974. The behavior of temperate cave bats in a subtropical environment. Ecology 55:943–958.

Miller, G. S., Jr., and G. M. Allen. 1928. The American bats of the genera *Myotis* and *Pizonyx*. Bull. U.S. National Mus. 144:1–128.

Myers, R. F. 1964. Ecology of three species of myotine bats in the Ozark Plateau. Unpubl. Ph.D. diss., University of Missouri, Columbia. 210 pp.

Richter, A. R., D. A. Seerley, J. B. Cope, and J. H. Keith. 1978. A newly discovered concentration of hibernating Indiana bats, *Myotis sodalis*, in southern Indiana. J. Mammal. 59:191.

Thomson, C. E. 1982. Myotis sodalis. Mammalian Species 163:1–5.

U.S. Fish and Wildlife Service. 1983. Recovery plan for the Indiana bat. Fish and Wildlife Reference Service, 1776 E. Jefferson Street, 4th Floor, Rockville, MD 20852.

U.S. Fish and Wildlife Service. 1989. Endangered and threatened wildlife and plants. U.S. Government Printing Office, Washington, D.C. 34 pp.

Prepared by: Stephen R. Humphrey, Florida Museum of Natural History, University of Florida, Gainesville, FL 32611.

Endangered

Gray Bat
Myotis grisescens
FAMILY VESPERTILIONIDAE
Order Chiroptera

TAXONOMY: Also known as gray myotis, cave bat, or Howell's bat. No subspecies have been named.

DESCRIPTION: The gray bat's unicolored dorsal fur clearly distinguishes it from similar species of the southeastern United States, all of which have bi- or tri-colored fur. Fur color is uniformly gray after the annual molt in mid-summer, but it may bleach to chestnut-brown or russet between molts (especially in May and June in adult females). The wing membrane connects to the foot at the ankle, rather than at the base of the toe as in other species of *Myotis*. The cartilaginous rod, or calcar, behind the ankle has no ridge of skin, or keel. Total length ranges from 80 to 96 mm and forearm length ranges from 40 to 46 mm. Gray bats can weigh from 7 to 16 g, but 8 to 10 g is most common.

POPULATION SIZE AND TREND: Accurately estimating the size of gray bat populations is very difficult (Tuttle 1979). The best measure of current population size may be the estimate of 1.6 million gray bats in the nine largest hibernacula (Brady et al. 1982). Current Florida populations are estimated to consist of approximately 10,000 individuals in summer and a few hundred in winter (Wenner 1984; J.A. Gore, unpubl. data).

Throughout its range the gray bat has declined in number by at least 50 percent since about 1965 (Brady et al. 1982). A census of 22 caves in Tennessee and Alabama showed a 54 percent decline in just six years (Tuttle 1979). Population size is probably still declining, but numerous efforts to protect colonies from disturbance are slowing the rapid decline (Tuttle 1987).

The Florida population of gray bats may be relatively stable now, but

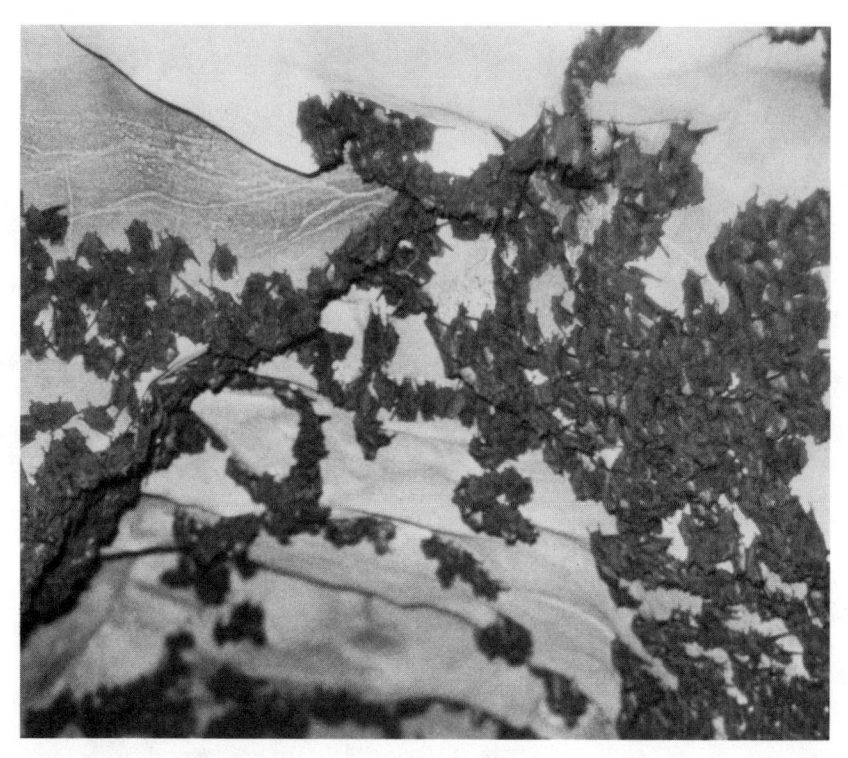

Gray bat, *Myotis grisescens*. (Photo by Stephen R. Humphrey)

only because the primary maternity colony (Judge's Cave) has been protected. This colony roosts over deep water along with a large colony of southeastern bats, so an accurate census of the gray bats is not practical. A lactating female gray bat and a newly volant juvenile were captured at the cave in 1987, and observations of emerging bats indicated that more than 50,000 *Myotis* spp. were present during the summer (J. A. Gore, unpubl. data). In 1970 when the total population of *Myotis* spp. was estimated to contain 50,000 to 100,000 *Myotis* spp., approximately 10,000 gray bats were thought to be present (Wenner 1984). Therefore, the current population size is assumed to be of that magnitude.

Historically, many thousands of gray bats hibernated in Florida in Old Indian Cave (Rice 1955; Lee and Tuttle 1970). After disturbance in the cave caused the bats to nearly disappear in the 1970s, the winter population in Florida has slowly increased to a few hundred individuals (O. G. Brock, personal communication).

Myotis grisescens 65

DISTRIBUTION AND HISTORY OF DISTRIBUTION: The gray bat occurs in portions of the midwestern and southeastern United States that have a karst topography with underlying limestone caverns. Populations are found mainly in Alabama, Tennessee, Kentucky, and Missouri, but they also occur in portions of adjacent states. In Florida major colonies have been found only in caves in Jackson County. Small colonies may exist in

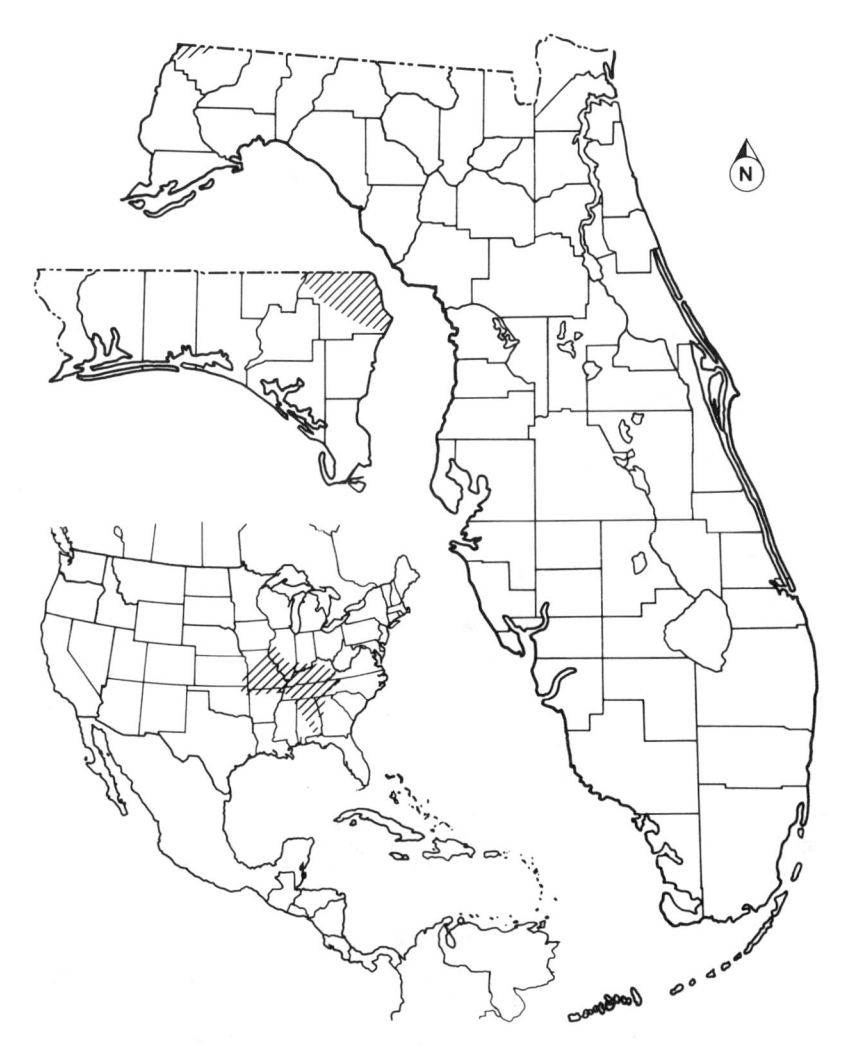

Distribution map of the gray bat, *Myotis grisescens*.

caves in adjacent counties, and occasionally individuals are found outside this restricted range (Humphrey and Tuttle 1978).

GEOGRAPHICAL STATUS: Because the gray bat inhabits only areas having suitable caves and nearby lakes or streams (Tuttle 1976), it likely has never been distributed evenly across its range. Its distribution has become more patchy as human disturbance has caused the bats to abandon many caves (Tuttle 1979). The colonies of gray bats in Florida represent the southernmost extent of the species' range. Although these colonies are located far south of much of the species' range, most of the gray bats that summer in Florida migrate each fall to northern Alabama and central Tennessee to hibernate in particular caves (Tuttle 1976).

HABITAT REQUIREMENTS AND HABITAT TREND: Gray bats are almost exclusively cave-dwelling and are narrowly restricted to particular cave environments. In summer they roost in warm caves (13–26°C), but they move to colder ones (6–11°C) to hibernate (Tuttle 1975, 1976). Because caves having stable, cold temperatures are rare, as much as 95 percent of the total population of gray bats is believed to hibernate in just nine caves (Tuttle 1979).

Available roosting and hibernating habitat has declined as caves have been flooded by man-made reservoirs, lighted for commercial tours, or sealed shut for various reasons (Tuttle 1979). Natural flooding and "cave-ins" may also result in the loss of cave habitat. The number of caves lost in these ways is not known; however, since 1967 at least three gray bat maternity caves in Florida have been filled or closed (Wenner 1984).

Gray bats forage primarily over water, especially over streams or pools bordered by forests (Tuttle 1976; LaVal et al. 1977). Foraging gray bats have been found more than 25 km from their cave roost (LaVal et al. 1977). Gray bats may use corridors of forest cover when traveling between roosts and foraging sites (Tuttle 1979). Availability of foraging habitat is not believed to be limiting populations of gray bats, but few quantitative data exist regarding the species' requirements for foraging habitat (Brady et al. 1982).

VULNERABILITY OF SPECIES AND HABITAT: The gray bat is highly specialized for particular cave habitats and is loyal to certain caves (Tuttle 1979). This bat is also sensitive to noise, lights, and other human disturbance (Tuttle 1979; Brady et al. 1982). Human intrusion into hibernacula may cause bats to expend their limited supply of stored fat, resulting

Myotis grisescens

in higher mortality levels. Disturbance of summer colonies can cause bats to abandon the cave or cause nonvolant young to drop to the cave floor.

Gray bats are potentially vulnerable to pesticide poisoning and loss of food sources due to pesticides, siltation, and other pollution. Forest clearing, stream impoundment, and other habitat changes also may affect gray bat populations. The impact of these potentially adverse effects, however, is currently outweighed by the losses caused by human disturbance of the cave habitat (Tuttle 1979).

CAUSES OF THREAT: Tuttle (1979) determined that loss of suitable caves for roosting was the greatest threat to the gray bat. Caves become unavailable to gray bats not only if entrances are blocked, but also if humans enter the cave frequently. Technological advances and an increasing human population probably have hastened the loss of caves through land development, commercialization for tourists, and disturbance by vandals, amateur cave explorers, and careless biologists.

RESPONSES TO HABITAT MODIFICATION: As noted above, gray bats are sensitive to modification of their normally stable cave habitat. In several instances even a gate erected to keep people out of a cave has caused gray bats to abandon the site (Tuttle 1977; Brady et al. 1982). The response of gray bats to modifications of their habitat outside the caves is less well understood. For example, gray bats use forests for cover and, sometimes, for foraging (LaVal et al. 1977; Tuttle 1979), but it is not known how their populations might respond to changes in forest cover. It is also not known whether impoundment of the streams over which the bats feed would significantly affect local populations.

DEMOGRAPHIC CHARACTERISTICS: Although gray bats copulate at their wintering caves just prior to the start of hibernation, fertilization is delayed until spring (Guthrie 1933). Pregnant females give birth to a single young in late May or early June. Neither males nor females breed until their second year (Guthrie 1933; Miller 1939; Tuttle 1976).

Gray bats have been known to survive 18 years in the wild (M. D. Tuttle, personal communication), and they are predicted to be able to live past 30 years (Stevenson and Tuttle 1981). However, mortality is high among juveniles and mean lifespan is probably less than five years.

KEY BEHAVIORS: In autumn gray bats migrate from relatively warm summer caves to colder caves, where they hibernate for about six months.

Bats that summer in Florida migrate as far as 500 km to hibernate in cold caves in central Tennessee and northern Alabama (Tuttle 1976). Relatively few gray bats, mostly adult males and yearlings, hibernate in Florida, in Old Indian Cave.

Gray bats return to Florida in late March. Pregnant females roost in "maternity" or "nursery" colonies in certain warm caves, while most males and yearling females roost in other nearby caves. The bats emerge each evening and follow corridors of trees to streams and lakes where they forage for flying insects (LaVal et al. 1977). The bats apparently defend feeding territories above the water.

CONSERVATION MEASURES TAKEN: In 1976 the gray bat was listed as endangered by the U.S. Fish and Wildlife Service. Since then private groups and government agencies have purchased or otherwise protected several important maternity and hibernation caves throughout the species' range (Tuttle 1987).

Judge's Cave, which harbors the largest maternity colony of gray bats in Florida, was purchased by the Nature Conservancy in 1982 for sale to the Florida Game and Fresh Water Fish Commission (Wenner 1984). Included in the purchase was a 14-ha corridor between the cave and the Chipola River. The commission has erected a fence around the cave entrance and monitors bat use of the cave.

Also in 1982, the commission designated Gerome's Cave as a "critical wildlife area," a legal designation that allows a site to be closed to entry. Since that time, this formerly important maternity cave has been posted with signs prohibiting entry.

Old Indian Cave in Florida Caverns State Park served as a bachelor cave for about 20,000 *Myotis* spp. and as a hibernaculum for a much larger number before human disturbance in the 1960s caused all but a few hundred bats to abandon the cave (Lee and Tuttle 1970). The cave's entrances were gated in 1970 to prevent public entry; unfortunately, the gates also apparently made the caves unsuitable for the bats. Surveys in 1974–75 found only two wintering gray bats (Humphrey and Tuttle 1978; Wenner 1984). Summer surveys in 1981 revealed fewer than five *Myotis* spp., and winter surveys that year found approximately 150 hibernating individuals (Wenner 1984).

The gate at the main entrance was removed in 1982 and the entrance fenced. As a result, bats appear to be returning to the cave. Fifty *Myotis* spp. were seen during a survey in summer 1983 and 98 in July 1984 (O.G. Brock, personal communication). On 17 June 1987, 1428 *Myotis* spp. were counted emerging from the cave (J.A. Gore, unpubl. data).

About 1000 *M. austroriparius* and 150 *M. grisescens* were hibernating there in February 1986 (O. G. Brock, personal communication), and several hundred *M. grisescens* were there in two subsequent winters (J. A. Gore, unpubl. data).

CONSERVATION MEASURES PROPOSED: Periodic surveys of bats at Old Indian Cave and Judge's Cave should be continued to monitor Florida's gray bat population. New techniques for more accurately censusing gray bats at the caves should be evaluated. Entry into the caves, even by biologists, however, must continue to be severely restricted (Brady et al. 1982). Maintenance of the gates and fences at the caves should, of course, be a regular part of management of the species.

Further protection for Gerome's Cave may be needed because, after being posted for >5 years, the cave is still used only irregularly by bats (J. A. Gore, unpubl. data). If disturbance is the reason the cave is not being used, efforts should be made to fence or gate the cave.

Protection of other secondary gray bat roosts may be difficult due to landowner objections or relatively unproductive because the roosts are marginally suitable for gray bats (Wenner 1984). This should not, however, preclude efforts to post the caves as closed to entry during the summer and to educate landowners and the local public about the bats.

If gray bat populations are found to be declining after the majority of suitable caves have been protected, the effects of pesticides, stream pollution, or other potential impacts should be investigated.

ACKNOWLEDGMENTS: Stephen R. Humphrey and Merlin D. Tuttle provided helpful comments on an earlier draft of the manuscript. They also wrote the account of the gray bat that appeared in the 1978 edition of this volume. Portions of that account, particularly the species and range descriptions, are repeated in the current account.

Literature Cited

Brady, J. T., T. H. Kunz, M. D. Tuttle, and D. E. Wilson. 1982. Gray bat recovery plan. U.S. Fish and Wildl. Serv., Denver, Colorado. 26 pp. + appendices.
Guthrie, M. J. 1933. The reproductive cycles of some cave bats. J. Mammal. 14:199–216.
Humphrey, S. R., and M. D. Tuttle. 1978. Gray bat. Pp. 1–3 *in* J. N. Layne, ed., Rare and endangered biota of Florida. Vol. 1. Mammals. University Presses of Florida, Gainesville. xx + 52 pp.

LaVal, R. K., R. L. Clawson, M. L. LaVal, and W. Caire. 1977. Foraging behavior and nocturnal activity patterns of Missouri bats, with special emphasis on the endangered species *Myotis grisescens* and *Myotis sodalis*. J. Mammal. 58:592–599.

Lee, D. S., and M. D. Tuttle. 1970. Old Indian Cave: Florida's first bat sanctuary. Florida Nat. 44:150–152.

Miller, R. E. 1939. The reproductive cycle in male bats of the species *Myotis lucifugus* and *Myotis grisescens*. J. Morph. 64:267–295.

Rice, D. W. 1955. Status of *Myotis grisescens* in Florida. J. Mammal. 36:289–290.

Stevenson, D. E., and M. D. Tuttle. 1981. Survivorship in the endangered gray bat (*Myotis grisescens*). J. Mammal. 62:244–257.

Tuttle, M. D. 1975. Population ecology of the gray bat (*Myotis grisescens*): Factors influencing early growth and development. Occas. Papers Museum Nat. Hist., Univ. of Kansas 36:1–24.

Tuttle, M. D. 1976. Population ecology of the gray bat (*Myotis grisescens*): Philopatry, timing and patterns of movement, weight loss during migration and seasonal adapative strategies. Occas. Papers Museum Nat. Hist., Univ. of Kansas 58:1–38.

Tuttle, M. D. 1977. Gating as a means of protecting cave dwelling bats. Pp. 77–82 *in* T. Aley and D. Rhoades, eds., National cave management symposium proceedings, 1976. Speleobooks, Albuquerque, New Mexico. 146 pp.

Tuttle, M. D. 1979. Status, causes of decline, and management of endangered gray bats. J. Wildl. Manage. 43:1–17.

Tuttle, M. D. 1987. Endangered gray bat benefits from protection. Endangered Species Tech. Bull. 21(3):4–5.

Wenner, A. S. 1984. Current status and management of gray bat caves in Jackson County, Florida. Florida Field Nat. 12:1–6.

Prepared by: Jeffery A. Gore, Florida Game and Fresh Water Fish Commission, 6938 Highway 2321, Panama City, FL 32409.

Endangered

Lower Keys Marsh Rabbit
Sylvilagus palustris hefneri
FAMILY LEPORIDAE
Order Lagomorpha

TAXONOMY: This subspecies also is known as the playboy bunny because its description was partially financed by the Playboy Foundation. Although Schwartz (1952) suggested that marsh rabbits on the Lower Keys might represent an undescribed subspecies, this population was only recently described (Lazell 1984) as a race distinct from *S. p. paludicola*, which inhabits the Upper Keys and southern Florida.

DESCRIPTION: This subspecies is generally similar to other races of the species. It has brownish dorsal pelage; the venter and underside of the tail are gray. The fur is darker than Upper Keys specimens, with no reddish tint to the upper pelage (some exceptions may occur on Big Pine Key). The sides of the face are grayer than the dorsal pelage. Measurements (in mm) reported for adults are: total length 361–405, tail 20–43, hind foot 77–86, ear from notch 44–49, condylobasal length 61.8–75.0, zygomatic breadth 31.7–36.0, molariform tooth row 12.8–13.8, frontonasal height at anterior tips of frontals 15.9–18.5, dental symphysis 16.3–17.4, cranium at posterior tips of zygomata 25.9–28.5, diastema 18.6–20.3, width of foramen magnum 8.9–10.0. The molariform tooth row is <80 percent of the length of the dental symphysis. The frontonasal region is convex with a high profile, and the cranium is broad. The dorsal surfaces of the parietal, frontal, and nasal bones are ornately sculptured. This subspecies is significantly different from Upper Keys rabbits in the following ratios: molariform row to dental symphysis, frontonasal height to molariform row, and molariform row to cranial width (Lazell 1984).

POPULATION SIZE AND TREND: Unpublished population estimates of 500 in 1976 and 259 in 1984 are cited by Howe (1988), who reported

Lower Keys marsh rabbit, *Sylvilagus palustris hefneri*. (Photo by Alan S. Maltz)

that the present population is between 200 and 400. These estimates suggest that the total population has been low for more than 10 years, with no obvious trend in overall population size or density. The race has apparently been extirpated from several localities (see below).

DISTRIBUTION AND HISTORY OF DISTRIBUTION: The subspecies is present on several keys from Big Pine to Boca Chica. The historical range may have included Key West (DePourtales 1877). Recent records include Big Pine, Hopkins, Sugarloaf, Welles, Saddlebunch, Geiger, and Boca Chica keys (Schwartz 1952; Layne 1974; Lazell 1984, personal communication; Howe 1988). It may have disappeared from Ramrod, Cudjoe, and the Torch keys in recent years (J.D. Lazell, personal communication). Howe (1988) reported that it no longer occurred on Geiger Key, but I observed one individual and fresh droppings near the western end of Geiger Key in 1987. It also occurs at sites on Boca Chica and Saddle-

Sylvilagus palustris hefneri 73

bunch keys not reported by Howe (1988). Howe's 1988 survey provided new and useful information but was limited in scope and duration. Further study is needed to more fully document the subspecies' distribution, especially on islands accessible only by water. The nearest record for *S. p.*

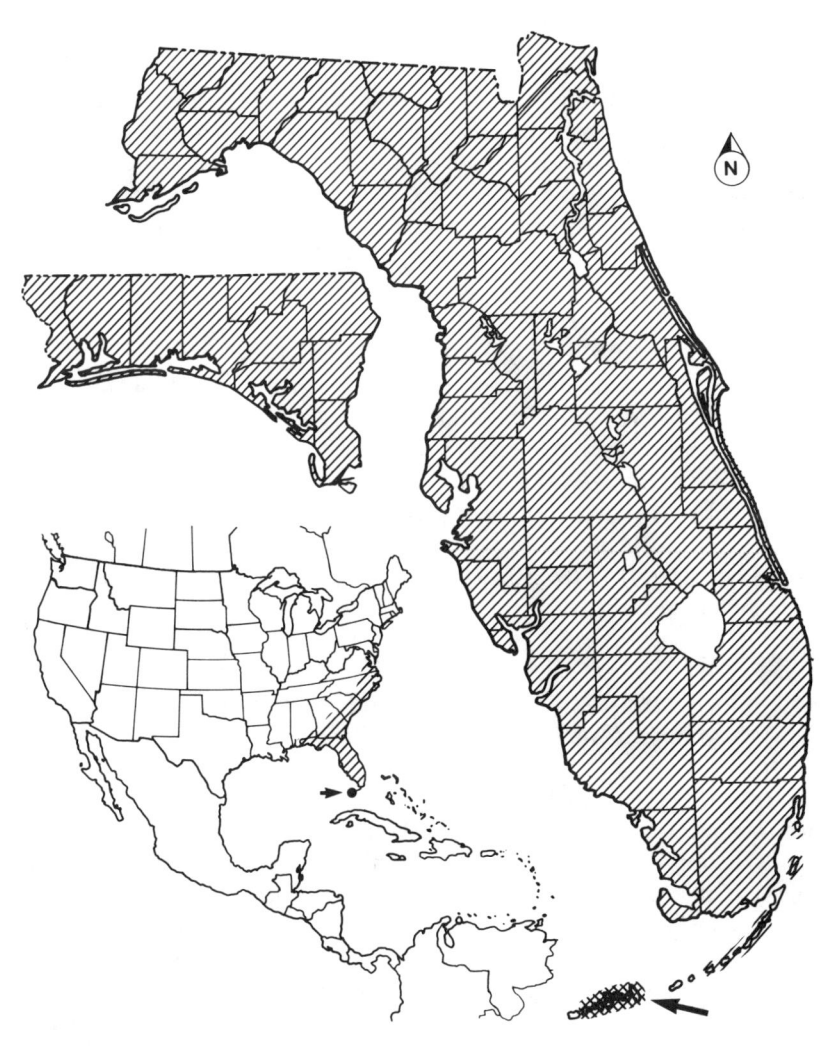

Distribution map of the Lower Keys marsh rabbit, *Sylvilagus palustris hefneri*. Hatching represents the species; crosshatching represents the subspecies.

paludicola is Long Key (Schwartz 1952). Thus there is an apparent hiatus of about 61 km between the two races.

GEOGRAPHICAL STATUS: This subspecies is endemic to the Lower Keys. Although present over most of its presumed historic range, the remaining populations are highly fragmented.

HABITAT REQUIREMENTS AND HABITAT TREND: This animal prefers marshes and adjacent low vegetative cover, especially sedges and grasses. Occasionally it is found in lesser numbers in areas farther upland such as grassy fields and tropical hammocks. More detail on plant associations in preferred habitat is provided by Howe (1988). Upland habitats in the Lower Keys not under state or federal ownership have been extensively developed.

VULNERABILITY OF SPECIES AND HABITAT: These rabbits are vulnerable to natural predators (Chapman and Wilner 1981) and, especially in the Keys, to predation by domesticated and feral pets. Highway traffic is a frequent cause of mortality. Hunting is not thought to be widespread but may be a problem at some localities. Vast areas of the preferred low habitat have been destroyed by dredge-and-fill operations to make them more suitable as commercial or residential sites.

CAUSES OF THREAT: Conversion of the preferred habitat to human uses poses the greatest threat.

RESPONSES TO HABITAT MODIFICATION: Habitat modifications that result in dense secondary growth near water could have positive impacts. Those that reduce cover near water are detrimental.

DEMOGRAPHIC CHARACTERISTICS: No information is available for this race. Marsh rabbits elsewhere in southern Florida breed throughout the year. The period of maximum production of young is December through June. Estimated potential productivity is 14–19 per female (Holler and Conaway 1979). Little is known about other aspects of demography.

KEY BEHAVIORS: Behavior of the Lower Keys marsh rabbit has not been described. In general, marsh rabbits are antisocial. They are primarily nocturnal but may be active during daylight. Vocalizations (squeals) are more common in marsh rabbits than in cottontails. Marsh rabbits swim well but usually do not enter water to escape predators. When foraging they typically walk rather than hop (Blair 1936; Chapman and Wilner 1981). They ordinarily do not live in burrows (Tomkins 1935) but occasionally do (J. D. Lazell, personal communication).

CONSERVATION MEASURES TAKEN: This animal is listed as endangered by the U.S. Fish and Wildlife Service and the Florida Game and Fresh Water Fish Commission.

CONSERVATION MEASURES PROPOSED: Distribution and abundance of the Lower Keys marsh rabbit should be better documented, and studies of the factors affecting its populations should be undertaken. Private lands occupied by this animal should be purchased by conservation agencies. Public lands occupied by this animal should be managed to protect the animals and their habitat.

ACKNOWLEDGMENTS: James D. Lazell generously provided unpublished observations. James N. Layne reviewed the manuscript and had many helpful comments.

Literature Cited

Blair, W. F. 1936. The Florida marsh rabbit. J. Mammal. 17:197–207.
Chapman, J. A., and G. R. Wilner. 1981. Sylvilagus palustris. Mammalian Species 153:1–3.
DePourtales, L. F. 1877. Hints on the origin of the flora and fauna of the Florida Keys. Amer. Nat. 11:137–144.
Holler, N. R., and C. H. Conaway. 1979. Reproduction of the marsh rabbit (*Sylvilagus palustris*) in south Florida. J. Mammal. 60:769–777.
Howe, S. E. 1988. Lower Keys marsh rabbit status survey. Report to U.S. Fish and Wildlife Service, 3100 University Boulevard, South Suite 120, Jacksonville. 8 pp.
Layne, J. N. 1974. The land mammals of south Florida. Pp. 386–413 *in* P. J. Gleason, ed., Environments of south Florida: Present and past. Miami Geol. Soc., Memoir 2:386–413.
Lazell, J. D., Jr. 1984. A new marsh rabbit (*Sylvilagus palustris*) from Florida's Lower Keys. J. Mammal. 65:26–33.
Schwartz, A. 1952. Land mammals of southern Florida and the Upper Florida Keys. Unpubl. Ph.D. diss., University of Michigan, Ann Arbor. 180 pp.
Tomkins, I. R. 1935. The marsh rabbit: an incomplete life history. J. Mammal. 16:201–205.

Prepared by: James L. Wolfe, Division of Biological Science, Emporia State University, Emporia, KS 66801.

Endangered

Choctawhatchee Beach Mouse
Peromyscus polionotus allophrys
FAMILY CRICETIDAE
Order Rodentia

TAXONOMY: No other names. Bowen (1968) investigated subspeciation in the Gulf coast forms of *P. polionotus* and described five new subspecies, of which *P. p. allophrys* was one. His treatment was accepted by Hall (1981), who included 16 subspecies. Bowen's treatment was based primarily on variation in pelage patterns, hair color values, and cranial measurements. Relationships among these factors are complex and additional data on genetic relationships of these taxa are needed. Prior to construction of the intracoastal ship canal, interchange might have occurred among *P. p. allophrys*, *P. p. albifrons*, and *P. p. sumneri*.

DESCRIPTION: The Choctawhatchee beach mouse is a small *Peromyscus*, polymorphic with respect to pelage coloration and patterns (Bowen 1968). This subspecies is distinctly more orange-brown or yellow-brown on the pigmented dorsum than are the other subspecies. In general, it is paler and shows more white than the inland forms. The pattern of pelage pigmentation at the rump either extends down the thighs or is squared, but it is not tapered as in *P. p. leucocephalus*, the palest of the subspecies (Bowen 1968). The mid-dorsal line is distinctly darker than the remainder of the pigmented area. Pigmentation on the head varies from extending along the dorsal surface of the nose to the tip in darker animals, to lighter individuals whose pigmentation ends posterior to the eyes with cheeks and rostral area pure white. White hairs are unpigmented to their base except immediately adjacent to the pigmented dorsum. The tail may show a dorsal stripe of varying length or this stripe may be absent. Means and extremes (in parentheses) for external measurements (in mm) of males ($n = 73$) and females ($n = 67$) respectively (age not specified; Bowen 1968) were: total length, 131, 135 (113–153); length of tail, 54,

Choctawhatchee beach mouse, *Peromyscus polionotus allophrys*. (Photo by George W. Folkerts)

55 (43–64); length of hind foot, 18.7, 18.6 (15.8–19.2); ear, 15.4, 15.4 (13.9–17.2). Hall (1981) described the skull of *P. polionotus* as resembling that of the smaller subspecies of *P. maniculatus* but with palatine slits relatively shorter and auditory bullae relatively larger.

POPULATION SIZE AND TREND: The 1979 population of this subspecies was estimated to exceed 515 individuals, >178 at Topsail Hill and >357 on Shell Island (Humphrey and Barbour 1981). The population of mice on Shell Island appeared to be secure in 1982 (Meyers 1983). As recently as January 1987, the population at Topsail Hill appeared to be low (Holler, unpubl. data); two trapping occasions (200 trap-nights each) yielded only two male mice (one adult and one subadult). Mouse sign was, however, seen at locations throughout the habitat zone and it is likely that the population was at a normal low. As noted below, beach mouse populations may exhibit large fluctuations. The best measure of their status is probably given by km of habitat occupied. Presently, the mice occur on about 6.5 km of beach dunes at Topsail Hill and 9.4 km on Shell Island. Status of a third population being established by translocation at Grayton Beach State Recreation Area is indefinite. Minimum effective population size for long-term survival of beach mouse populations has not been estimated.

DISTRIBUTION AND HISTORY OF DISTRIBUTION: Bowen (1968) described the distribution of the Choctawhatchee beach mouse as the coastal dunes between the entrance to Choctawhatchee Bay, Okaloosa County, and St. Andrew Bay, Bay County, Florida. Bowen (1968) noted that by

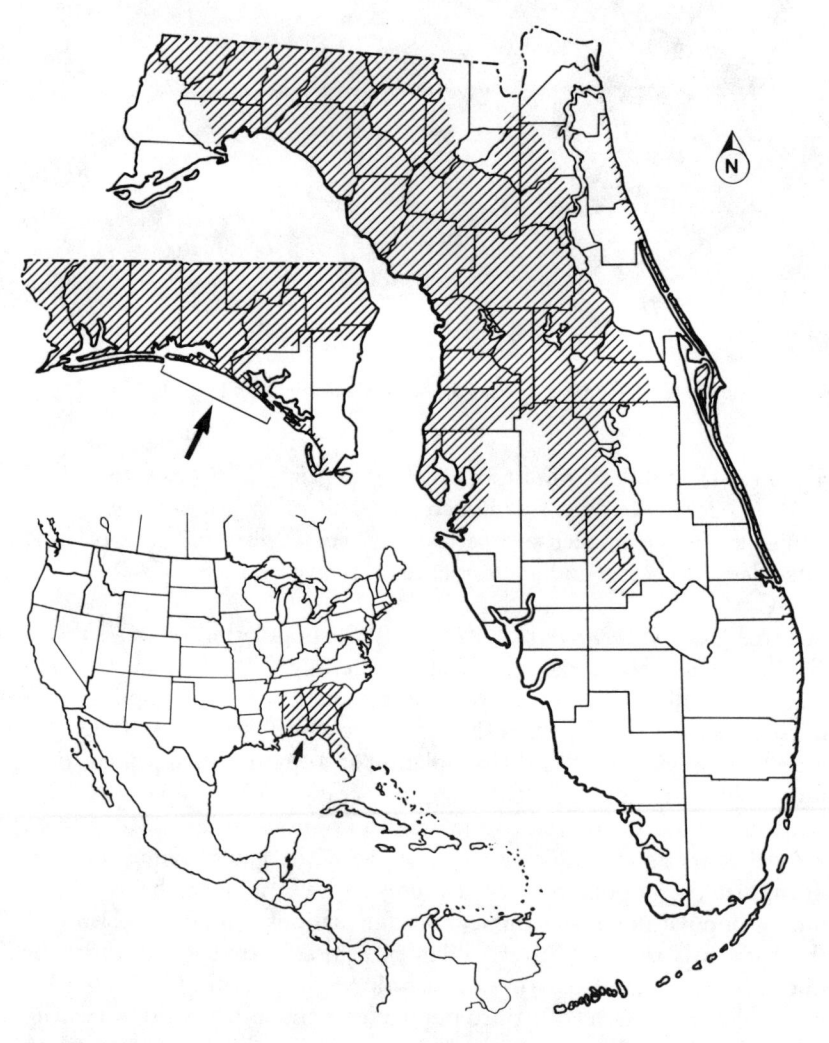

Distribution map of the Choctawhatchee beach mouse, *Peromyscus polionotus allophrys*. Hatching represents the species; crosshatching and bracket represent the subspecies.

the end of his study more than two-thirds of the habitat for this subspecies had been lost as a result of coastal development. Humphrey and Barbour (1981) trapped all remnants of suitable habitat in 1979, including dunes on the western end of Shell Island, just to the east of the previously reported range. They captured beach mice only along 7.9 km of habitat at Topsail Hill near Destin and on Shell Island (9.4 km of suitable habitat). The latter locality represented an extension of the known range for the subspecies. Shell Island was formerly connected with the mainland. Holler (unpubl. data) confirmed the presence of Choctawhatchee mice on most of Shell Island as recently as December 1987. The only other areas of apparently suitable habitat are at Grayton Beach (approximately 4.3 km) and east of Seagrove Beach (approximately 4.0 km). These areas were unoccupied in 1979; a translocation program during 1987 and 1988 has resulted in a low but persistent population at Grayton Beach.

GEOGRAPHICAL STATUS: The Choctawhatchee beach mouse is endemic to Florida. Once present in more or less continuous populations along the coastal dunes between the entrances to Choctawhatchee and St. Andrew Bay, it remains as only two (possibly three) disjunct populations.

HABITAT REQUIREMENTS AND HABITAT TREND: Habitat for this mouse consists of the sand dunes along the Gulf coast beaches. Optimal habitat consists of the primary and secondary dunes vegetated primarily by sea oats (*Uniola paniculata*), beach grass (*Panicum amarum*), and bluestem (*Andropogon maritimus*). Other locally common plants include *Panicum repens*, beach morning glory (*Ipomoea stolonifera*), pennywort (*Hydrocotyle bonariencis*), and, on the frontal slopes, sea rocket (*Cakile constricta*). Additionally, the mice are known to occupy the older scrub dunes found adjacent to and immediately inland to the optimal habitat. These dunes are characterized by scrubby oaks (*Quercus virginiana* and *Q. myrtifolia*), dwarfed magnolia (*Magnolia grandiflora*), and rosemary (*Ceratiola ericoides*). Extine and Stout (1987) found greater populations of *P. p. niveiventris* on the interior dunes than on the primary dunes at Merritt Island on the Atlantic coast of Florida, a pattern not found in Alabama beach mice (*P. p. ammobates*; Hill 1989).

The two remaining natural populations occupy areas of quite different habitat. Topsail Hill has a continuous and quite high system of frontal dunes, densely vegetated on the beach side with coarse grasses. A broad area of interior dunes with scrub vegetation approaches the crest of these frontal dunes from the landward side. Shell Island lacks the interior scrub dunes. Habitat there consists essentially of the primary and in some areas secondary dunes vegetated with coarse grasses; density of these grasses and height of the dune system are greater on the western than on the eastern end of the island.

Four areas of critical habitat totaling 20.2 km have been defined for the Choctawhatchee beach mouse (U.S. Fish and Wildlife Service 1987). Zone 1 (5.6 km) is in the Topsail Hill area east of Destin and is currently occupied. Zone 2 (2.6 km) includes Grayton Beach State Recreation Area and adjacent privately owned land to the east; it is occupied by a translocated population (status indefinite; see below). Zone 3 (1.6 km) consists of the mainland portion of St. Andrews State Recreation Area and is not presently occupied. Zone 4 (10.5 km) is on Shell Island; at least 9 km are occupied.

Diet of the Choctawhatchee beach mouse and of other beach subspecies has not been adequately determined. The food of *P. polionotus* appears to be primarily seeds of grasses and herbs; insects may be eaten when available (Golley 1962). In Georgia stomachs of mice collected in March contained 50 percent seeds and 50 percent arthropods. Blair (1951) noted that mouse trails on Santa Rosa Island (*P. p. leucocephalus*) indicated frequent visits to bluestem and to windrows of sea oat seeds. Beach grass is also a prolific seed producer and may be used. I noticed several locations where mice (based on tracks) had fed upon fruits of sea rocket in April. We also have identified beetle (Coleoptera) parts in fecal pellets of Alabama beach mice (*P. p. ammobates*; A. Appel, Auburn University, personal communication).

VULNERABILITY OF SPECIES AND HABITAT: The Choctawhatchee beach mouse is vulnerable to habitat loss (Bowen 1968; Ehrhart 1978*a*), direct loss of populations during tropical storms (Meyers 1983), predation, especially by domestic cats (*Felis catus*; Bowen 1968; Holliman 1983), genetic isolation of small populations (U.S. Fish and Wildlife Service 1987), and competition with house mice (*Mus musculus*; Briese and Smith 1973; Humphrey and Barbour 1981); but see Caldwell (1964), Caldwell and Gentry (1965*a*), and Gentry (1966) for data that indicate competition with inland subspecies of *P. polionotus* may vary depending on food supply and relative population densities. Habitat loss results primarily from coastal development (Bowen 1968); however, tropical storms can also cause extensive losses. Hurricane Elena in 1985 destroyed an estimated 75 percent of primary dune habitat at Fort Morgan in Alabama (U.S. Fish and Wildlife Service 1986).

CAUSES OF THREAT: Habitat loss from real estate development is the single most important factor causing the endangered status of this subspecies (Bowen 1968). One major area of occupied critical habitat (Topsail Hill) and a small area adjacent to Grayton Beach State Recreation Area are privately owned and still subject to development. Other areas of critical habitat, including that occupied by mice on Shell Island, are in public ownership. Shell Island is particularly vulnerable to storm damage because of the narrow configuration of the beach mouse habitat and the lack of interior dunes. The length of this habitat (approximately 9.4 km occupied) provides some protection against loss of the entire population. Predation by house cats would probably be most severe at critical-habitat zones 2 and 3 if mice were reestablished there. House mice have been present at zone 3 (Humphrey and Barbour 1981), although they were not found in 1982 (Meyers 1983).

RESPONSES TO HABITAT MODIFICATION: Development of the dune habitat is normally accompanied by loss of beach mouse populations (Bowen 1968) either as a direct result of habitat loss or indirectly as a result of predation (Bowen 1968; Holliman 1983) or competition with house mice (Humphrey and Barbour 1981). Beach mice can be expected to respond positively to restoration of dune habitat following storm damage. Populations of Alabama beach mice are using dune habitat rebuilt from damage incurred during Hurricane Elena at Fort Morgan (Hill 1989). Data are lacking, however, on response of beach mouse populations to specific modifications or manipulations of their habitat (but see *P. p. phasma* account for some information).

DEMOGRAPHIC CHARACTERISTICS: Demographic data are limited for *P. p. allophrys* and must be augmented with data published for other subspecies. Laboratory-reared female *P. polionotus* (several subspecies) reach sexual maturity (i.e., first estrus) at 30 days (Clark 1938). Wild-trapped male and female *P. p. leucocephalus* were judged to be sexually active at not more than 6 or 7 weeks (Blair 1951). Gestation is reported as 24 days (Smith 1939; Laffoday 1957; both in Layne 1968); 28 (25–31) days if conception is at postpartum estrus (Laffoday 1957, cited in Layne 1968). Litter size averages 3.1–4.1, extremes 1–8 (Caldwell and Gentry 1965*b*; Kaufman and Kaufman 1982, 1987). Litter size has been positively correlated with female size (Caldwell and Gentry 1965*b*; Kaufman and Kaufman 1987). Littering interval in the laboratory is 26 days (Bowen 1968) and postpartum estrus may occur (Golley 1962). Blair (1951) reported much breeding activity in November, December, and early January, with reduced activity in May and early June for *P. p. leucocephalus*. I found extensive breeding in *P. p. trissyllepsis* in fall and spring, 1986–89 (unpubl. data). Hill (1989) found that Alabama beach mice breed throughout the year, but that breeding peaked in fall. Inland forms are reported to have fall and spring breeding peaks with reduced reproduction in winter and summer (Golley 1962; Caldwell and Gentry 1965*b*). The monthly survival rate for Alabama beach mice was 0.68, and 87 percent of this population survived four months or less (Hill 1989). Blair (1951) indicated a fairly rapid turnover; only 19.5 percent of mice resident in early January survived the 4-month period until early May. Densities in his study of *P. p. leucocephalus* ranged from 2.05 to 3.5/ha. Seasonal fluctuations in population size may be great (Bowen 1968); the mean number of *P. p. niveiventris* known to be alive in a 1.44-ha study site on the Atlantic coast varied from 19.75 to 93.75 during a 13-month period (Extine and Stout 1987).

KEY BEHAVIORS: No data have been reported specifically for *P. p. allophrys*; the following is based on reports for several other subspecies. *P. polionotus* appear to be primarily monogamous based on trapping (Blair 1951) and genetic (Foltz 1981) data. They tend to pair rather permanently, and the male and female may associate during feeding activities (Blair 1951). Antagonism was strong between adult females and considerably weaker between adult males.

P. polionotus is nocturnal (Wolfe and Esher 1978) with activity highest on moonless or cloudy nights (Blair 1951). These mice inhabit burrows (Hayne 1936). Ehrhart (1978*b*:9) notes that in beach forms the burrow entrance is usually on the sloping side of a dune and that burrows typically have three main parts: "1) the entrance tunnel, usually descending obliquely for some distance and then continuing straight into the bank, 2) a nest chamber, formed at the end of the level portion of the entrance tunnel at a depth of .6 to .9 m (23 ft), and 3) an 'escape tunnel,' which rises steeply from the nest chamber to within about 2.5 cm (1 in) of the surface." The mice also use burrows of ghost crabs (*Ocypode*; Blair 1951).

Adult beach mice, after establishing residence, tend to remain in one area for the duration of their life. Most immigrants are juvenile mice. Blair (1951) found that juveniles (*P. p. leucocephalus*) moved an average of 431 m prior to establishing residence. Home ranges of males and females did not differ in Blair's study. They were larger in the spring than fall and in more open secondary habitat than in densely vegetated primary dunes. Mean sizes varied from 0.8 ha in the dune area in the fall to 4.3 ha in secondary habitat in the spring. Mean linear distances moved along the dunes by Alabama beach mice were greatest in the fall (males = 188 m; females = 164 m) and least in the spring (males = 86 m; females = 85 m; Hill 1989).

CONSERVATION MEASURES TAKEN: The Choctawhatchee beach mouse was listed by the U.S. government as endangered pursuant to the Endangered Species Act of 1973 on 6 June 1985 (50 FR 23872). It is also listed as endangered in Title 39-27.002 of the Florida Administrative Code, so it has both federal and state protection. Critical habitat (20.2 km) has been designated of which 14 km are in public ownership. On publicly owned land this mouse is given consideration in all management decisions. The Florida Game and Fresh Water Fish Commission has initiated, in cooperation with the Alabama Cooperative Fish and Wildlife Research Unit and the Florida Department of Natural Resources, a program to reestablish this subspecies at Grayton Beach State Recreation Area (Holler and Mason 1987). Fourteen pairs of wild-trapped mice were translocated from Shell Island to two disjunct blocks of habitat at Grayton Beach during 1987 and 1988; current status at Grayton Beach is indefinite but small populations (augmented with 20 pairs of captive-reared mice) persist at both release sites. A captive breeding colony has been established at Auburn University to provide a source of mice for reestablishment should natural populations be lost. A slide and tape show has been developed by the Alabama Cooperative Fish and Wildlife Research Unit, for the U.S. Fish and Wildlife Service, to educate the public regarding beach mice and their need for protection. Additionally, an educational brochure is being prepared by the U.S. Fish and Wildlife Service.

CONSERVATION MEASURES PROPOSED: Full potential recovery of the Choctawhatchee beach mouse would result in self-sustaining populations at each of the five areas of suitable habitat. These are the only remaining areas suitable for sustained habitation by this subspecies within its native range. Establishment of populations in each area might justify downlisting of the subspecies to threatened status. Because of the extensive loss of habitat it will probably never be possible to remove the subspecies from protection under the Endangered Species Act of 1973 (U.S. Fish and Wildlife Service 1987).

The recovery plan (U.S. Fish and Wildlife Service 1987) identifies three principal thrusts for recovery of this and two other subspecies of endangered beach mice: maintenance and restoration of suitable habitat, development of reestablishment programs, and education of the general public regarding beach mice and their requirements for protection. The recovery plan provides detailed recommendations in each of these areas.

The single most important measure that could be undertaken to benefit this beach mouse would be the protection of its habitat in critical habitat zone 1, the eastern portion of zone 2, and the area east of Seagrove Beach. These areas are currently under private ownership and subject to development. Their preservation might be achieved by protective easements, cooperative agreements, land exchanges, fee title acquisition, or other means (U.S. Fish and Wildlife Service 1987) and should be aggressively pursued.

Reestablishment of beach mice at Grayton Beach State Recreation Area is underway. This program should be continued until a self-sustaining population of mice is established. Habitat at St. Andrews State Recreation Area (zone 3) should be surveyed to determine suitability and current status of domestic cats and house mice. Reestablishment of beach mice in this area should be attempted after habitat restoration and reduction of predators and house mice, if warranted. A captive breeding colony composed of at least 30 breeding pairs of progeny from wild-trapped mice is proposed for maintenance at Auburn University. Genetic studies are also needed to determine the degree of inbreeding within each population, the need for interchange of genetic material among populations, and the interrelatedness of the various subspecies.

Literature Cited

Blair, W. F. 1951. Population structure, social behavior, and environmental relations in a natural population of the beach mouse *Peromyscus polionotus leucocephalus*. Contrib. Lab. Vert. Biol., Univ. of Michigan 48:1–47.

Bowen, W. W. 1968. Variation and evolution of Gulf coast populations of beach mice, *Peromyscus polionotus*. Bull. Florida State Mus., Biol. Sci. 12(1):1–91.

Briese, L. A., and M. H. Smith. 1973. Competition between *Mus musculus* and *Peromyscus polionotus*. J. Mammal. 54:968–969.

Caldwell, L. D. 1964. An investigation of competition in natural populations of mice. J. Mammal. 45:12–30.

Caldwell, L. D., and J. B. Gentry. 1965a. Interactions of *Peromyscus* and *Mus* in a one-acre field enclosure. Ecology 46:189–192.

Caldwell, L. D., and J. B. Gentry. 1965b. Natality in *Peromyscus polionotus* populations. Amer. Midland Nat. 74:168–175.

Clark, F. H. 1938. Age of sexual maturity in mice of the genus *Peromyscus*. J. Mammal. 19:230–234.

Ehrhart, L. M. 1978a. Choctawhatchee beach mouse. Pp. 18–19 *in* J. N. Layne, ed., Rare and endangered biota of Florida. Vol. 1. Mammals. University Presses of Florida, Gainesville. xx + 52 pp.

Ehrhart, L. M. 1978b. Pallid beach mouse. Pp. 8–10 *in* J. N. Layne, ed., Rare and endangered biota of Florida. Vol. 1. Mammals. University Presses of Florida, Gainesville. xx + 52 pp.

Extine, D. D., and I. J. Stout. 1987. Dispersion and habitat occupancy of the beach mouse, *Peromyscus polionotus niveiventris*. J. Mammal. 68:297–304.

Foltz, D. W. 1981. Genetic evidence for long-term monogamy in a small rodent. *Peromyscus polionotus*. Amer. Nat. 117:665–675.

Gentry, J. B. 1966. Invasion of a one-year abandoned field by *Peromyscus polionotus* and *Mus musculus*. J. Mammal. 47:431–439.

Golley, F. B. 1962. Mammals of Georgia. University of Georgia Press, Athens. 218 pp.

Hall, E. R. 1981. The mammals of North America. John Wiley and Sons, New York. 1181 + 90 pp.

Hayne, D. W. 1936. Burrowing habits of *Peromyscus polionotus*. J. Mammal. 17: 420–421.

Hill, E. A. 1989. Population dynamics, habitat, and distribution of the Alabama beach mouse. M.S. thesis, Auburn University. 53 pp.

Holler, N. R., and D. W. Mason. 1987. Reestablishment of Perdido Key and Choctawhatchee beach mice into areas of unoccupied critical habitat. J. Alabama Acad. Sci. 58:66.

Holliman, D. C. 1983. Status and habitat of Alabama Gulf coast beach mice *Peromyscus polionotus ammobates* and *P. p. trissyllepsis*. Northeast Gulf Sci. 6:121–129.

Humphrey, S. R., and D. B. Barbour. 1981. Status and habitat of three subspecies of *Peromyscus polionotus* in Florida. J. Mammal. 62:840–844.

Kaufman, D. W., and G. A. Kaufman. 1982. Secondary sex ratio in *Peromyscus polionotus*. Trans. Kansas Acad. Sci. 85:216–217.

Kaufman, D. W., and G. A. Kaufman. 1987. Reproduction by *Peromyscus polionotus*: Number, size, and survival of offspring. J. Mammal. 68:275–280.

Layne, J. N. 1968. Ontogeny. Pp. 148–253 *in* J. A. King, ed., Biology of *Peromyscus* (Rodentia). Special Publ. 2. Amer. Soc. Mammal. 593 pp.

Meyers, J. M. 1983. Status, microhabitat, and management recommendations for *Peromyscus polionotus* on Gulf coast beaches. Unpubl. report to U.S. Fish and Wildlife Service, Atlanta, Georgia. 29 pp.

U.S. Fish and Wildlife Service. 1986. Hurricanes damage southeastern beach mouse habitat. Endangered Species Tech. Bull. 11(1):7–8.

U.S. Fish and Wildlife Service. 1987. Recovery plan for the Alabama beach mouse (*Peromyscus polionotus ammobates*), Perdido Key beach mouse (*P. p. trissyllepsis*), and Choctawhatchee beach mouse (*P. p. allophrys*). U.S. Fish and Wildlife Service, Atlanta, Georgia. 45 pp.

Wolfe, J. L., and R. J. Esher. 1978. The behavior of a burrowing mouse (*Peromyscus polionotus*) in a residential maze. J. Mississippi Acad. Sci. 23:100–109.

Prepared by: Nicholas R. Holler, U.S. Fish and Wildlife Service, Alabama Cooperative Fish and Wildlife Research Unit,* 331 Funchess Hall, Auburn University, AL 36849.

*Cooperators: U. S. Fish and Wildlife Service, Game and Fish Division of the Alabama Department of Conservation and Natural Resources, Wildlife Management Institute, Auburn University (Alabama Agricultural Experiment Station—Contribution #15-881494P, Department of Fisheries and Allied Aquacultures, Department of Zoology and Wildlife Science).

Endangered

St. Andrew Beach Mouse
Peromyscus polionotus peninsularis
FAMILY CRICETIDAE
Order Rodentia

TAXONOMY: Of the 16 currently recognized subspecies of the oldfield mouse, *Peromyscus polionotus*, eight have very pale pelage and are adapted to living in dunes. *P. p. peninsularis* is the easternmost of the five dune-adapted forms (beach mice) that occur along the northern coast of the Gulf of Mexico. It was given its trinomial name by Howell (1939), who had trapped the type specimen on St. Andrew Point Peninsula in Bay County. The present geographic range of the subspecies extends in dune habitats from St. Joseph Point in Gulf County to Panama City at the entrance of St. Andrew Bay in Bay County (Hall 1981).

The subspecific assignments of the various populations of beach mice were made on the basis of geographic variation in pelage and skeletal measurements. The results of breeding experiments by Sumner (1930, 1932), Bowen (1968), and others show that such variation has a genetic basis. Nevertheless, subspecific divisions in the genus *Peromyscus* have not been found to be concordant with geographic variation in chromosomal types (Baker et al. 1983), allozymes (Robbins et al. 1985), or mitochondrial DNA clones (Avise et al. 1983). When there is concordance among these various types of data, it seems to be attributable simply to distance (Schnell and Selander 1981). There is substantial morphological and pelage variation within individual subspecies, especially in *P. p. leucocephalus* (see Bowen 1968 for examples).

DESCRIPTION: *P. polionotus* is one of the smallest of the 55 species in its genus. Sumner (1930, 1932) and Bowen (1968) give careful analyses of morphological and pelage variation in the Gulf coast beach populations and the adjacent darker inland populations. In general the beach populations have proportionately larger ears and feet than the inland populations

St. Andrew beach mouse, *Peromyscus polionotus peninsularis*. (Photo by A. Ken Womble)

do. In *P. p. peninsularis* the white fur of the underparts and sides is extensive, and the back is nearly as pale buffy as that of the palest form, *P. p. leucocephalus*, which occurs on Santa Rosa Island. On the basis of an examination of 40 specimens in the U.S. National Museum, Bowen (1968) stated that the pattern of color of the tips of the hair on the rump in *P. p. peninsularis* is either tapered, as in *leucocephalus*, or square, as in some *allophrys*. He found no intermediates among the specimens, and he concluded that there are two distinct morphs in the population. Labels with these specimens gave measurements made on freshly killed animals. The averages were: head and body length 75 mm, tail 52 mm, hind foot length 18.5 mm, male ear length 16.5 mm, female ear length 15 mm.

DISTRIBUTION AND HISTORY OF DISTRIBUTION: The stated geographic range of *P. p. peninsularis* is from the St. Joseph spit in Gulf County to the entrance of St. Andrew Bay in Bay County. Since 1950, real estate development in both counties has caused the destruction of beach-mouse

Peromyscus polionotus peninsularis

habitat in many areas. The recreational use of sand dunes by pedestrians and vehicles has resulted in serious erosion of the remaining dunes. In the fall of 1985, Hurricanes Elena and Kate scoured the dune systems of the area, causing huge blowouts in the high dunes of the St. Joseph spit.

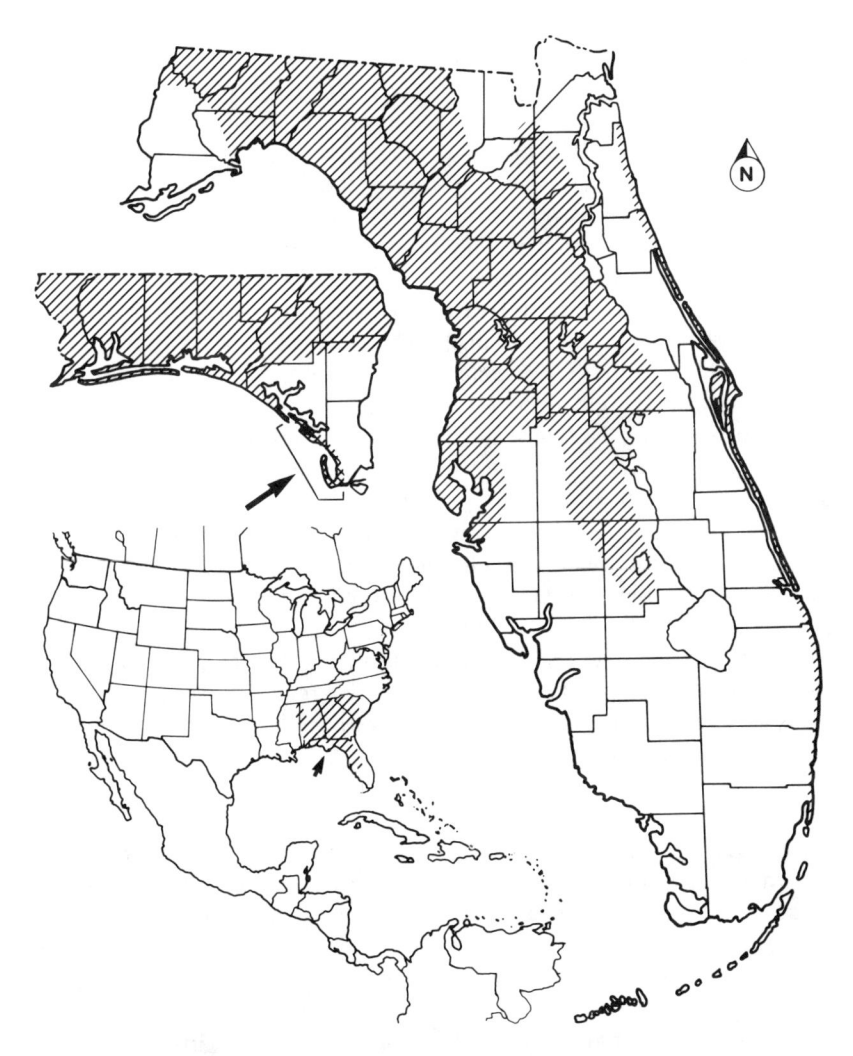

Distribution map of the St. Andrew beach mouse, *Peromyscus polionotus peninsularis*. Hatching represents the species; crosshatching and bracket represent the subspecies.

The only substantial population of beach mice remaining is in an 8-km strip of high dunes at the northern end of the spit, in St. Joseph Peninsula State Park. There, when conditions are favorable, one can find tracks and the entrances of burrows in both the front dunes and back dunes. The other locality where *P. p. peninsularis* persists is Crooked Island East in Bay County. Here signs of beach-mouse activity are much harder to find, and the population appears to be small.

POPULATION SIZE AND TREND: On the basis of surveys in fall 1986 and spring 1987, D. Martin and F. James estimated that there were approximately 500 mice in the more easterly population within the St. Joseph Peninsula State Park. Tracks of beach mice were seen outside the park on the St. Joseph spit in only one place, 1 mi south of the park entrance. None were seen in former localities such as Port St. Joe and Mexico Beach. In Bay County all recent records of *P. p. peninsularis* are from Crooked Island East, on Tyndall Air Force Base. There, in February and March 1986, J. Short, C. Petrick, and F. James live trapped several mice in low dunes and well-vegetated back dunes. They estimated that there were approximately 150 mice at this site. In the spring of 1987, however, C. Petrick trapped numerous house mice (*Mus musculus*) at the same site, but only a single beach mouse. Petrick was concerned that competition with *Mus* may have caused a crash in the beach-mouse population. Also, trial exercises with military hovercraft had damaged the dunes. J. A. Gore trapped several animals there in 1989, so the population persists.

GEOGRAPHICAL STATUS: The St. Andrew beach mouse is a subspecies of the oldfield mouse, *P. polionotus*, a widespread species in the southeastern United States. It persists only in two highly isolated subpopulations in Gulf and Bay counties. The former area is within St. Joseph Peninsula State Park; the latter is on Tyndall Air Force Base.

HABITAT REQUIREMENTS AND HABITAT TREND: *P. p. peninsularis* occurs in well-developed, high front dunes of the St. Joseph Peninsula State Park, where the major plant cover is sea oats (*Uniola paniculata*). It also occurs in the older and even higher back dunes, where burrows are often found at the base of blowouts of sand held up by roots of live oak (*Quercus geminata*) shrubs. Such areas also support sea oats and rosemary (*Ceratiola ericoides*). On Crooked Island beach mice occur in the low front dunes (3–5 m high) and in even lower back dunes covered with bunch grass (*Andropogon*) and beach grass (*Panicum*).

VULNERABILITY OF SPECIES AND HABITAT: See the species account on the Choctawhatchee beach mouse, *Peromyscus polionotus allophrys*.

CAUSES OF THREAT: See the species account on the Choctawhatchee beach mouse, *Peromyscus polionotus allophrys*.

RESPONSES TO HABITAT MODIFICATION: Some drift fences have been placed in the dune area of Crooked Island East on Tyndall Air Force Base, Bay County, to help maintain the dune system. These dunes are low and degraded in comparison with the dune system on the St. Joseph spit.

DEMOGRAPHIC CHARACTERISTICS: See the species account on the Choctawhatchee beach mouse, *Peromyscus polionotus allophrys*.

KEY BEHAVIORS: The only Gulf coast population for which the behavior and ecology have been studied in the wild is *P. p. leucocephalus* (Blair 1951). See the species account on the Choctawhatchee beach mouse, *Peromyscus polionotus allophrys*.

CONSERVATION MEASURES TAKEN: Despite the fact that the genus *Peromyscus* has been the subject of intensive research by mammalogists and that the genetics of *P. polionotus* had been studied carefully, *P. p. peninsularis* has been neglected. It was not collected by Bowen (1968), Selander (1970), Selander et al. (1971), or Avise et al. (1979) in their major surveys of variation in this species. The only remaining substantial population of the St. Andrew beach mouse is the one within the St. Joseph Peninsula State Park in Gulf County. This population, although protected by the Florida Department of Natural Resources, is completely isolated. The status of the only other population, on Tyndall Air Force Base, is tenuous. Although that area is protected from public use by the U.S. Air Force, military exercises have damaged the beach-mouse habitat. Because of the fragile nature of this habitat and the isolation of these two remaining populations, the St. Andrew beach mouse has been listed as endangered by the Florida Game and Fresh Water Fish Commission. It is presently under review for listing by the U.S. Fish and Wildlife Service.

CONSERVATION MEASURES PROPOSED: The western subpopulation of the St. Andrew beach mouse is in urgent need of management. The primary need is for measures to encourage greater development of the dune system. The extent of genetic differentiation in this population is unknown. Its genetic heterozygosity is probably low. A management program should include genetic studies of both populations, with a view toward the possibility of introducing mice into the western subpopulation, after the establishment of a more stable dune system. See the species account on the Choctawhatchee beach mouse, *Peromyscus polionotus allophrys*, for related comments.

Literature Cited

Avise, J. C., M. H. Smith, and R. K. Selander. 1979. Biochemical polymorphism and systematics in the genus *Peromyscus*. VII. Geographic differentiation in members of the *truei* and *maniculatus* species groups. J. Mammal. 60:177–192.

Avise, J. C., J. F. Shapira, S. W. Daniel, C. F. Aquadro, and R. A. Lansman. 1983. Mitochondrial DNA differentiation during the speciation process in *Peromyscus*. Mol. Biol. Evol. 1:38–56.

Baker, R. J., L. W. Robbins, F. B. Strange, Jr., and E. C. Birney. 1983. Chromosomal evidence for a major subdivision in *Peromyscus leucopus*. J. Mammal. 64:356–359.

Blair, W. F. 1951. Population structure, social behavior, and environmental relations in a natural population of the beach mouse *Peromyscus polionotus leucocephalus*. Contrib. Lab. Vert. Biol., Univ. of Michigan 48:1–47.

Bowen, W. W. 1968. Variation and evolution of Gulf coast populations of beach mice, *Peromyscus polionotus*. Bull. Florida State Mus., Biol. Sci. 12:1–91.

Hall, E. R. 1981. The mammals of North America. John Wiley and Sons, New York. 1181 + 90 pp.

Howell, A. H. 1939. Descriptions of five new mammals from Florida. J. Mammal. 20:363–365.

Robbins, L. W., M. H. Smith, M. C. Wooten, and R. K. Selander. 1985. Biochemical polymorphism and its relationship to chromosomal and morphological variation in *Peromyscus leucopus* and *Peromyscus gossypinus*. J. Mammal. 66:498–510.

Schnell, G. D., and R. K. Selander. 1981. Environmental and morphological correlates of genetic variation in mammals. Pp. 60–69 *in* M. H. Smith and J. Joule, eds., Mammalian population genetics. University of Georgia Press, Athens.

Selander, R. K. 1970. Biochemical polymorphism in populations of the house mouse and old-field mouse. Symp. Zool. Soc. Lond. 26:73–91.

Selander, R. K., M. H. Smith, S. Y. Yang, W. E. Johnson, and J. B. Gentry. 1971. Biochemical polymorphism and systematics in the genus *Peromyscus*. I. Variation of the old-field mouse (*Peromyscus polionotus*). Stud. Genet. 6:49–90.

Sumner, F. B. 1930. Genetic and distributional studies of three subspecies of *Peromyscus*. J. Genetics 23:257–376.

Sumner, F. B. 1932. Genetic, distributional, and evolutionary studies of the subspecies of deer mice (*Peromyscus*). Bibliogr. Genet. 9:1–106.

Prepared by: Frances C. James, Department of Biological Science, Florida State University, Tallahassee, FL 32306.

Endangered

Anastasia Island Beach Mouse
Peromyscus polionotus phasma
FAMILY CRICETIDAE
Order Rodentia

TAXONOMY: This beach mouse was named as *Peromyscus phasma* by Bangs (1898) based on 29 specimens from the southern tip of Anastasia Island. The population was treated as a subspecies of *P. polionotus* by Osgood (1909), who examined 54 specimens, all from the type locality. This taxonomic arrangement has not been reviewed since then; Bowen (1968) dealt only with Gulf coastal subspecies of *P. polionotus*. R. D. Ivey (personal communication) assigned specimens from the barrier beach north of St. Augustine Inlet to the subspecies *phasma* on the basis of the white spot above the eye.

This population is quite distinctive genetically (Selander et al. 1971; Ramsey 1973), differing from other subspecies of *P. polionotus* to a degree normally found among separate species (M.H. Smith, personal communication). In view of the poor survival status of this population, a reevaluation of geographic variation is urgently needed, so the nature of the genetic entity at risk can be clearly recognized.

DESCRIPTION: This is a small mouse whose whitish color closely matches the color of the beach sand on which it lives. The Anastasia Island subspecies is paler than *niveiventris* from the southeastern Atlantic coast and darker than *decoloratus* from south of Matanzas Inlet. Hairs of the upper parts are pinkish buff, with a grayish tinge in the middle of the back; the nose, a spot above the eye, and a spot at the base of the ear are pure white and very conspicuous; the underparts are pure white to the bases of the hairs; the feet and legs are completely white; and the tail is completely white or may have traces of dusky on its upper side (Bangs 1898; Osgood 1909). The skull is similar to that of *niveiventris*. Average measurements of 11 adults are: total length 138.7 mm, tail length 53.5 mm, hind

Anastasia Island beach mouse, *Peromyscus polionotus phasma*. (Photo by Philip A. Frank)

foot length 18.7 mm, and length of ear from notch 13.9 mm (Bangs 1898). The comparable measurements on 17 adults in the University of Florida mammal collection are 141.0, 57.4, 19.3, and 14.0 mm.

POPULATION SIZE AND TREND: Bangs (1898:199) said the Anastasia Island beach mouse "fairly swarmed on the sand hills at the lower end of the island." No other indications of numbers have been reported. Today, substantial populations remain at each end of Anastasia Island, and very small populations are scattered along its length (Humphrey et al. 1987). Based on the apparent reduction in distribution, it is reasonable to infer that the total population has declined greatly.

DISTRIBUTION AND HISTORY OF DISTRIBUTION: Published literature records the Anastasia Island beach mouse from only two sites: the type locality at Point Romo, Anastasia Island, St. Johns County, and on the coastal dunes of the peninsular barrier beach extending from St. Augustine Inlet north to the border between Duval and St. Johns counties (Ivey 1949). Ivey noted that the population did not extend farther north into Duval County. Hall (1981) also indicated marginal records of *phasma*

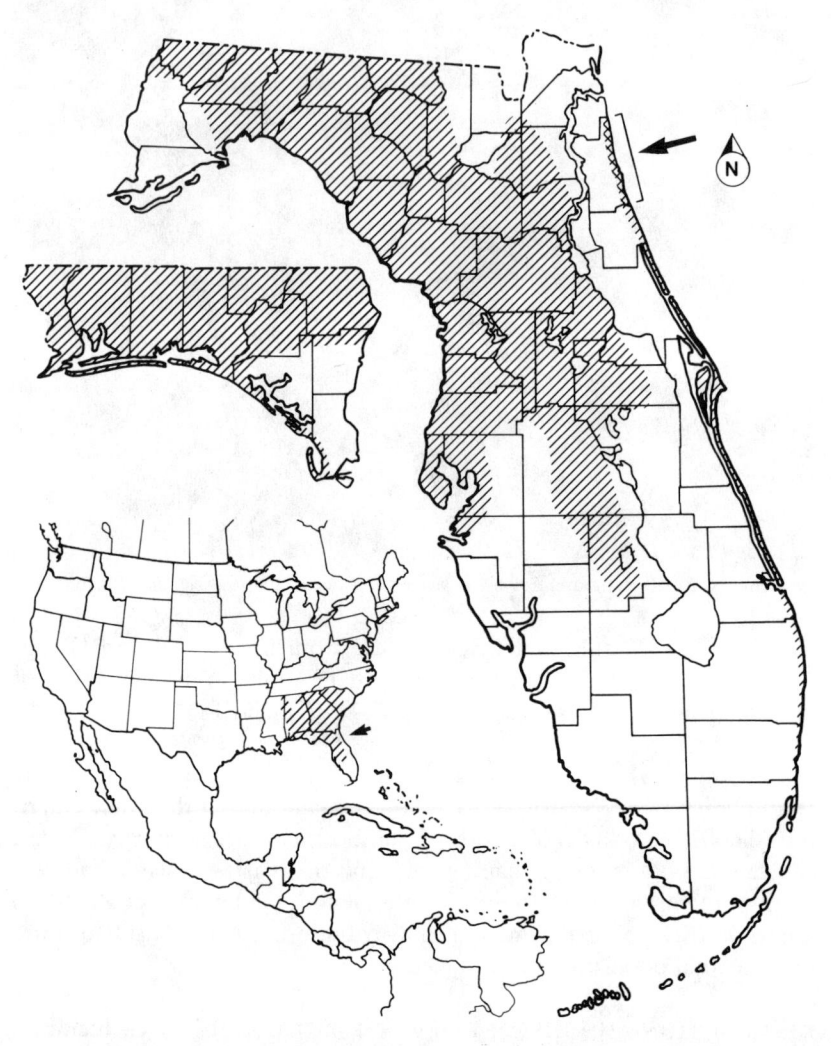

Distribution map of the Anastasia Island beach mouse, *Peromyscus polionotus phasma*. Hatching represents the species; crosshatching and bracket represent the subspecies.

from the border of St. Johns and Duval counties, presumably based on the report of Ivey (1949). R.D. Ivey (personal communication) actually found beach mice all along the coast from St. Augustine Inlet (Vilano Beach) as far north as Mickler Pier (now Mickler Landing). Therefore this subspecies can be expected to occur along the entire coast of St. Johns County (except south of Matanzas Inlet). Much of the dune habitat along this coast has been developed, however, and no longer is suitable for beach mice. Some undeveloped habitat remains on Anastasia Island and between Ponte Vedra Beach and South Ponte Vedra Beach in St. Johns County, but the northern barrier beach has very narrow, poor-quality habitat apparently now devoid of beach mice (Humphrey et al. 1987).

Although Bangs (1898) reported that beach mice were absent from the beaches north of St. Augustine, he referred to failure to catch animals at Burnside Beach, a place-name no longer in use. It is clear from Ivey's (1949) discussion that this locality was near the northern end (perhaps near Mayport or Jacksonville Beach) of the peninsula he studied, and that beach mice occurred only on the southern half.

In a survey in May and June 1986, Humphrey et al. (1987) captured Anastasia Island beach mice in Fort Matanzas National Monument, two sites just north of the monument, a site just north of Versaggi Road in St. Augustine Beach, and Anastasia State Recreation Area (including Conch Island, originally the northernmost of several sand bars in St. Augustine Inlet, but now connected to Anastasia Island by dredge spoil and alongshore deposition). Beach mice were not taken at a few sites from which captures were anticipated on the basis of habitat characteristics. No beach mice were captured on the barrier beach north of St. Augustine Inlet, including four sites in Guana River State Park, between South Ponte Vedra Beach and Mickler Landing.

GEOGRAPHICAL STATUS: This taxon is endemic to a small geographic area and a single habitat.

HABITAT REQUIREMENTS AND HABITAT TREND: Beach mice are restricted to sand dunes vegetated mainly by sea oats (*Uniola paniculata*) and dune panic grass (*Panicum amarulum*), and to the adjoining scrub, characterized by oaks (*Quercus* sp.) and sand pine (*Pinus clausa*) or palmetto (*Serenoa repens*). Although this restriction ordinarily limits the range of beach mice to the 25- to 500-ft-wide strip of coastal dunes as stated by Ivey (1959), Pournelle and Barrington (1953) reported *P. p. phasma* in scrub up to 1800 ft from the dunes. All such areas on Anastasia Island now have been developed for human residence.

Although the historic range in the northern half of St. Johns County appears to be no longer occupied, populations remain distributed along the length of Anastasia Island. However, much of the former habitat has been converted to lawn or concrete associated with development of houses and condominiums. As a result, the remaining habitat is much reduced, and most remaining populations are small. Significant areas of good habitat, densely occupied by beach mice, remain along the lengths of both Anastasia State Recreation Area and Fort Matanzas National Monument, at either end of Anastasia Island.

VULNERABILITY OF SPECIES AND HABITAT: The survival status of the Anastasia Island beach mouse is precarious. Its range appears to be reduced and fragmented. The subspecies persists in large, presumably viable populations on public land at each end of Anastasia Island (Anastasia State Recreation Area and Fort Matanzas National Monument) and in small, marginal populations on numerous parcels of privately owned land. These latter populations are threatened by habitat conversion. The population on the northern end of the island is vulnerable to competition by house mice (*Mus musculus*) and predation by house cats (*Felis catus*) (Humphrey et al. 1987). The population appearing to be least at risk is at Fort Matanzas National Monument, where house mice have not been recorded and house cats are uncommon. Although some uncertainty exists about the threat of exotic rodents and cats to native beach mice, enough is known to justify management action.

CAUSES OF THREAT: Habitat loss appears to have caused substantial reduction of the distribution of this animal. Florida's Coastal Setback Law, which prohibits construction of buildings within certain distances of the mean high-tide line, has not eliminated this threat, because beach mice occur well behind this limit.

A new threat is that house mice have colonized the dune grassland on which beach mice depend. The inference that these two species compete strongly is speculative, but Humphrey and Barbour (1981) presented prima facie evidence for competitive exclusion of other subspecies of beach mice by house mice. The situation on Anastasia Island is unprecedented, because for the first time the two species have been found to co-occur locally. A related issue is that activity of house cats is widespread on Anastasia Island. The effect of these two exotic species on survival of beach mouse populations may be quite important. Either a competitor or a predator alone can eliminate another species, and the effects of a competitor and a predator together could be additive. The two processes operate differently—predation stimulates reproduction of most prey species, whereas competition suppresses reproduction of most species. These biological threats can be expected to affect the fragmented populations even if no new direct loss of habitat occurs.

RESPONSES TO HABITAT MODIFICATION: Because the influence of commensal house mice and house cats extends considerably beyond the limits of human land use, beach mice are extremely sensitive to habitat modification. The survival of beach mouse populations appears to depend on the wilderness quality of their habitat. Preserves designed for them may need to be several square kilometers in size.

DEMOGRAPHIC CHARACTERISTICS: See the species account for the Choctawhatchee beach mouse, *Peromyscus polionotus allophrys*.

KEY BEHAVIORS: See the species account for the Choctawhatchee beach mouse, *Peromyscus polionotus allophrys*.

CONSERVATION MEASURES TAKEN: This subspecies is listed as endangered by the U.S. Fish and Wildlife Service (1989). The best undeveloped habitat is already under public ownership, in Anastasia State Recreation Area and Fort Matanzas National Monument. This taxon was introduced onto a small barrier island off Folly Island, South Carolina, in 1969 and 1971, and onto Devaux Bank, at the mouth of Stono River Inlet, South Carolina, in 1970 (Pelton et al. 1979). The former population (38 released) apparently did not survive; it suffered severe predation by owls, intense wounding of females (presumably from intraspecies fights), and the presence of at least a few house mice. Habitat at the latter site was much more extensive, and this population (18 released) survived at least through 1972, at which time it showed signs of inbreeding depression (Ramsey 1973).

CONSERVATION MEASURES PROPOSED: It is questionable whether the small beach mouse populations remaining on privately owned land are viable, and whether they contribute to the genetic diversity of the two remaining large populations by dispersal. Little opportunity remains for habitat acquisition. Instead, conservation action should focus on habitat management, to ensure that the habitat already acquired is secure from threats other than development. Management of the dune grassland on public lands should strive to prevent colonization or activity of house mice and house cats. Information is needed on the nature of the house mouse and cat populations, however, before management techniques can be recommended with certainty. How to discourage survival of house mice is especially problematic. Systematic control of free-ranging house cats appears to be relatively straightforward. Because of the sensitivity of suburban cat owners, the best method is to live-trap and take captured cats to the local animal shelter.

Although effects on beach mice are unknown, signs of use of the dunes as a public toilet and physical damage of dune vegetation by foot traffic are obvious, especially at Anastasia State Recreation Area, where foot traffic is permitted. Providing sanitary facilities and limiting the number of beach visitors could prevent deterioration of habitat for beach mice.

So much of the coastline once occupied by this subspecies has been lost to human uses that the taxon probably will remain endangered or threatened indefinitely. Regular monitoring of the populations on public lands should be conducted, to allow detection of trends or catastrophes and effects of management practices. Wildlife managers should consider regularly interchanging individuals from the two populations on public land, to reduce the risk that genetic diversity might decline because of inbreeding. If possible, a third population should be established on public land by introducing *P. p. phasma* from Anastasia Island to the suitable habitat at Guana River State Park, within the historical distribution, because addition of a third population would substantially increase the probability of survival of the taxon.

ACKNOWLEDGMENTS: We are grateful to William H. Kern, Jr., and Mark E. Ludlow for essential help with the field work; good information on status is the first step in evaluating the need for and design of conservation action. We thank Mark E. Ludlow and I. Jack Stout for helpful review of the manuscript.

Literature Cited

Bangs, O. 1898. The land mammals of peninsular Florida and the coastal region of Georgia. Boston Soc. Nat. Hist. Proc. 28:157–235.

Bowen, W. W. 1968. Variation and evolution of Gulf coast populations of beach mice, Peromyscus polionotus. Bull. Florida State Museum, Biol. Sci. 12:1–91.

Hall, E. R. 1981. The mammals of North America. 2d ed. John Wiley and Sons, New York. 2:601–1181 + 90 pp.

Humphrey, S. R., and D. B. Barbour. 1981. Status and habitat of three subspecies of *Peromyscus polionotus* in Florida. J. Mammal. 62:840–844.

Humphrey, S. R., W. H. Kern, Jr., and M. E. Ludlow. 1987. Status survey of seven Florida mammals. Unpubl. final report to U.S. Fish and Wildlife Serv., 3100 University Boulevard, South Suite 120, Jacksonville. 39 pp.

Ivey, R. D. 1949. Life history notes on three mice from the Florida east coast. J. Mammal. 30:157–162.

Ivey, R. D. 1959. The mammals of Palm Valley, Florida. J. Mammal. 40:585–591.

Osgood, W. H. 1909. Revision of the mice of the American genus *Peromyscus*. North Amer. Fauna 28:1–285.

Pelton, M. R., F. W. Kinard, Jr., E. Pivorum, A. E. Sanders, and M. Smith. 1979. Status report: The mammals. Pp. 88–92 *in* D. M. Forsythe and W. B. Ezell, Jr., eds., Proc. First South Carolina Endangered Species Symp., South Carolina Wildlife and Marine Res. Dept., Charleston. 196 pp.

Pournelle, G. H., and B. A. Barrington. 1953. Notes on mammals of Anastasia Island, St. Johns County, Florida. J. Mammal. 34:133–135.

Ramsey, P. R. 1973. Spatial and temporal variation in genetic structure of insular and mainland populations of *Peromyscus polionotus*. Unpubl. Ph.D. diss., University of Georgia, Athens. 103 pp.

Selander, R. K., M. H. Smith, S. Y. Yang, W. E. Johnson, and J. B. Gentry. 1971. Biochemical polymorphism and systematics in the genus *Peromyscus*. I. Variation in the old-field mouse (*Peromyscus polionotus*). Univ. Texas Studies in Genetics 6:49–90.

U.S. Fish and Wildlife Service. 1989. Endangered and threatened wildlife and plants; endangered status for the Anastasia Island beach mouse and threatened status for the southeastern beach mouse. Federal Register 54:20598–20602.

Prepared by: Stephen R. Humphrey, Florida Museum of Natural History, University of Florida, Gainesville, FL 32611; Philip A. Frank, Department of Wildlife and Range Sciences, University of Florida, Gainesville, FL 32611.

Endangered

Perdido Key Beach Mouse
Peromyscus polionotus trissyllepsis
FAMILY CRICETIDAE
Order Rodentia

TAXONOMY: No other names. See the species account on the Choctawhatchee beach mouse. Prior to construction of the intracoastal ship canal, interchange might have occurred between *P. p. trissyllepsis* and *P. p. polionotus* because a substantial land connection existed between Perdido Key and the mainland. Dune habitat is separated from occupied areas to the north, however, by a band of poorly drained soils.

DESCRIPTION: The Perdido Key beach mouse is a small *Peromyscus*, dimorphic with respect to color and patterns of head and rump pelage (Bowen 1968). The dorsal pigmented pelage is paler than in *P. p. ammobates* and *P. p. allophrys*. The patterns of pigmentation at the rump are either squared or squared superimposed on tapered; pigmentation does not extend down the thighs as in some *P. p. allophrys*. The mid-dorsal stripe is indistinct. Pigmentation on the head is reduced, with white patches extending dorsally posterior to the ears to the mid-dorsal stripe. White hairs are unpigmented to their base; the tail stripe is lacking (Bowen 1968). Means and extremes (in parentheses) for external measurements (in mm) of males ($n = 8$) and females ($n = 7$), respectively (age not specified; Bowen 1968), were: total length, 122, 127 (115–139); length of tail, 49, 50 (45–54); length of hind foot, 17.2, 17.8 (16.2–18.2); ear 14.2, 14.5 (13.4–15). Hall (1981) described the skull of *P. polionotus* as resembling that of the smaller subspecies of *P. maniculatus* but with palatine slits relatively shorter and auditory bullae relatively larger.

POPULATION SIZE AND TREND: The Perdido Key beach mouse is the most endangered of the five subspecies listed as endangered by the U.S.

Perdido Key beach mouse, *Peromyscus polionotus trissyllepsis*. (Photo by George W. Folkerts)

government. The 1979 population of this subspecies was estimated to be 78 individuals, 52 on Gulf Islands National Seashore at the eastern end of Perdido Key in Florida and 26 at Gulf State Park in Alabama on the western tip of Perdido Key (Humphrey and Barbour 1981). In 1982 Holliman (1983) captured only one mouse at Gulf State Park (110 trap-nights). Meyers (1983) captured 13 individuals at Gulf State Park (400 trap-nights) in 1982 but failed to capture mice on Gulf Islands National Seashore. He suggested that the mice were extirpated from that area during Hurricane Frederic in 1979. Both Humphrey and Barbour (1981) and Meyers (1983) found no evidence of beach mice in the central portion of the key. Thus, the only known population of this subspecies in April 1986 was at Gulf State Park in Alabama. That population has been through several severe genetic bottlenecks. Estimates obtained in 1986 and 1987 (Holler et al. 1989), however, indicate that it is still viable (30 individuals in April 1986; >100 individuals in April and November 1987). A second population has been established by translocation from Gulf State Park, Alabama, to Gulf Islands National Seashore, Florida (see below; Holler et al. 1989). That population occupies 11 km of habitat and probably exceeds 100 individuals. Minimum effective population size for long-term survival of beach mouse populations has not been estimated.

DISTRIBUTION AND HISTORY OF DISTRIBUTION: Bowen (1968) described the distribution of the Perdido Key beach mouse as the coastal dunes between Perdido Bay, Alabama, and Pensacola Bay, Florida. Humphrey and Barbour (1981), trapping in 1979, found the distribution to be restricted to 10.4 km of habitat at Gulf Islands National Seashore on

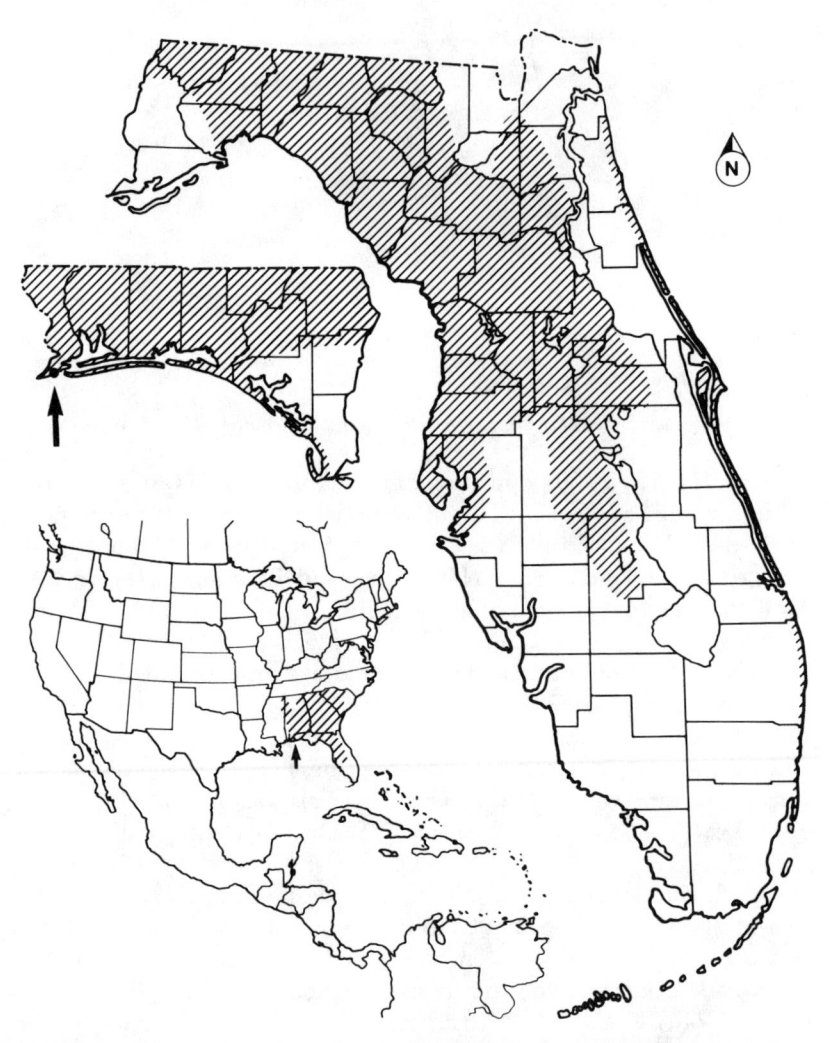

Distribution map of the Perdido Key beach mouse, *Peromyscus polionotus trissyllepsis*. Hatching represents the species; the dot represents the subspecies.

the eastern end of Perdido Key and to 2.6 km at Gulf State Park, Alabama, on the western end. Meyers (1983) reported loss of the Gulf Islands National Seashore population. Current known distribution is restricted to 1.9 km of habitat at Gulf State Park, Alabama, and 11 km at Gulf Islands National Seashore occupied by a recently reestablished population (see below).

GEOGRAPHICAL STATUS: The Perdido Key beach mouse is native only to the coast along Perdido Key in Florida and Alabama. Once present in more or less continuous populations along the coastal dunes between the entrances to Perdido and Pensacola bays, it remains as two disjunct populations.

HABITAT REQUIREMENTS AND HABITAT TREND: Habitat for this mouse consists of the sand dunes along the Gulf coast beaches. See the species account on the Choctawhatchee beach mouse. No specific requirements are known for this subspecies. The remaining natural population occupies an area characterized by a primarily continuous line of medium-height, frontal dunes with no or very few secondary dunes lying inland; scrub dunes are lacking. The frontal dunes are vegetated with coarse grasses (primarily sea oats, *Uniola paniculata*, and bluestem, *Andropogon maritimus*) at moderate density. Gulf Islands National Seashore, occupied by the translocated population, has a broken line of primary dunes and secondary dunes vegetated with coarse grasses; scrub dunes are lacking.

Three areas of critical habitat totaling 16.4 km (Mike Dawson, U.S. Fish and Wildlife Service, personal communication) have been defined for the Perdido Key beach mouse (U.S. Fish and Wildlife Service 1987). Zone 1 (1.9 km) is at Gulf State Park, Alabama, and is occupied. Zone 2 (2.3 km) is at Perdido Key State Recreation Area, Florida, and is reportedly not occupied, although mouse sign (species undetermined; possibly *Mus musculus*) has been observed there recently. Zone 3 (12.2 km) is on Gulf Islands National Seashore and until recently was unoccupied; it has been repopulated by translocation from the Gulf State Park population.

See the species account on the Choctawhatchee beach mouse for information on diet.

VULNERABILITY OF SPECIES AND HABITAT: See the species account on the Choctawhatchee beach mouse. The major threat to this subspecies is habitat loss and direct mortality of mice as a result of tropical storms. This threat has been heightened as a result of habitat fragmentation that precludes natural reestablishment of lost populations from contiguous areas. All critical habitat is in public ownership. Zones 1 and 3, however, because of their narrow configuration and lack of high inland scrub dunes, are very susceptible to storm damage. Additional threats to the existing population at zone 1 may result from predation by domestic cats (*Felis catus*) and red fox (*Vulpes vulpes*). This area is small, bordered at the eastern end by condominium developments, and has a high fox population. Zone 2 has a well-developed system of scrub dunes; however, it is small and bordered on both ends by residential development.

CAUSES OF THREAT: Habitat loss as a result of real estate development and storm damage has resulted in the endangered status of this subspecies. Vulnerability of remaining habitat and populations to storm damage is the greatest present threat. Predation and potential for competition with house mice are additional threats in critical habitat zones 1 and 2.

RESPONSES TO HABITAT MODIFICATION: See the species account on the Choctawhatchee beach mouse.

DEMOGRAPHIC CHARACTERISTICS: Demographic data are lacking for *P. p. trissyllepsis* and must be inferred from data published for other subspecies; see the species account on the Choctawhatchee beach mouse. Recent trapping data (Holler et al. 1989) indicates continued viability of this subspecies. The population at Gulf State Park increased from an estimated 30 individuals to >100 individuals between April 1986 and April 1987 and stabilized at that level through 1989. Furthermore, reproductively active females (pregnant, lactating, or both) were numerous in spring and fall.

KEY BEHAVIORS: No data have been reported specifically for *P. p. trissyllepsis*; see the species account on the Choctawhatchee beach mouse for reports on other subspecies.

CONSERVATION MEASURES TAKEN: The Perdido Key beach mouse was listed by the U.S. government as endangered pursuant to the Endangered Species Act of 1973 on 6 June 1985 (50 FR 23872). It is also listed as endangered in Title 39-27.002 of the Florida Administrative Code and it is included as a protected species under Alabama's nongame regulation 87-GF-7. Thus, this subspecies has both federal and state protection throughout its range. Critical habitat (16.4 km) has been designated and is in public ownership; the beach mouse is given consideration in all management decisions involving these lands. The Florida Game and Fresh Water Fish Commission, in cooperation with the National Park Service, U.S. Department of the Interior, and the Alabama Cooperative Fish and Wildlife Research Unit, has initiated a program to reestablish this subspecies at Gulf Islands National Seashore. Seventeen pairs of mice were translocated from Gulf State Park to the seashore between fall 1986 and spring 1989. Trapping along 6.2 km and observation of sign in the remainder of habitat at the seashore in 1987 and 1988 has shown that the population is well established (Holler et al. 1989). All habitat (approximately 11 km) is occupied and the population is believed to exceed 100 individuals. A captive breeding colony is being established at Auburn University to provide a source of mice should the natural populations be lost. A slide and tape show has been developed by the Alabama Cooperative Fish and Wildlife Research Unit for the U.S. Fish and Wildlife Service to educate the public regarding beach mice and their need for protection. Additionally, an educational brochure is being prepared by the U.S. Fish and Wildlife Service.

CONSERVATION MEASURES PROPOSED: Full potential recovery of the Perdido Key beach mouse would result in self-sustaining populations at each of the three remaining areas of suitable habitat. These are the only areas remaining suitable for sustained habitation by this subspecies within its native range. Establishment of populations in each area might justify downlisting of the subspecies to threatened status. Because of the extensive loss of habitat it will probably never be possible to remove the subspecies from protection under the Endangered Species Act of 1973 (U.S. Fish and Wildlife Service 1987).

The recovery plan (U.S. Fish and Wildlife Service 1987) identifies three principal thrusts for recovery of this and two other subspecies of endangered beach mice: maintenance and restoration of suitable habitat, development of reestablishment programs, and education of the general public regarding beach mice and their requirements for protection. The recovery plan provides detailed recommendations in each of these areas.

The most critical need for recovery of this subspecies at the time of its listing was reestablishment of the mice at a second location. This has been accomplished. The population must now be monitored to determine long-term success and to evaluate impacts of a major beach renourishment project at the site (Department of the Navy 1987). Habitat at

Perdido Key State Recreation Area (zone 2) should be surveyed; recently we observed mouse sign there. Live trapping should be conducted to determine which species (*P. p. trissyllepsis, Mus musculus,* or both) are present. If beach mice are absent, a translocation program (including control of house mice, if warranted) should be undertaken. A captive colony composed of at least 30 breeding pairs of progeny from wild-trapped mice is proposed for maintenance at Auburn University. Genetic studies are also needed to determine the degree of inbreeding within each population, the need for interchange of genetic materials among populations (assuming relocation efforts are successful), and the interrelatedness of the various subspecies.

ACKNOWLEDGMENTS: The author of the species account in the 1978 edition was Donald W. Linzey.

Literature Cited

Bowen, W. W. 1968. Variation and evolution of Gulf coast populations of beach mice, *Peromyscus polionotus*. Bull. Florida State Mus., Biol. Sci. 12(1):1–91.

Department of the Navy. 1987. Final environmental impact statement. United States Navy, Southern Division, Naval Facilities Engineering Command, Charleston, South Carolina.

Hall, E. R. 1981. The mammals of North America. John Wiley and Sons, New York. 1181 + 90 pp.

Holler, N. R., D. W. Mason, R. M. Dawson, T. Simons, and M. C. Wooten. 1989. Reestablishment of the Perdido Key beach mouse (*Peromyscus polionotus trissyllepsis*) on Gulf Islands National Seashore. Conserv. Biol. 3:397–404.

Holliman, D. C. 1983. Status and habitat of Alabama Gulf coast beach mice *Peromyscus polionotus ammobates* and *P. p. trissyllepsis*. Northeast Gulf Sci. 6:121–129.

Humphrey, S. R., and D. B. Barbour. 1981. Status and habitat of three subspecies of *Peromyscus polionotus* in Florida. J. Mammal. 62:840–844.

Meyers, J. M. 1983. Status, microhabitat, and management recommendations for *Peromyscus polionotus* on Gulf coast beaches. Unpubl. report to U.S. Fish and Wildlife Service, Atlanta, Georgia. 29 pp.

U.S. Fish and Wildlife Service. 1987. Recovery plan for the Alabama beach mouse (*Peromyscus polionotus ammobates*), Perdido Key beach mouse (*P. p. trissyllepsis*), and Choctawhatchee beach mouse (*P. p. allophrys*). U.S. Fish and Wildlife Service, Atlanta, Georgia. 45 pp.

Prepared by: Nicholas R. Holler, U.S. Fish and Wildlife Service, Alabama Cooperative Fish and Wildlife Research Unit,* 331 Funchess Hall, Auburn University, AL 36849.

*Cooperators: U.S. Fish and Wildlife Service, Game and Fish Division of the Alabama Department of Conservation and Natural Resources, Wildlife Management Institute, Auburn University (Alabama Agricultural Experiment Station—Contribution #15-881494P, Department of Fisheries and Allied Aquacultures, Department of Zoology and Wildlife Science).

Endangered

Key Largo Cotton Mouse
Peromyscus gossypinus allapaticola
FAMILY CRICETIDAE
Order Rodentia

TAXONOMY: This subspecies was described from a sample of 21 specimens, which was compared with 44 specimens of three other subspecies from the mainland of Florida (Schwartz 1952*a*). The Key Largo form was more reddish in color of the dorsal fur and larger in most external and cranial measurements, including total length, tail length, length of ear from notch, greatest length of skull, basilar length, palatal length, length of palatine slits, and length of diastema. The name was derived from the Seminole Indian term *allapattah* for the tropical dry deciduous hammocks of southern Florida.

DESCRIPTION: The Key Largo cotton mouse is distinctly larger and more reddish in color than other subspecies of cotton mice from peninsular Florida. Average measurements for 12 adults of this subspecies are: total length 179 mm, tail length 77 mm, ear from notch 18 mm, greatest skull length 28.3 mm, zygomatic breadth 14.1 mm, and length of the maxillary tooth row 4.0 mm. Its fur is dark hazel dorsally, grading into grizzled medium brown on the sides, and white ventrally with a cinnamon-buff wash over the throat and chest. The tail is bicolored, brown above and white below (Schwartz 1952*a*).

POPULATION SIZE AND TREND: An average of 21.2 Key Largo cotton mice per ha of forest was estimated in the dry season of 1986 (Humphrey 1988*a*). Three grids near housing subdivisions averaged 15.5 cotton mice per ha, and three distant from such development averaged 26.9 cotton mice per ha. This difference is unexplained. Elsewhere in the species' range, densities ranging from 1.8 to >250 per ha have been reported (Smith 1982; Smith et al. 1984; Young and Stout 1986). Even the lowest

Peromyscus gossypinus allapaticola 111

Key Largo cotton mouse,
Peromyscus gossypinus allapaticola.
(Photo by Stephen R. Humphrey)

densities found on Key Largo, however, are higher than those typical in most parts of Florida. Extrapolation over the 851 ha of undeveloped upland on northern Key Largo suggests that a population of about 18,000 cotton mice is attainable (Humphrey 1988*a*).

DISTRIBUTION AND HISTORY OF DISTRIBUTION: The past known or possible distribution of this subspecies included all the tropical hardwood forest on Key Largo, from the northern end south at least as far as Planter, near Tavernier, on Plantation Key (Osgood 1909). (Key Largo and Plantation Key appear on many maps to be contiguous, but they are separated by a narrow, shallow marine channel.) In a later study, this subspecies was not trapped in appropriate habitat on Plantation Key, Elliot Key, and Upper and Lower Matecumbe keys (Schwartz 1952*b*). The type locality is 19.3 km northeast of Rock Harbor (Schwartz 1952*a*). No

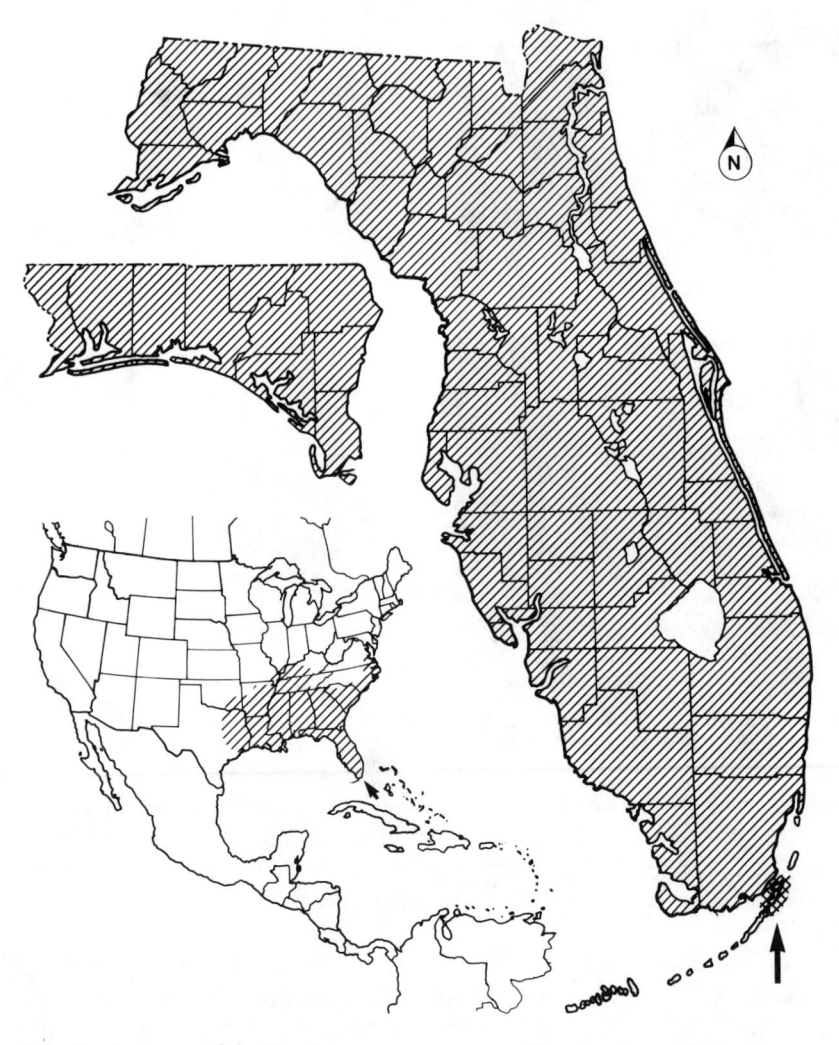

Distribution map of the Key Largo cotton mouse, *Peromyscus gossypinus allapaticola*. Hatching represents the species; the dots represent the subspecies.

modern hiatus occurs in the range of cotton mice in southern Florida, although this subspecies is absent from several islands north of Key Largo (Schwartz 1952*b*).

All recent reports (Schwartz 1952*b*; Barbour and Humphrey 1982) show that the subspecies is now restricted to the northern half of Key Largo, north of the point where U.S. Highway 1 enters Key Largo. Efforts to find populations in apparently suitable but isolated fragments of habitat to the south (1.3 km north of Key Largo Elementary School, in John Pennekamp State Park, and between highway mile markers 96.8 and 97.2 south of the town of Rock Harbor) have been unsuccessful (Goodyear 1985; S. R. Humphrey, unpubl. observation).

In 1970, 14 individuals of this subspecies from Key Largo were introduced on Lignumvitae Key, Monroe County, Florida (Brown and Williams 1971). The habitat there (approximately 90 ha) appears to be suitable. A specimen was taken there by Jennie Parks in the early 1980s, but none were found during trapping in 1987–90 (P. W. Wells, personal communication).

GEOGRAPHICAL STATUS: Present and past distribution of the Key Largo cotton mouse is restricted to the Upper Keys of Florida, so this is an endemic subspecies of a species that is widely distributed in the southeastern United States.

HABITAT REQUIREMENTS AND HABITAT TREND: The Key Largo cotton mouse occurs in deciduous forest in the life zone designated "dry tropical forest" (Holdridge 1967). The forest is taxonomically and ecologically similar to that found on the coasts of Caribbean islands and Central America. Composition of this forest community on northern Key Largo has been described as follows (Hilsenbeck 1976). It includes 48 species of trees, 6 shrubs, 10 vines, 8 epiphytes, 3 herbs, 2 grasses, and 2 ferns. An area of 0.4 ha included 3703 trees of 34 species. Seven dominant species of trees accounted for 61 percent of the basal area, 69 percent of the importance value, and 79 percent of the density in this area.

Key Largo cotton mice find shelter underground and use entrance holes that are about 3–5 cm in diameter (Goodyear 1985). Most such holes are next to the bases of trees, but some occur in the forest floor away from tree bases. Often the entrances of holes are partly covered by leaves or pieces of bark. Cotton mice also are captured in woodrat holes, nests, or runways.

Early indications that this subspecies is most abundant in older hardwood forest and less numerous in young stands (Brown 1978; Barbour and Humphrey 1982) appear to be incorrect. Humphrey (1988*a*) was unable to confirm this pattern and also found cotton mice to occupy *Salicornia* coastal strand adjacent to forest (unpubl. observation). Goodyear (1985) found cotton mice to occupy the entire sere from recently burned forest (dominated by bracken fern and tree seedlings) to maturing forest. The features of the environment that constitute optimal habitat for this animal are unknown.

The type locality had deep leaf litter and abundant dead and decaying logs. Where animals were absent farther southwest on Key Largo, "the Key Largo limestone comes almost to the surface, trees (*Lysiloma bahamensis*) are dense, and there is good shrubby growth between them" (Schwartz 1952*b*:94). A pineland once occurred on the southern end of Key Largo, but only a few of the pineland-related species persist there (Alexander 1953). This is good evidence that Key Largo has been free of naturally occurring fires for many decades. Given the habitat requirements of the Key Largo cotton mouse, this portion of the key probably has been unoccupied for a long time. Where fires occur repeatedly, both the living plants and the peaty soil are removed. A bare limestone surface remains, so succession to the climax association apparently most favorable to this subspecies probably is very slow.

Much of the hardwood forest on Key Largo was cleared for agriculture (mostly pineapples) in the late 19th century. Selective harvest of mahogany (*Swietenia mahogani*) for furniture and lignumvitae (*Guaiacum sanctum*) for ships also occurred. In 1906 a severe pineapple blight occurred, and by 1915 production had ended. Many fields underwent subsequent succession to young hardwood forests. In the 1930s, some lime groves were established, and some of these trees can still be found alive in successional forest. Only a few clearings for agriculture have occurred in recent decades, but sizable portions of the forest have been burned. Although people have lived on Key Largo for more than a century, land clearing for residential and commercial development accelerated greatly after Word War II. On southern Key Largo, which receives fresh water from the mainland by pipeline, only early successional fragments of the hardwood forest remain among intensive real estate development.

An important fact for restoration of cotton mouse habitat is that any upland site in an early successional stage, if adjacent to a mature forest serving as a source of seeds and allowed to undergo succession, should mature to form good habitat for the subspecies. Most species of trees of the tropical dry deciduous forest are adapted for dispersal of seeds either by wind or by frugivorous birds. The white-crowned pigeon (*Columba leucocephala*) is common and forages on the fruits of many species in the forest, dispersing seeds widely while foraging, flying, or perching. Consequently this species speeds succession and is a major force in the resilience of the forest to disturbances that fall short of permanent conversion of the habitat.

Occupied habitat now occurs only on the northern end of the key. The total area of upland on northern Key Largo that could revert to forest habitat after succession is estimated at 851 ha (Siemon et al. 1986; Humphrey 1988*b*).

VULNERABILITY OF SPECIES AND HABITAT: Although the habitat of this subspecies is quite vulnerable to conversion to human uses, the subspecies appears to be relatively secure from other types of threats. If the forest is not developed, cotton mice can survive in the range of seral stages from recent burns to maturing forest. See the species account on the Key Largo woodrat (*Neotoma floridana smalli*) for a detailed discussion of vulnerability. That black rats (*Rattus rattus*) seem not to be a problem for Key Largo cotton mice is in curious contrast to their implication in the extinction of two other subspecies of cotton mice (*P. g. restrictus* and *P. g. anastasae*); perhaps the numerous, large woodrats on Key Largo prevent residence of black rats through agonistic behavior.

CAUSES OF THREAT: See the species account on the Key Largo woodrat, *Neotoma floridana smalli*.

RESPONSES TO HABITAT MODIFICATION: Considering the long history of modification of the forests on northern Key Largo and the potential for invasion by exotic competitors and predators, the cotton mouse there seems remarkably resilient. However, details and limits of its responses are unknown. The relatively low densities of cotton mice near human developments may have a cause-and-effect relationship, but none has been demonstrated. Possible causes include pesticides, young age of forest stands, competing black rats, and predatory house cats (Humphrey 1988*b*).

DEMOGRAPHIC CHARACTERISTICS: No thorough study of reproduction of the Key Largo cotton mice has been done; composite data from three studies are somewhat contradictory, perhaps because the pattern varies among years. Schwartz (1952*b*) concluded that litters are probably born year-round, as assumed for the mainland form, *P. g. palmarius*. No pregnant females were captured. Two females taken on 6 and 7 February 1951 were lactating. Immatures with gray pelage were captured on 28

December 1950. In contrast, Barbour and Humphrey (1982) concluded that breeding is highly seasonal. Among the 15 adult females examined in June 1979, none was pregnant or lactating, and six had perforate vulvae, indicating recent copulation. Similarly, most of the adult males (14 of 20) had descended testes in June. No juvenile animals were captured in June. Goodyear (1985) captured juveniles from mid-May to mid-June with a peak in late May.

On tree islands in the Florida Everglades, the timing of reproduction in *P. g. palmarius* varies from year to year. Usually breeding occurs during the wet season and is curtailed during the dry season, and the year's peak population density occurs in mid- to late winter. Breeding occurred in the dry season during one year, however, and no breeding occurred in another, causing a gradual population decline without the usual peak (Smith 1982).

Some information on maximum longevity is available for the cotton mouse subspecies *gossypinus* in a different habitat in northern Florida (Pearson 1953). Three of 68 individuals trapped lived at least 1 year and 9 months. The subspecies *gossypinus* has an average litter size of 3.7 with a 1:1 sex ratio at birth (Pournelle 1952). The earliest record of a fertile mating of a female cotton mouse (subspecies *gossypinus*, in captivity) was for one 73 days old (Pournelle 1952). Captive-born young matured with no ill effects when separated from their mothers at 20–25 days of age (Pournelle 1952).

KEY BEHAVIORS: If the Key Largo cotton mouse resembles other subspecies of cotton mice, its activity is nocturnal and both subterranean and arboreal, but not terrestrial. Cotton mice nest underground in cavities made by tree roots, and they forage in the forest canopy, traveling from tree to tree throughout the home range. Cotton mice probably spend relatively little time on the forest floor, where most of their potential predators live. When released at trapping sites, individuals run directly to small, inconspicuous holes in the forest floor, under logs or roots (Barbour and Humphrey 1982). The diet of the Key Largo cotton mouse has not been documented, but other members of its species and genus are omnivorous.

Data on activity range of another subspecies (*gossypinus*) in a different habitat (Pearson 1953) show that population size was correlated with the amount of food produced by the forest. Home ranges (as reflected by the average distance between points of release and subsequent recapture) were smaller during high population density than during lower density. The average distance moved by males was larger than for females, indicating larger activity ranges. Activity ranges of the two sexes overlapped extensively, but territoriality appeared to be strong within each sex. Activity ranges of males overlapped somewhat, but after a resident male was removed, the ranges of neighboring males extended into the vacant area. Activity ranges of females did not overlap at all.

CONSERVATION MEASURES TAKEN: The Key Largo cotton mouse is listed as endangered by the Florida Game and Fresh Water Fish Commission (1990) and the U.S. Fish and Wildlife Service (1989). See the species account on the Key Largo woodrat, *Neotoma floridana smalli*.

CONSERVATION MEASURES PROPOSED: See the species account on the Key Largo woodrat, *Neotoma floridana smalli*.

ACKNOWLEDGMENTS: I thank James N. Layne and I. Jack Stout for helping me to improve the manuscript.

Literature Cited

Alexander, T. R. 1953. Plant succession on Key Largo, Florida, involving *Pinus caribea* and *Quercus virginiana*. Quart. J. Florida Acad. Sci. 16:133–138.

Barbour, D. B., and S. R. Humphrey. 1982. Status and habitat of the Key Largo woodrat and cotton mouse (*Neotoma floridana smalli* and *Peromyscus gossypinus allapaticola*). J. Mammal. 63:144–148.

Brown, L. N. 1978. Key Largo cotton mouse. Pp. 10–11 *in* J. N. Layne, ed., Inventory of rare and endangered biota of Florida. Vol. 1. Mammals. University Presses of Florida, Gainesville. xx + 52 pp.

Brown, L. N., and R. L. Williams. 1971. The Key Largo woodrat (*Neotoma floridana snalli*) (sic) and cotton mouse (*Peromyscus gossypinus allapaticola*) on Lignum Vitae Key, Florida. Florida Nat. 44:95–96.

Florida Game and Fresh Water Fish Commission. 1990. Official lists of endangered and potentially endangered fauna and flora in Florida. Tallahassee, Florida. 23 pp.

Goodyear, N. C. 1985. Results of a study of Key Largo woodrats and cotton mice: Phase I, spring and summer 1985. Unpubl. report to U.S. Fish and Wildlife Serv., 3100 University Boulevard, South Suite 120, Jacksonville. 76 pp.

Hersh, S. L. 1978. Ecology of the Key Largo woodrat. Unpubl. M.S. thesis, University of Miami, Coral Gables. 106 pp.

Hersh, S. L. 1981. Ecology of the Key Largo woodrat (*Neotoma floridana smalli*). J. Mammal. 62:201–206.

Hilsenbeck, C. E. 1976. A comparison of forest sampling methods in hammock vegetation. Unpubl. M.S. thesis, University of Miami, Coral Gables. 106 pp.

Holdridge, L. R. 1967. Life zone ecology. Tropical Sci. Center, San Jose, Costa Rica. 206 pp.
Humphrey, S. R. 1988a. Density estimates of the Key Largo woodrat and cotton mouse (*Neotoma floridana smalli* and *Peromyscus gossypinus allapaticola*), using the nested grid approach. J. Mammal. 69:524–531.
Humphrey, S. R. 1988b. The habitat conservation plan. Conservation Biol. 2: 240–244.
Osgood, W. H. 1909. Revision of the mice of the American genus *Peromyscus*. North American Fauna 28:1–285.
Pearson, P. G. 1953. A field study of *Peromyscus* populations in Gulf Hammock, Florida. Ecology 34:199–207.
Pournelle, G. H. 1952. Reproduction and early post-natal development of the cotton mouse, *Peromyscus gossypinus gossypinus*. J. Mammal. 33:1–20.
Schwartz, A. 1952a. Three new mammals from southern Florida. J. Mammal. 33:381–385.
Schwartz, A. 1952b. The land mammals of southern Florida and the Upper Florida Keys. Unpubl. Ph.D. diss., University of Michigan, Ann Arbor. 180 pp.
Siemon, Larsen, and Purdy. 1986. North Key Largo, Florida, Habitat Conservation Plan. Unpubl. report to Gov. Bob Graham. 128 pp., 2 appendices.
Smith, A. T. 1982. Population and reproductive trends of *Peromyscus gossypinus* in the Everglades of south Florida. Mammalia 46:467–475.
Smith, M. W., W. R. Teska, and M. H. Smith. 1984. Food as a limiting factor and selective agent for genic heterozygosity in the cotton mouse *Peromyscus gossypinus*. Amer. Midland Nat. 112:110–118.
U.S. Fish and Wildlife Service. 1983. Final land protection plan, Crocodile Lake National Wildlife Refuge, Monroe County, Florida. U.S. Fish and Wildlife Serv., Atlanta, Georgia. 22 pp., 3 appendices, 2 maps.
U.S. Fish and Wildlife Service. 1984. Endangered and threatened wildlife and plants; determination of endangered status for the Key Largo woodrat and Key Largo cotton mouse. Federal Register 49:34504–34510.
U.S. Fish and Wildlife Service. 1985. Environmental impact statement on North Key Largo (Florida) habitat conservation plan and endangered species permit. Federal Register 50:14299–14301.
U.S. Fish and Wildlife Service. 1989. Endangered and threatened wildlife and plants. U.S. Government Printing Office, Washington, D.C. 34 pp.
Young, B. L., and J. Stout. 1986. Effects of extra food on small rodents in a south temperate zone habitat: Demographic responses. Canadian J. Zool. 64: 1211–1217.

Prepared by: Stephen R. Humphrey, Florida Museum of Natural History, University of Florida, Gainesville, FL 32611.

Endangered

Key Largo Woodrat
Neotoma floridana smalli
FAMILY CRICETIDAE
Order Rodentia

TAXONOMY This subspecies was described from a sample of 46 specimens from Key Largo, Monroe County, Florida, which was compared with 110 specimens of eastern woodrats from elsewhere in Florida (Sherman 1955). The sphenopalatine vacuity in island specimens was found to be narrower and positioned farther posterior than in mainland woodrats. Measuring this trait as the ratio of the distance from the front of the incisor to the vacuity to skull length, Sherman was able to separate 84 percent of the Key Largo specimens from mainland ones. Sherman also was able to distinguish all but two mainland specimens from island ones based on shape and position of the vacuity. Schwartz and Odum (1957) compared *N. f. smalli* with samples from the rest of the species' range and judged subspecific distinction to be appropriate. The subspecies was named for the botanist John Small, who first recorded its presence on Key Largo (Small 1923).

DESCRIPTION: The Key Largo woodrat is a medium-sized rodent. Its fur is sepia dorsally, cinnamon on the sides, and cream or white below; the feet are white. The tail is shorter than the head and body, sparsely haired, and not sharply bicolored. The subspecies is distinguishable from mainland conspecifics only by the skull features described above. Males are slightly larger than females. The normal range of measurements of adults is: total length 320–420 mm, length of the head and body 170–220 mm, length of the hind foot 34–40 mm, and condylobasal skull length 38–48 mm (Sherman 1955). This subspecies is slightly smaller than conspecifics from peninsular Florida, but overlap in size of the two is considerable. Body size changes clinally across the range of the eastern woodrat, and this subspecies represents the small extreme of the spectrum (Schwartz and Odum 1957).

Key Largo woodrat, *Neotoma floridana smalli*. (Photo by Barry W. Mansell)

POPULATION SIZE AND TREND: The first density estimate available for Key Largo woodrats was 2.2 per ha from a single study area selected for its high density of woodrat nests (Hersh 1981). Barbour and Humphrey (1982) extrapolated a lower value to the 475 ha of maturing forest in which they found woodrat houses in 1979, suggesting a total population of 654 woodrats. Goodyear (1985), however, found that woodrat numbers are poorly correlated with the abundance of houses and that woodrats also occur in locations where houses are absent. Subsequently, an average of 7.6 Key Largo woodrats per ha of forest was estimated on six study areas in the dry season of 1986 (Humphrey 1988*a*). Three grids near housing subdivisions averaged 3.1 woodrats per ha, and three distant from such development averaged 12.2 woodrats per ha. This difference is unexplained. The new data show that woodrat density on northern Key Largo can reach about 7 times the level found by Hersh and about three times the maximum reported for other subspecies of the eastern woodrat (Fitch and Rainey 1956). Extrapolation over the 851 ha of undeveloped upland on northern Key Largo suggests that a population of about 6500 woodrats is attainable (Humphrey 1988a).

DISTRIBUTION AND HISTORY OF DISTRIBUTION: The past known or possible distribution of the native range of this subspecies included all

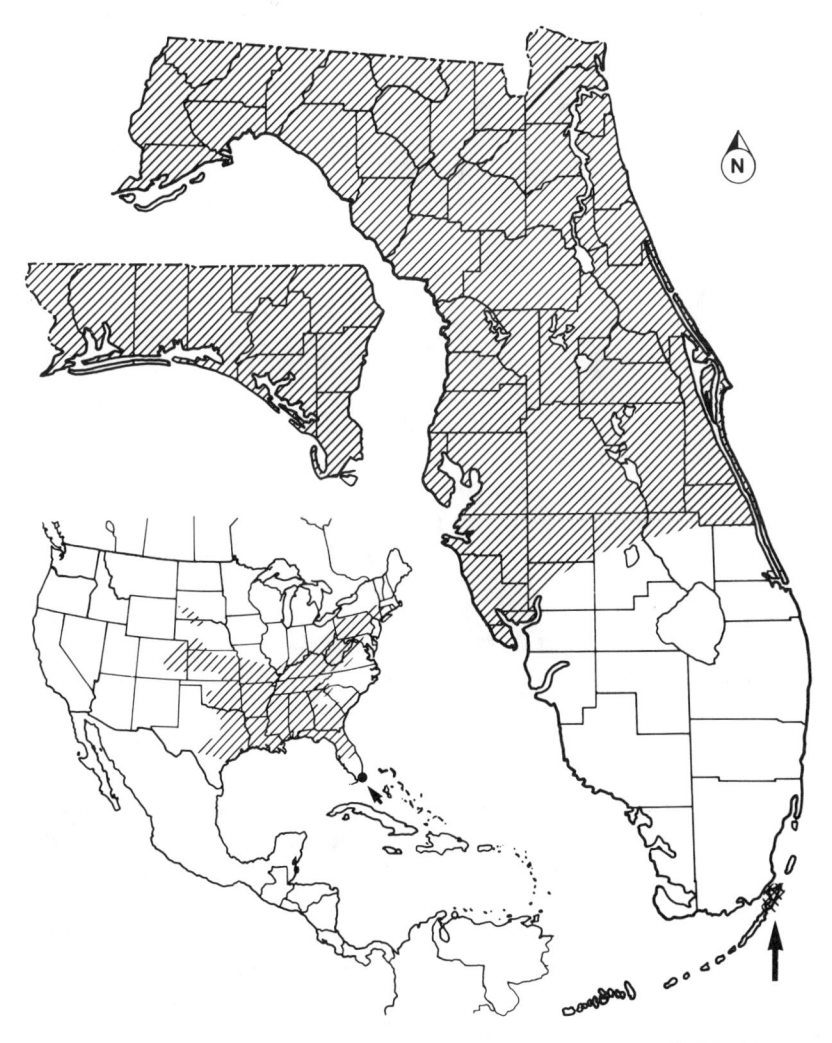

Distribution map of the Key Largo woodrat, *Neotoma floridana smalli*. Hatching represents the species; crosshatching represents the subspecies.

the tropical hardwood forest on Key Largo, from the northern end south at least as far as Rock Harbor (Schwartz 1952). The earliest report on population status (Schwartz 1952) indicated that woodrats were most common on the northern end of the key. The present known or possible distribution of this subspecies (Barbour and Humphrey 1982; Goodyear 1985) includes approximately 851 ha (Siemon et al. 1986) of forest or afforestable upland on the northern half of Key Largo, Monroe County, Florida. The actual distribution is smaller than this, however, because not all of the upland is in suitable habitat at present. All of this distribution is north of the point where U.S. Highway 1 enters Key Largo. Efforts to find populations in apparently suitable but isolated fragments of habitat to the south (1.3 km north of Key Largo Elementary School, in John Pennekamp State Park, and between highway mile markers 96.8 and 97.2 south of the town of Rock Harbor) have been unsuccessful (Goodyear 1985; S. R. Humphrey, unpubl. observation).

A population was established (Brown and Williams 1971) by introduction to Lignumvitae Key, Monroe County, Florida, which supports approximately 90 ha of forest habitat (Barbour and Humphrey 1982). Based on the abundance of woodrat houses, this population grew substantially after its introduction. An abrupt decline in the number of active houses occurred in 1987, however, and none have been seen in recent years (P.W. Wells, personal communication). Furthermore, trapping in 1990 did not produce any captures, indicating that this population may have been lost to unknown causes (J. G. Duquesnel, personal communication).

GEOGRAPHICAL STATUS: This subspecies is endemic to Key Largo, Monroe County, Florida, and it is distinct from its mainland relatives (Sherman 1955; Schwartz and Odum 1957). The nearest modern records of the peninsular subspecies (*N. f. floridana*) are about 260 km from Key Largo—from Murdock, Charlotte County (Sherman 1944), and near Vero Beach, Indian River County (Layne 1974). There is credible evidence, however, that the eastern woodrat once ranged southward along the Atlantic coastal ridge to a point about 35 km north of Key Largo. A subfossil specimen of *N. floridana* was found in a solution pit in Nichol's Hammock near Princeton, Dade County, in a stratum judged to be post-Columbian (Hershfeld 1968). This record suggests that the presettlement distribution of woodrats may have been nearly continuous along the southeastern coast of Florida, unlike the present disjunct distribution.

HABITAT REQUIREMENTS AND HABITAT TREND: Woodrats seek shelter underground, constructing a burrow system that includes a nest chamber and one or more entrances usually covered by a structure of sticks and other debris. Well-defined runs often lead into the forest from burrow entrances. Where the forest soil is shallow, the burrow may be so short that it is contained within the mound of sticks. Some burrows lead into cracks in the coral bedrock. Where the maturing forest soil has a deep humus layer, the animals use extensive subterranean tunnels in which the nest chamber may be far from the entrance mound (Schwartz 1952). That the burrows can be extensive is exemplified by a radio-collared individual located underground many meters from the nearest hole or sign of activity (S. R. Humphrey, unpubl. observation). Because depth of the soil is correlated with age of the forest stand, depth of the stratum within which tunnels can be made may be a significant feature of habitat quality. Goodyear (1985) described the occurrence of burrow-entrance holes in a variety of situations, including in nests, at the bases of trees, under exposed tree roots or mounds of roots, under fallen trees, and under rocks (especially at the edges of solution holes). The structures of sticks often are called woodrat houses or nests, but often they contain no nests and instead may constitute mazes to confuse predators at the entrance to the burrow. Although the presence of woodrat houses is a clear indication that animals are present, animals may live where no house is present; Goodyear (1985) found about 30 percent of captures not associated with stick structures. Sign of presence may be limited to a few twigs placed over a hole, or simply holes next to tree bases, exposed roots, or rocks.

The Key Largo woodrat occurs in deciduous forest in the life zone designated "dry tropical forest" (Holdridge 1967). The woodrat has a narrower range of occupied habitats than the Key Largo cotton mouse; woodrats are usually absent from deforested and oldfield sites (Goodyear 1985) and uncommon in young forest (Humphrey 1988*a*). However, the assumption that abundance of houses of this subspecies is directly related to the maturity of forest stands (Brown 1978; Hersh 1978; Barbour and Humphrey 1982) appears to be incorrect (Goodyear 1985), because the density of houses may not be a reliable index to the density of individuals. The specific features of the Key Largo forest that constitute optimal habitat for woodrats are unclear, possibly hindering effective management. A promising lead is that numerous observers (Goodyear 1985; P. E. Moler, personal communication; S. R. Humphrey, unpubl. observation) think that piles of rubble (including junk, building materials, tree-trimming refuse, and coral rock from agricultural and road-building operations) provide sites preferred by woodrats for their burrows and stick piles. Observations reported by Goodyear (1985) even indicate that woodrats will colonize modest distances away from forest habitat where such structures are available. As a consequence, the density of sign or animals often is very high where trash is piled adjacent to roads, in utility rights-of-way, and in forest tracts in which surface rock has been piled or windrowed by heavy machinery. Based on this observation, placement of artificial substrates for burrow sites has been proposed and attempted as mitigation for development activities on northern Key Largo.

See the account of the Key Largo cotton mouse (*Peromyscus gossypinus allapaticola*) for a more detailed description and history of this habitat. Occupied habitat now occurs only on the northern end of the key. Except for a small parcel in and adjacent to John Pennekamp Coral Reef State Park, no older forest remains on the southern two-thirds of the key. Furthermore, all the forest has been cut over to some degree, and no fully mature hammock occurs anywhere on the island. Judging from soil depth and tree diameter, the area between the northeast 1/4 of Sec. 29, T 60 S, R 39 E and the northern end of the key includes several large tracts of forest that have been much less changed than the rest, perhaps by virtue of having never been farmed. Considering only forested land, this area was estimated to include approximately 475 ha of suitable habitat (Barbour and Humphrey 1982). The total area of upland on northern Key Largo that could revert to forest habitat after succession is estimated at 851 ha (Siemon et al. 1986; Humphrey 1988*b*).

VULNERABILITY OF SPECIES AND HABITAT: Although the habitat of this subspecies is quite vulnerable to conversion to human uses, the subspecies appears to be relatively secure from other types of threats. If the forest is not developed, woodrats can survive in forest of a variety of seral ages.

Although house cats (*Felis catus*) are abundant in the mangrove habitat of northern Key Largo, none were captured during removal of raccoons and opossums from rodent-trapping grids, not even near subdivisions (Humphrey 1988*a*). The only potential competitor of the Key Largo woodrat is the black rat (*Rattus rattus*), which in autumn may occur at densities on the order of 2 per ha (Hersh 1981). However, black rats do not breed in the hardwood forest. Instead, they are a nonbreeding, floating population of dispersers from mangrove swamp, human residences, and garbage dumps (Hersh 1978). Goodyear (1985) captured only 3 individual black rats along with 233 captures of woodrats and cotton mice, at only 2 of 45 trapping sites. Humphrey (unpubl. observation) had only 20 captures of 10 black rats in 2 of 6 grids along with 1407 captures of woodrats and cotton mice in 1986. Perhaps the numerous, large woodrats on Key Largo prevent residence of black rats through agonistic behavior.

With one exception, incidents of predation are undocumented. By far the most abundant predatory animals occurring on northern Key Largo are raccoons (*Procyon lotor*). Raccoons quickly learn to check mammalogists' traplines when an odiferous bait such as peanut butter is used, and they remove woodrats from the traps and eat them. This behavior is less pronounced with a less attractive bait such as oatmeal, but removal of raccoons while live trapping Key Largo rodents is advisable in any case. Other less common predators there include bobcats (*Felis rufus*), diamondback rattlesnakes (*Crotalus adamanteus*), and eastern indigo snakes (*Drymarchon corais*).

CAUSES OF THREAT: This subspecies is threatened by continued conversion of the remaining habitat. A decade ago, its future looked bleak. A total of 4089 housing units was reported as approved or under construction in this area (U.S. Fish and Wildlife Service 1984). In 1981, with funds loaned by the Farmer's Home Administration, the Florida Keys Aqueduct Authority completed a branch pipeline from the junction of U.S. Highway 1 and C-905 up northern Key Largo to the Ocean Reef Club development. A restrictive hookup policy was adopted, including most of the proposed Crocodile Lake National Wildlife Refuge (U.S. Fish and Wildlife Service 1984), but the woodrat had no official status during that consultation, so hookups were not precluded from most habitat outside of the proposed refuge. A loan was proposed to the Florida Keys Electric Cooperative by the Rural Electrification Administration to finance an electric substation and system expansion allowing approximately 6000 more electric drops on northern Key Largo. On 27 October 1983, the USFWS regional director in Atlanta, Georgia, issued a biological opinion that the proposal would jeopardize the subspecies by facilitating development. "No-electric-hookup areas" were proposed, but this issue was not resolved (U.S. Fish and Wildlife Service 1984). As of 1989, little pressure had occurred for water or electric hookups in hammock areas.

RESPONSES TO HABITAT MODIFICATION: The Key Largo woodrat has survived modification of the forests and exotic competitors and predators, but details and limits of its responses are unknown. The low densities of woodrats near human developments might be caused by pesticides, young age of forest stands, competing black rats, or predatory house cats, but no relationship has been demonstrated (Humphrey 1988*b*).

DEMOGRAPHIC CHARACTERISTICS: Body mass at sexual maturity is approximately 200 g for males and 160 g for females (Hersh 1981). Lactating females were captured mainly from April to September with a peak in August; a few others were captured in December (Hersh 1981). Recruitment of juveniles occurred from July through March, with a major peak in August–October and a minor peak in February (Hersh 1981). Hersh noted little seasonal change in woodrat density from August to May.

Increase of the introduced population on Lignumvitae Key from 19 in 1970 (Brown and Williams 1971) to a crude estimate of 85 in 1979 (Barbour and Humphrey 1982) suggests very rapid population growth. This demonstrates that introduction of the subspecies into suitable but unoccupied habitat can be an effective management action. However, the subsequent apparent loss of this population indicates substantial variation in population levels, such that small populations may not be sustainable for very long.

KEY BEHAVIORS: The Key Largo woodrat is nocturnal. If it resembles other subspecies of the eastern woodrat, it forages mostly in the forest canopy. Some information on diet of this subspecies is available from food remains on nest structures and from laboratory preference tests (Hersh 1978). Food left on nest mounds included remains of a slug, a cicada wing, a freshly gnawed shell of a tree snail (*Liguus*), four fruits and two leaves of pigeon plum (*Coccoloba diversifolia*), one leaflet of blackbead (*Pithecellobium guadalupense*), fruit and two leaves of princewood (*Exostema caribaeum*), one stalk of a poisonwood (*Metopium toxiferum*) fruit, one seed of mahogany (*Swietenia mahogani*), and two leaves of bustic (*Dipholis salicifolia*). Laboratory animals completely ate the leaves of paradise tree (*Simarouba glauca*), strangler fig (*Ficus aurea*), and wild lime (*Zanthoxylum fagara*). They partially ate the leaves of gumbo-limbo (*Bursera simaruba*), lignumvitae (*Guaiacum sanctum*), and wild coffee (*Psychotria undata*). They did not eat leaves of pigeon plum, Jamaica dogwood (*Piscidia piscipula*), or mahogany.

Managers of Lignumvitae Key State Botanical Site expressed concern over predation by the introduced woodrats on indigenous tree snails (*Liguus fasciatus*), and they considered trapping out the woodrats for that reason. At least three considerations cast doubt on the wisdom of such an action. (1) On Key Largo, woodrats and tree snails have coexisted for years. (2) The evidence for predation is inconclusive, consisting of rodent-gnawed shells observed on woodrat houses and the forest floor (Schwartz 1952; Hersh 1978). Conceivably these could have been taken in the ordinary course of "packrat" behavior and chewed to gain mineral nutrients.

(3) The only study of the tree snails on Lignumvitae Key deals at length with predation but makes no mention of woodrats as agents of mortality (Tuskes 1981).

Average home range calculated by the minimum-complex-polygon method for four females and six males captured four or more times was 2.37 ha, with no difference between the sexes (Hersh 1981).

Nests of the Key Largo woodrat are occupied by numerous species of arthropods (R. E. Woodruff, unpubl. observation). Perhaps 10 or 12 taxa appear to be true commensals or parasites, associated only with woodrats. Taxa whose scientific descriptions have been published include the minute Key Largo woodrat dung beetle (*Ataenius brevicollis*), the green Key Largo woodrat dung beetle (*Onthophagus orpheus orpheus*), a blind weevil (*Caecossonus dentipes*), a blind staphylinid (*Cubanotyphlus largo*), a blind wingless enicocephalid bug, and an isopod. Neither species of dung beetle is associated exclusively with this subspecies of woodrat, and both occur in nests of woodrats in peninsular Florida (Woodruff 1973). A single flea (*Orchopeas howardi*) was reported from one woodrat (Schwartz 1952).

CONSERVATION MEASURES TAKEN: The Key Largo woodrat is listed as endangered by the Florida Game and Fresh Water Fish Commission (1990) and the U.S. Fish and Wildlife Service (1989). Protection of the introduced population on Lignumvitae Key by public ownership and management as a state botanical site by the Florida Department of Natural Resources is mooted by the apparent extirpation of this population. Recent conservation action focused partly on this rodent has included planning and some land acquisition for Crocodile Lake National Wildlife Refuge, land acquisition under Florida's Conservation and Recreation Lands program, and preparation of a Habitat Conservation Plan (HCP).

Most of the habitat occupied by this subspecies is contained within proposed federal and state land acquisition projects. Progress on acquisition has been slow and the outcome is uncertain. About half of the potential habitat (approximately 486 ha of upland) is included in the proposed Crocodile Lake National Wildlife Refuge (U.S. Fish and Wildlife Service 1983). Most of the older forest lies outside the proposed federal boundaries and is included in the North Key Largo State Preserve planned by the Florida Department of Natural Resources.

The draft HCP (Siemon et al. 1986; summarized in part by Humphrey 1988*b*) has not been implemented by the Monroe County Commission and may never be. Nonetheless, it catalyzed a rough consensus on future land use. The purpose of the plan was to conserve and enhance the habitats of several endangered species while accommodating some development (U.S. Fish and Wildlife Service 1985). Probably its greatest effects were to: (1) raise land prices in some areas and lower them in others by designating the land for either development or conservation; (2) legitimizing the primacy of compensation over regulatory taking for lands to be conserved; and (3) focusing the attention of conservation agencies on the need to act. The HCP called for conservation of 84 percent of the undeveloped lands and transfer of those development rights to less sensitive parts of the county; development of the remaining 16 percent in nodes adjacent to preexisting developments and subject to further restriction or mitigation; and user fees and impact fees to fund development infrastructure and biological research and management. The HCP identified no source of funding for purchasing land for conservation, but north Key Largo has become a high-priority project of the Conservation and Recreation Lands program of the Florida Department of Natural Resources. Several purchases of lands designated for development have been made instead by conservation agencies, so the ultimate use of the landscape will differ from that envisioned in the HCP.

CONSERVATION MEASURES PROPOSED: The occurrence of high densities of Key Largo woodrats in undeveloped habitat indicates that, if substantial tracts can be protected from development and other threats, perpetuation of the population should be an attainable goal. If a substantial refuge is established with suitable management, delisting of the subspecies could be considered.

Research is needed to determine the cause of comparatively low densities of Key Largo woodrats documented near human developments (Humphrey 1988*a*, 1988*b*) and to devise a way to reduce this effect. Evidence that woodrats have been lost from the Miami Ridge and southern portions of Key Largo suggests that vigilance will be necessary to be certain that habitat conserved on northern Key Largo is secure from other threats. Loss of the population introduced onto Lignumvitae Key indicates that the variation in population density in space and time on Key Largo, and its causes, should be characterized by long-term study. Research also is needed on ways to restore and repopulate degraded habitat. Attention should be paid to the population status of the white-crowned pigeon, whose consumption and dispersal of forest seeds are essential for the succession of disturbed sites to tropical hardwood forest.

ACKNOWLEDGMENTS: I am grateful to James N. Layne and I. Jack Stout for helpful review of the manuscript.

Literature Cited

Alexander, T. R. 1953. Plant succession on Key Largo, Florida, involving *Pinus caribea* and *Quercus virginiana*. Quart. J. Florida Acad. Sci. 16:133–138.

Barbour, D. B., and S. R. Humphrey. 1982. Status and habitat of the Key Largo woodrat and cotton mouse (*Neotoma floridana smalli* and *Peromyscus gossypinus allapaticola*). J. Mammal. 63:144–148.

Brown, L. N. 1978. Key Largo woodrat. Pp. 11–12 *in* J. N. Layne, ed., Inventory of rare and endangered biota of Florida. Vol. 1. Mammals. University Presses of Florida, Gainesville. xx + 52 pp.

Brown, L. N., and R. L. Williams. 1971. The Key Largo woodrat (*Neotoma floridana snalli*) (sic) and cotton mouse (*Peromyscus gossypinus allapaticola*) on Lignum Vitae Key, Florida. Florida Nat. 44:95–96.

Fitch, H. S., and D. G. Rainey. 1956. Ecological observations on the woodrat, *Neotoma floridana*. Univ. Kansas Publ., Mus. Nat. Hist. 8:499–533.

Florida Game and Fresh Water Fish Commission. 1990. Official lists of endangered and potentially endangered fauna and flora in Florida. Tallahassee, Florida. 23 pp.

Goodyear, N. C. 1985. Results of a study of Key Largo woodrats and cotton mice: Phase I, spring and summer 1985. Unpubl. report to U.S. Fish and Wildlife Service, 3100 University Boulevard, South Suite 120, Jacksonville. 76 pp.

Hersh, S. L. 1978. Ecology of the Key Largo woodrat (*Neotoma floridana smalli*). Unpubl. M.S. thesis, University of Miami, Coral Gables. 106 pp.

Hersh, S. L. 1981. Ecology of the Key Largo woodrat (*Neotoma floridana smalli*). J. Mammal. 62:201–206.

Hershfeld, S. E. 1968. Vertebrate fauna of Nichol's Hammock, a natural trap. Quart. J. Florida Acad. Sci. 31:177–189.

Holdridge, L. R. 1967. Life zone ecology. Tropical Sci. Center, San Jose, Costa Rica. 206 pp.

Humphrey, S. R. 1988*a*. Density estimates of the Key Largo woodrat and cotton mouse (*Neotoma floridana smalli* and *Peromyscus gossypinus allapaticola*), using the nested grid approach. J. Mammal. 69:524–531.

Humphrey, S. R. 1988*b*. The habitat conservation plan. Conservation Biol. 2:240–244.

Layne, J. N. 1974. The land mammals of south Florida. Pp. 386–413 *in* P. J. Gleason, ed., Environments of south Florida: Present and past. Miami Geol. Soc., Memoir 2:1–452.

Schwartz, A., 1952. The land mammals of southern Florida and the upper Florida Keys. Unpubl. Ph.D. diss., University of Michigan, Ann Arbor. 180 pp.

Schwartz, A., and E. P. Odum. 1957. The woodrats of the eastern United States. J. Mammal. 38:197–206.

Sherman, H. B. 1944. Recent literature and some new distribution records concerning Florida mammals. Proc. Florida Acad. Sci. 7:199–202.

Sherman, H. B. 1955. Description of a new race of woodrat from Key Largo, Florida. J. Mammal. 36:113–120.

Siemon, Larsen, and Purdy. 1986. North Key Largo, Florida, Habitat Conservation Plan. Unpubl. report to Gov. Bob Graham. 128 pp., 2 appendices.

Small, J. K. 1923. Green deserts and dead gardens. J. New York Botanical Garden 24(286):215.

Tuskes, P. M. 1981. Population structure and biology of *Liguus* tree snails on Lignumvitae Key, Florida. Nautilus 95:162–169.

U.S. Fish and Wildlife Service. 1983. Final land protection plan, Crocodile Lake National Wildlife Refuge, Monroe County, Florida. U.S. Fish and Wildlife Serv., Atlanta, Georgia. 22 pp., 3 appendices, 2 maps.

U.S. Fish and Wildlife Service. 1984. Endangered and threatened wildlife and plants; determination of endangered status for the Key Largo woodrat and Key Largo cotton mouse. Federal Register 49:34504–34510.

U.S. Fish and Wildlife Service. 1985. Environmental impact statement on North Key Largo (Florida) habitat conservation plan and endangered species permit. Federal Register 50:14299–14301.

U.S. Fish and Wildlife Service. 1989. Endangered and threatened wildlife and plants. U.S. Government Printing Office, Washington, D.C. 34 pp.

Woodruff, R. E. 1973. The scarab beetles of Florida. Arthropods of Florida. Vol. 8. Florida Dept. Agric. and Consumer Services, Division of Plant Industry, Gainesville. 220 pp.

Prepared by: Stephen R. Humphrey, Florida Museum of Natural History, University of Florida, Gainesville, FL 32611.

Endangered

Florida Saltmarsh Vole
Microtus pennsylvanicus dukecampbelli
FAMILY SIGMODONTIDAE
Order Rodentia

TAXONOMY: *M. p. dukecampbelli* was described by Woods et al. (1982) as a Pleistocene relict living in a coastal salt marsh near Cedar Key. The common name for most forms of *M. pennsylvanicus* is meadow mouse or meadow vole. The Florida subspecies of this taxon appears to be restricted to coastal saltmarsh habitats, however, and the common name (Florida saltmarsh vole) is more descriptive and preferable. In the list of the Florida Game and Fresh Water Fish Commission (1990), one common name for this species is given as Duke's saltmarsh vole in honor of the late Dr. Howard W. "Duke" Campbell, Jr., for whom the form is named. Since the saltmarsh vole is an endemic mammal of Florida, it is preferable to use the common name to stress its unique status within the state.

DESCRIPTION: The Florida saltmarsh vole is distinctly larger than most other races of *M. pennsylvanicus*. Diagnostic skull measurements are: greatest length of skull 31.17 mm, condylozygomatic length 24.65 mm, incisive foramen length 5.69 mm, diastema length 9.22 mm, cranial breadth 11.71 mm, and zygomatic breadth 16.89 mm. All of these measurements are larger than other forms of *M. pennsylvanicus*, as are measurements for total length (186.10 mm), tail length (48.80 mm), and hind foot length (23.36 mm). The rostrum is shorter than most other forms (6.52 mm), and the pelage is much darker in coloration. A comparison of *M. p. dukecampbelli* with other forms of *M. pennsylvanicus* from throughout the range of the species indicates that it is most distinct from the Georgia population of *M. p. pennsylvanicus*. It is most similar to *M. p. nigrans* from coastal salt marshes in Maryland, and to the insular forms *M. p. copelandi* from Grand Manan Island (New Brunswick) and *M. p. magdalenensis* from Magdalen Island (Quebec).

Florida saltmarsh vole, *Microtus pennsylvanicus dukecampbelli*. (Photo by Charles A. Woods)

POPULATION SIZE AND TREND: The Florida saltmarsh vole is difficult to trap and may be cyclical in population size. In a survey in June 1979 no saltmarsh voles were captured at the type locality at the same time that 55 rice rats (*Oryzomys palustris*), 10 cotton rats (*Sigmodon hispidus*), and 4 cotton mice (*Peromyscus gossypinus*) were captured. In March 1980 at the same location, 53 *Oryzomys*, 2 *Sigmodon*, and no *Peromyscus* were captured along with 9 *M. p. dukecampbelli*. The 9 individual saltmarsh voles were captured a total of 36 times over the 6-day trapping period in 1980, indicating that they are not difficult to trap when their numbers are high (or when numbers of *Sigmodon* and *Peromyscus* are low). This information is important in evaluating a later survey of the status of the Florida saltmarsh vole at the type locality during 1987 and 1988. Over a 7-day period between 29 December 1987 and 4 January 1988, trapping in the Island Field Marsh type locality for a total of 343 trap-nights did not produce any saltmarsh voles. During the same period 8 *Sigmodon hispidus*, 3 *Peromyscus gossypinus*, and only 1 *Oryzomys palustris* were captured. Capture locations as well as the relative abundance of the three species in the 1987–88 survey differed dramatically from the results of the 1979–80 surveys (Woods 1988).

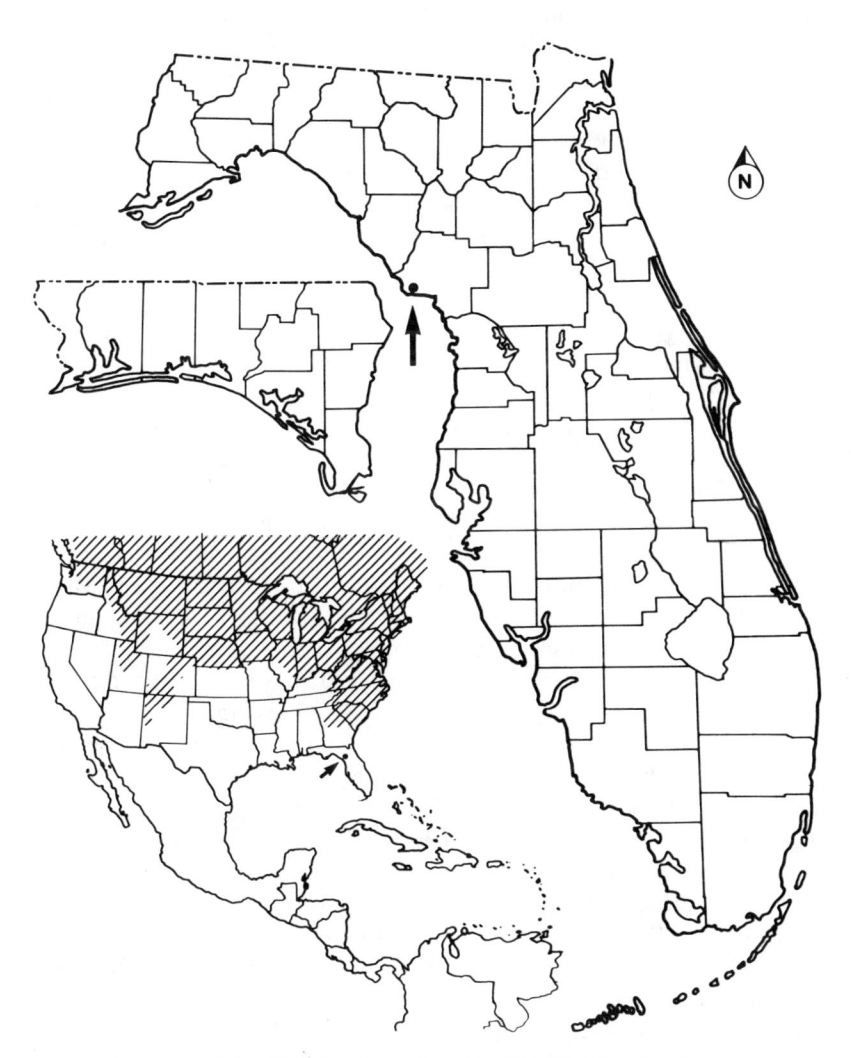

Distribution map of the Florida saltmarsh vole, *Microtus pennsylvanicus dukecampbelli*. Hatching represents the species; the dot represents the subspecies.

A subsequent survey in May and June 1988 over a 13-day period (1025 trap-nights) produced a very different pattern in species composition and habitat use. No *Sigmodon hispidus* were captured, while 38 *Oryzomys palustris*, 23 *Peromyscus gossypinus*, and 1 *M. p. dukecampbelli* were captured. Based on these data it appears that the small-mammal community in the habitat of the Florida saltmarsh vole near Cedar Key is characterized by shifting densities of species that use different habitats at different times. *M. p. pennsylvanicus* exists as one of four small mammals that compete with each other for limited resources and space in the saltmarsh community. The most recent estimate of its population size is not encouraging, with only one male Florida saltmarsh vole captured between December 1987 and June 1988. More recent efforts to trap saltmarsh voles at the type locality and in adjacent locations along the west coast of Florida have also been unsuccessful (M. Bentzien, personal communication).

DISTRIBUTION AND HISTORY OF DISTRIBUTION: The Florida saltmarsh vole is known only from the type locality at Island Field Marsh along the shore of Waccasassa Bay, Levy County, Florida. Attempts to capture saltmarsh voles in adjacent areas of coastal saltmarsh habitat north and south of the type locality, as well as in interior pastures east of Cedar Key, have not been successful.

Microtus pennsylvanicus is known to have occurred in the Levy County region during the late Pleistocene, where fossil specimens have been collected at Devil's Den near Williston (Arata 1961; Martin 1968; Martin and Webb 1974), as well as from 10 km southwest of Gainesville in Alachua County, 13 km west of Dunnellon in Citrus County, and near Gulf Hammock in Levy County (Woods et al. 1982). The ages of these fossil sites range between 30,000 and 8000 yBP. It is probable that *M. pennsylvanicus* was broadly distributed along the Gulf coast of Florida between 5000 and 10,000 yBP when sea levels were somewhat lower and the climate cooler. Blackwelder et al. (1979) estimated that 10,000 yBP sea levels were 25 m lower than present levels in the Gulf of Mexico, which would extend suitable terrestrial habitats 100 km west of the present coastline at the type locality of the Florida saltmarsh vole along the shore of Waccasassa Bay. The area of expanded habitat 10,000 yBP would have been the most extensive between Clearwater and the mouth of the Apalachicola River, at the center of the extensive Gulf Coast Savanna Corridor discussed by Webb (1977).

The favored habitat of the ancestor of the Florida saltmarsh vole may have been along water courses and wet areas in patches of bluestem prairie and savanna in northern and central Florida that existed until about 5000 yBP. Forest habitat returned as pine expanded and eliminated the prairie plants between 6710 and 5000 yBP, with black gum, sweetgum, ironwood, and other mesic-forest species becoming established by 5000 yBP (Watts 1980), eliminating suitable upland habitat for grassland species such as *M. pennsylvanicus*. When the grasslands disappeared in central and northern Florida, voles apparently became isolated in saltmarsh habitats. Therefore, *M. p. dukecampbelli* probably has been disjunct from *M. p. pennsylvanicus* to the north and west for at least 5000 years, and possibly longer if Watts (1980) is correct in his interpretation of late Quaternary vegetative history of the southeastern United States. The disjunct relict population of *M. pennsylvanicus* along the Florida Gulf coast, which may now be restricted to the Cedar Key area, is analogous to a relict population of the prairie vole (*M. ochrogaster*) that occurred along the Gulf coast of Texas and Louisiana (Bailey 1900; Lowery 1974). This form, named *M. p. ludovicianus*, was abundant at several locations early in the present century but is probably now extinct (Lowery 1974).

GEOGRAPHICAL STATUS: This subspecies is endemic to the Gulf coast area of central Florida and is known only from the type locality on the shore of Waccasassa Bay near Cedar Key.

HABITAT REQUIREMENTS AND HABITAT TREND: The habitat requirements of the Florida saltmarsh vole within the saltmarsh community are complex and poorly understood. In the 1982 survey *M. p. dukecampbelli* appeared to be restricted to areas near the edge of patches of black rush (*Juncus roemerianus*) and in patches of seashore saltgrass (*Distichlis spicata*), and appeared to avoid areas dominated by smooth cordgrass (*Spartina alterniflora*). The only Florida saltmarsh vole captured in the 1988 survey was in the mixed *Distichlis/Spartina* microhabitat area of the Island Field salt marsh.

VULNERABILITY OF SPECIES AND HABITAT: The coastal saltmarsh habitats of the Florida Gulf coast are very vulnerable to flooding because they are low-lying. The waters of the adjacent Gulf of Mexico are shallow and strong west winds can increase the flooding effects of normal tidal cycles by driving water into the marshes. How a largely terrestrial form like a vole can survive in a habitat that regularly floods and is occasionally devastated by high winds and water that completely submerge its habitat is a matter of great concern. Harris (1953) investigated *M. p. nigrans* in tidal marshes in Maryland and determined that they survived by using houses and feeding platforms of muskrats (*Ondatra zibethicus macrodon*).

There are no *Ondatra* in Florida, and the round-tailed muskrat (*Neofiber alleni*) is not found in the Waccasassa Bay area, which may partially explain why *Microtus* is more common in tidal marshes in Maryland than it is in Florida. The lack of suitable vegetative cover at times of high water and the extremely high wind-driven tidal floods that inundate the coast near Cedar Key serve as continued hazards to the survival of the Florida saltmarsh vole. Fisler (1961) studied the California vole (*Microtus californicus*), which survives in areas of marsh in the San Francisco Bay area that are totally submerged by tides. He observed that the California vole climbs into the vegetation to stay above the water level. When the vegetation was covered by water, the voles swam (even staying within their home range), and he observed them swimming through rough water and against strong tidal currents.

Microtus p. dukecampbelli, like *M. californicus*, may be able to remain in their natural habitat and successfully maintain adult populations within the tidal salt marshes even under the most extreme conditions of high tide and strong west winds. In Florida, survival during high tides and at times of extremely high water associated with hurricanes and wind-driven water off the shallow Gulf of Mexico directly off Cedar Key might be facilitated in one or more of the following ways: (1) saltmarsh voles might be able to move to high land adjacent to Island Field Marsh; (2) they could make use of the tallest vegetation within the marsh to stay above the water and get some protection from the high winds; (3) they might be able to survive brief periods of exposure while swimming about; (4) they might have a broad distribution throughout the salt marshes of the Levy County area, and rely on recolonization of Island Field Marsh after a severe storm has wiped out the resident population there. All of these factors may contribute to the survival of *M. p. dukecampbelli* in this relict population.

CAUSES OF THREAT: The main threat to the survival of *M. p. dukecampbelli* is catastrophic weather such as a hurricane. An example of such a storm in recent times is Hurricane Elena, which sat off the coast of Cedar Key for 24 hours in August 1985, possibly causing the extremely low population level of the Florida saltmarsh vole in 1988. There is also a threat of competition from one or more of the other small rodents of the salt marsh, which fluctuate in population density and distribution within the microhabitats of the salt marsh. The species has been able to survive these dangers in the past and was still present in moderate numbers as late as March 1980. Finally, a sea-level rise of 1–2 m by the year 2100 as a result of global warming is expected to greatly diminish the extent of salt marsh everywhere (Park et al. 1988).

Most factors that may enable the saltmarsh vole to survive require a broad area of suitable habitat to serve as refugium for animals displaced by storms or temporary competition from another species. It is imperative, therefore, that the upland adjacent to the type locality and large areas of coastal salt marsh in the Cedar Key region be preserved as a sanctuary for this subspecies if there is going to be any hope of the subspecies surviving into the future. The extinction of the relict population of the Gulf coast prairie vole in Louisiana and Texas is an unhappy reminder of how vulnerable the Florida saltmarsh vole is.

RESPONSES TO HABITAT MODIFICATION: Unknown.

DEMOGRAPHIC CHARACTERISTICS: *Microtus pennsylvanicus* as a species has a very high reproductive potential. The average distance between capture for the saltmarsh vole at the type locality was 32 m, which is half the distance for *Oryzomys* on the same site. During the period of maximum known density of saltmarsh voles at Waccasassa Bay in 1980, there were five male and nine female animals. Reproductive activity appears to extend throughout the year but is most pronounced during February and March. Thus the known demographic characteristics of the form indicates that it is a species of great reproductive potential and that it should be able to recover from severe reductions in population levels such as the one that characterized the saltmarsh vole in the Waccasassa Bay area in 1987 and 1988.

KEY BEHAVIORS: Little is known of the behavior of this subspecies within its restricted area of distribution.

CONSERVATION MEASURES TAKEN: The Florida saltmarsh vole is listed as a species of special concern by the Florida Game and Freshwater Fish Commission (1990). The U.S. Fish and Wildlife Service (1990) has published a notice of intent to list the Florida saltmarsh vole as endangered.

CONSERVATION MEASURES PROPOSED: A careful and systematic search for additional populations of *M. p. dukecampbelli* should be undertaken along the Gulf coast of Florida, especially between the Suwannee and Withlacoochee rivers. The saltmarsh habitat and adjacent uplands of the type locality should be protected by public ownership, and more extensive areas also should be protected as a permanent reservoir of suitable habitat for the saltmarsh vole. Because of the ease of maintaining this species in captivity, and the natural tendency of *Microtus* to recolonize habitats where it has been extirpated, serious consideration should be given to establishing a captive breeding center. Surplus individuals from this facility could be reintroduced at regular intervals into the area of the established population, and new populations could be established on public lands to provide insurance against a disaster at the type locality.

ACKNOWLEDGMENTS: The assistance of Leslie Hay-Brown during the 1987–88 survey was invaluable and is appreciated. The work of William Post in the Cedar Key region was the impetus for the original discovery of this form.

Literature Cited

Arata, A. A. 1961. Meadow vole (*Microtus pennsylvanicus*) from the Quaternary of Florida. Quart. J. Florida Acad. Sci. 24:117–121.

Bailey, V. 1900. Revision of the American voles of the genus *Microtus*. North American Fauna 17:1–88.

Blackwelder, B. W., O. H. Pilkey, and J. D. Howard. 1979. Late Wisconsinan sea levels on the southeast U.S. Atlantic shelf based on in-place shoreline indicators. Science 204:618–620.

Fisler, G. F. 1961. Behavior of salt-marsh *Microtus* during winter high tides. J. Mammal. 42:37–43.

Florida Game and Fresh Water Fish Commission. 1990. Official lists of endangered and potentially endangered fauna and flora in Florida. Tallahassee, Florida. 23 pp.

Harris, V. T. 1953. Ecological relationships of meadow voles and rice rats in tidal marshes. J. Mammal. 34:479–487.

Lowery, G. H., Jr. 1974. The mammals of Louisiana and its adjacent waters. Louisiana State University Press, Baton Rouge. 565 pp.

Martin, R. A. 1968. Late Pleistocene distribution of *Microtus pennsylvanicus*. J. Mammal. 49:265–271.

Martin, R. A., and S. D. Webb. 1974. Late Pleistocene mammals from Devil's Den Fauna, Levy County. Pp. 114–145 *in* S. D. Webb, ed., Pleistocene mammals of Florida, University Presses of Florida, Gainesville. 270 pp.

Park, R. A., M. S. Trehan, P. W. Mausel, and R. C. Howe. 1988. The effects of sea level rise on U.S. coastal wetlands. Holcomb Research Institute Rep. 142:1–60.

U.S. Fish and Wildlife Service. 1990. Endangered and threatened wildlife and plants; proposed endangered status for the Florida salt marsh vole. Federal Register 55:13576–13578.

Watts, W. A. 1980. The late Quaternary vegetation history of the southeastern United States. Ann. Rev. Ecol. Syst. 11:387–409.
Webb, S. D. 1977. Evolution of savanna vertebrates in the New World. Part 1: North America. Ann. Rev. Ecol. Syst. 11:387–409.
Woods, C. A. 1988. Final report on cooperative agreement No. 14-16-0009-1544-Work Order No. 15. Cooperative Fish and Wildlife Research Unit, University of Florida, Gainesville. 6 pp.
Woods, C. A., W. Post, and C. W. Kilpatrick. 1982. *Microtus pennsylvanicus* (Rodentia: Muridae) in Florida: A Pleistocene relict in a coastal saltmarsh. Bull. Florida State Mus., Biol. Sci. 28:25–52.

Prepared by: Charles A. Woods, Florida Museum of Natural History, University of Florida, Gainesville, FL 32611.

Endangered

Right Whale
Eubalaena glacialis
FAMILY BALAENIDAE
Order Cetacea

TAXONOMY: The right whale was described by Muller (1776). There is disagreement as to the inclusion of the right whale with the bowhead whale in the genus *Balaena* because the two are anatomically distinct (Honacki et al. 1982; Cummings 1985). Most recent authors prefer the genus *Eubalaena*. As Cummings (1985) pointed out, there is insufficient material available for a proper study. However, recent advances in the area of biochemical genetics may provide some insight in the future. Right whales of the southern hemisphere are considered a separate species (*E. australis*). Other authors have considered the right whale to be a single species (*E. glacialis*) with several subspecific designations. Here again, the question is open because of lack of material.

DESCRIPTION: The right whale is a large, generally uniformly dark colored whale. White belly patches of variable size, shape, and location may occur. Right whales reach a length of about 17 m, with females slightly larger than males. No dorsal fin is present. Right whales are rotund, and maximum girth may approach 60 percent of body length (Leatherwood and Reeves 1983). Mass may reach 100 tn (Klumov 1962; Leatherwood and Reeves 1983). The head is nearly one-third of the body length. Blowholes are paired and the jaws are greatly arched to accommodate the baleen, which may reach a length of 2.6–2.8 m. The number of plates of baleen in each of the two rows is usually between 205 and 270. Aside from the size of the animal and the arched jaws, the most characteristic external feature is the callosities (Kraus et al. 1989) or excrescences on the head. Callosities vary in color but are usually yellowish or whitish. See Payne et al. (1983) for a description of the external features of the southern right whale. Callosities may be used for individual identification

Eubalaena glacialis

Right whale, *Eubalaena glacialis*. (Photo by Sea World of Florida. Copyright © 1990 Sea World, Inc. All rights reserved. Reproduced by permission.)

and appear to be sexually dimorphic in the southern right whale (Payne and Dorsey 1983). The skull may reach a length of 5.2 m and weigh 100 kg. The seven cervical vertebrae are fused. The vertebral formula is: C7, T14–15, L10–11, Ca25, total = 56–58 (Cummings 1985). See True (1904), Allen (1908), and Omura et al. (1969) for a detailed description of the skeleton.

POPULATION SIZE AND TREND: There is no good estimate of the preexploitation population size because whaling records are incomplete (Brownell et al. 1986). The preexploitation size of the southern hemisphere stock is thought to have exceeded 100,000 (Leatherwood and Reeves 1983) and 100,000–300,000 worldwide (Scott 1985). The most recent analyses (National Marine Fisheries Service 1991) suggest a worldwide preexploitation population of about 100,000 animals. The current world population estimate is about 2300 animals, including 400? [sic] in the North Pacific, >350 in the North Atlantic and >1540 in the Southern Ocean (National Marine Fisheries Service 1991). Current population trends are unclear; numbers appear to be increasing in the southern hemisphere but not in the northern hemisphere.

DISTRIBUTION AND HISTORY OF DISTRIBUTION: The right whale is a coastal animal generally found at latitudes less than 50° in winter and greater than 40° in spring, summer, and fall (Leatherwood and Reeves 1983). In the North Pacific they have been sighted as far as 60°N and in the North Atlantic as far as 75°N (Cummings 1985). The current distribution, given the reduced status of the populations, is similar to the historic.

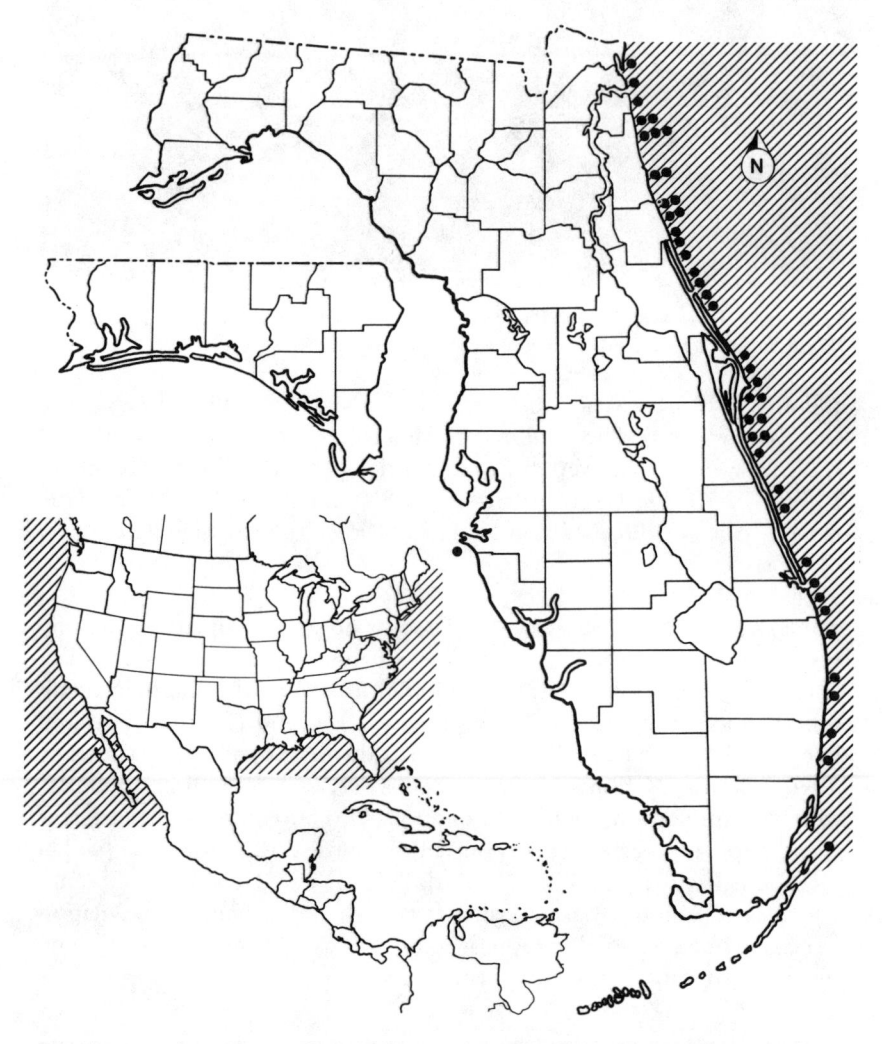

Distribution map of the right whale, *Eubalaena glacialis*. Hatching represents the species; the dots represent sightings and strandings.

Eubalaena glacialis

GEOGRAPHICAL STATUS: The Southern Ocean, North Pacific, and North Atlantic stocks are considered (by some) as subspecies (see Taxonomy). The Southern Ocean stock is currently the largest. North Pacific stocks (east and west) are the lowest in numbers, perhaps 200-250 (Leatherwood and Reeves 1983). In the North Atlantic, the eastern stock is virtually extinct, whereas the western stock numbers several hundred and currently ranges from eastern Canada to southeastern Florida (Winn et al. 1986; Kraus et al. 1989), with two records from the Gulf of Mexico (Schmidly 1981).

HABITAT REQUIREMENTS AND HABITAT TREND: The habitat requirements of the right whale are poorly understood. They are coastal/nearshore where, during spring, summer, and fall, they feed extensively on copepods and euphausids at the northern end of their range (Prescott et al. 1980; Reeves and Brownell 1982; Leatherwood and Reeves 1983; Kraus et al. 1989). Oceanographic factors relating to primary productivity are important in determining abundance of copepods and euphausids. Right whales migrate in nearshore waters to wintering/calving grounds in lower latitudes. In the western North Atlantic the still unidentified calving ground appears to be off the coasts of Georgia and northeast Florida (Kraus et al. 1986, 1989).

VULNERABILITY OF SPECIES AND HABITAT: The right whale earned its common name because it was the "right" or "correct" whale to take. It swam slowly near shore, making it easy to catch, and floated when dead. In contrast, the "wrong" whales (fin, blue, etc.) swam fast and sank when dead. The right whale and its coastal habitat are extremely vulnerable (Prescott et al. 1980; Anonymous 1986).

CAUSES OF THREAT: Whaling was responsible for the original decline of the species. Resulting changes in ecosystem equilibrium (e.g., less food because other copepod/euphausid feeders took advantage of increased resource availability) may be hindering recovery (Prescott et al. 1980). These changes are, however, speculative. In the western North Atlantic, coastal pollution could affect food distribution and abundance. More important, fishing nets and vessel traffic may be the most critical factors impeding recovery in this area. Observations of individuals have shown that many bear propeller wounds from large vessels and several whales have been killed in this manner (Reeves et al. 1978; Reeves and Mitchell 1986; Aguar Video Productions 1989; Kraus et al. 1989).

RESPONSES TO HABITAT MODIFICATION: Unknown.

DEMOGRAPHIC CHARACTERISTICS: Male right whales probably become sexually mature at 14.5–15.5 m, and females at 15–16 m. This may correspond to an age of about 10 years. Payne (1986) suggested that southern right whales in Argentina may reach sexual maturity at 7 years of age. Kraus et al. (1989) report that two known females from the western North Atlantic population gave birth at 5 and 7 years of age, respectively. The calving interval is about 2 years. Gestation is about 12 months followed by a lactation period of 6–7 months. Birth length is 4.5–6.0 m. Twins are rare. Longevity is unknown (Reeves and Brownell 1982; Cummings 1985; Brownell et al. 1986). Recruitment rate in Argentina is thought to be 4–7 percent (Scott 1985).

KEY BEHAVIORS: Right whales are often seen skim feeding at the surface of the water as they follow patches of copepods (Cummings 1985). Right whales generally prefer shallow, sheltered coastal waters for calving. In the western North Atlantic, wintering right whales often are seen just outside the surfline along the eastern coast of Florida. They may be accompanied by bottlenose dolphins. These whales are quite often reported as stranding or being stranded.

CONSERVATION MEASURES TAKEN: The right whale is protected by the Marine Mammal Protection Act of 1972 (National Marine Fisheries Service 1988), the Endangered Species Act of 1973 (U.S. Fish and Wildlife Service 1989), the International Convention for the Regulation of Whaling, 1946 (International Whaling Commission 1988), and the Convention on International Trade in Endangered Species (CITES 1989). The species is listed as endangered by the Florida Game and Fresh Water Fish Commission (1990). The first draft of the federal recovery plan for the northern right whale was issued in February 1990 (National Marine Fisheries Service 1990). At the local level, there is an ongoing public awareness campaign to ensure that boat and ship operators are aware of and on the lookout for right whales (e.g., Aguar Video Productions 1989).

CONSERVATION MEASURES PROPOSED: Wherever possible, sanctuaries should be established to limit or eliminate the impact of coastal vessel traffic and fishing gear on the right whale. Public education must continue. In the western North Atlantic, the still ill-defined calving grounds and the feeding grounds deserve special attention. All proposals are reviewed in the 1986 Workshop (Anonymous 1986). Kraus et al. (1989) update proposed conservation measures with recommendations for specific areas.

ACKNOWLEDGMENTS: This is Sea World of Florida Technical Contribution No. 8912-F. The authors of the cetaceans account in the 1978 edition were David K. Caldwell and Melba C. Caldwell.

Literature Cited

Aguar Video Productions. 1989. Right whale. Videotape, Aguar Video Productions, Athens, Georgia.
Allen, J. A. 1908. The North Atlantic right whale and its near allies. Bull. Amer. Mus. Nat. Hist. 24:277–329.
Anonymous. 1986. Report of the southeastern U.S. right whale workshop, Jekyll I, Georgia. 18–20 February 1986. The Georgia Conservancy, Savannah, Georgia. 41 pp.
Brownell, R. L., Jr., P. B. Best, and J. H. Prescott, eds. 1986. Right whales: Past and present status. Rep. Int. Whal. Commn. (Special Issue 10). 289 pp.
CITES. 1989. Convention on International Trade in Endangered Species of Wild Fauna and Flora. Regulations and appendices published by the Federal Wildlife Permit Office, U.S. Fish and Wildlife Service, Washington, D.C.
Cummings, W. C. 1985. Right whales—*Eubalaena glacialis* and *Eubalaena australis*. Pp. 275–304 *in* S. H. Ridgway and R. Harrison, eds., Handbook of marine mammals. Vol. 3. The sirenians and baleen whales. Academic Press, London. 362 pp.
Florida Game and Fresh Water Fish Commission. 1990. Official lists of endangered and potentially endangered fauna and flora in Florida. Tallahassee. 23 pp.
Honacki, J. H., K. E. Kinman, and J. W. Koeppl, eds. 1982. Mammal species of the world. A taxonomic and geographic reference. Allen Press and Assoc. of Systematics Collections, Lawrence, Kansas. 694 pp.
International Whaling Commission. 1988. International convention for the regulation of whaling, 1946. Schedule. As amended by the Commission at the 40th Annual Meeting, May/June 1988, and replacing that dated November 1987. International Whaling Commission, Cambridge, United Kingdom. 27 pp.
Klumov, S. K. 1962. The northern right whales in the Pacific Ocean. Tr. Inst. Okeanol. 58:202–297.
Kraus, S. D., M. J. Crone, and A. R. Knowlton. 1989. The North Atlantic right whale. Pp. 684–698 *in* W. J. Chandler, ed., Audubon Wildlife Report 1989/ 1990. Academic Press, San Diego. 817 pp.
Kraus, S. D., J. H. Prescott, and A. R. Knowlton. 1986. Wintering right whales (*Eubalaena glacialis*) along the southeastern coast of the United States, 1984–1986. Unpubl. report, New England Aquarium, Boston, Massachusetts. 14 pp.

Kraus, S. D., J. H. Prescott, A. R. Knowlton, and G. S. Stone. 1986. Migration and calving of right whales (*Eubalaena glacialis*) in the western North Atlantic. Pp. 139–144 *in* R. L. Brownell, Jr., P. B. Best, and J. H. Prescott, eds., Right whales: Past and present status. Rep. Int. Whal. Commn. (Special Issue 10), Cambridge. 289 pp.

Layne, J. N. 1965. Observations on marine mammals in Florida waters. Bull. Florida State Mus., Biol. Sci. 9:131–181.

Leatherwood, S., and R. R. Reeves. 1983. The Sierra Club handbook of whales and dolphins. Sierra Club Books, San Francisco. 302 pp.

Mitchell, E. D., V. M. Kozicki, and R. R. Reeves. 1986. Sightings of right whales, *Eubalaena glacialis*, on the Scotian shelf, 1966–1972. Pp. 83–107 *in* R. L. Brownell, Jr., P. B. Best, and J. H. Prescott, eds., Right whales: Past and present status. Rep. Int. Whal. Commn. (Special Issue 10), Cambridge. 289 pp.

Moore, J. C. 1953. Distribution of marine mammals to Florida waters. Amer. Midland Nat. 49:117–158.

Muller. 1776. Zool. Danicae Prodr., p. 7.

National Marine Fisheries Service. 1988. Marine Mammal Protection Act of 1972. Annual Report 1987/88. USDOC/NOAA/NMFS, Washington, D.C. 68 pp.

National Marine Fisheries Service. 1990. Draft national recovery plan for the northern right whale (*Eubalaena glacialis*). USDOC/NOAA/NMFS, Washington, D.C. 77 pp.

National Marine Fisheries Service. 1991. Endangered Whales: Update. USDOC, NOAA, National Marine Fisheries Service, Silver Spring, Maryland. 52 pp.

Odell, D. K. 1991. A review of the southeastern United States marine mammal stranding network: 1978–1987. Pp. 19–23 *in* J. E. Reynolds III and D. K. Odell, eds., Marine Mammal Strandings in the United States. Proceedings of the Second Marine Mammal Stranding Workshop, Miami, Florida, December 3–5, 1987. NOAA Technical Report NMFS 98, USDOC, NOAA, National Marine Fisheries Service, Seattle, Washington. 157 pp.

Omura, H., S. Ohsumi, T. Nemoto, K. Nasu, and T. Kasuya. 1969. Black right whales in the North Pacific. Sci. Rep. Whales. Res. Inst. 21:1–78.

Payne, R. 1986. Long term behavioral studies of the southern right whale, *Eubalaena australis*. Pp. 161–168 *in* R. L. Brownell, Jr., P. B. Best, and J. H. Prescott, eds., Right whales: Past and present status. Rep. Int. Whal. Commn. (Special Issue 10), Cambridge. 289 pp.

Payne, R., and E. M. Dorsey. 1983. Sexual dimorphism and aggressive use of callosities in right whales (*Eubalaena australis*). Pp. 295–329 *in* R. Payne, ed., Communication and behavior of whales. AAAS Selected Symposium 76, Amer. Assoc. for the Advancement of Science, Washington, D.C. 643 pp.

Payne, R., O. Brazier, E. M. Dorsey, J. S. Perkins, V. J. Rowntree, and A. Titus. 1983. External features in southern right whales (*Eubalaena australis*) and their use in identifying individuals. Pp. 371–445 *in* R. Payne, ed., Communication and behavior of whales. AAAS Selected Symposium 76. Amer. Assoc. for the Advancement of Science, Washington, D.C. 643 pp.

Prescott, J. E., S. D. Kraus, and J. R. Gilbert, eds. 1980. East coast/Gulf coast cetacean and pinniped research workshop. 25–26 September 1979, Boston, Massachusetts. Unpubl. final report, contract MM1533558-2, U.S. Marine Mammal Commission, Washington, D.C. NTIS PB80-160 104. 142 pp.

Reeves, R. R., J. G. Mead, and S. Katona. 1978. The right whale, *Eubalaena glacialis*, in the western North Atlantic. Rep. Int. Whal. Commn. 28:303–312.

Reeves, R. R., and R. L. Brownell, Jr. 1982. Baleen whales (*Eubalaena glacialis* and allies). Pp. 415–444 *in* J. A. Chapman and G. A. Feldhammer, eds., Wild mammals of North America. Biology, management, and economics. Johns Hopkins University Press, Baltimore, Maryland. 1147 pp.

Reeves, R. R., and E. D. Mitchell. 1986. The Long Island, New York, right whale fishery: 1650–1924. Pp. 201–220 *in* R. L. Brownell, Jr., P. B. Best, and J. H. Prescott, eds., Right whales: Past and present status. Rep. Int. Whal. Commn. (Special Issue 10), Cambridge. 289 pp.

Schmidly, D. J. 1981. Marine mammals of the southeastern United States coast and the Gulf of Mexico. FWS/OBS-80/41. U.S. Fish and Wildlife Service, Washington, D.C. 165 pp.

Scott, G. P., ed. 1985. Report of the working group on NEFC/SEFC marine mammal research. Results of the meeting held 8–9 January 1985. NOAA Tech. Memorandum NMFS-SEFC-168, National Marine Fisheries Service, Washington, D.C. 27 pp.

True, F. W. 1904. The whalebone whales of the western North Atlantic compared with those occurring in European waters with some observations on the species of the North Pacific. Smithson. Contrib. Knowl. 33:1–332.

U.S. Fish and Wildlife Service. 1989. Endangered and threatened wildlife and plants; animal notice of review. Federal Register 54:554–579.

Winn, H. E., C. A. Price, and P. W. Sorensen. 1986. The distributional biology of the right whale (*Eubalaena glacialis*) in the western North Atlantic. Pp. 129–138 *in* R. L. Brownell, Jr., P. B. Best, and J. H. Prescott, eds., Right whales: Past and present status. Rep. Int. Whal. Commn. (Special Issue 10), Cambridge. 289 pp.

Prepared by: Daniel K. Odell, Sea World of Florida, 7007 Sea World Drive, Orlando, FL 32821.

Endangered

Sei Whale
Balaenoptera borealis
FAMILY BALAENOPTERIDAE
Order Cetacea

TAXONOMY: *Balaenoptera borealis* was described by Lesson (1828). The type locality is Germany (BRD), Schleswig-Holstein, Lubeck Bay, near Gromitz (Honacki et al. 1982). Gambell (1985) and Horwood (1987) review the taxonomy of the sei whale. Sei whales in the northern hemisphere are slightly smaller than those in the southern hemisphere, and the two morphs have been given subspecific status. The northern hemisphere morph is *B. b. borealis*; the one from the southern hemisphere is *B. b. schleglii* (Gambell 1985; Horwood 1987). Gambell (1985) also lists a number of common names for *B. borealis*.

DESCRIPTION: The sei whale reaches a length of 20 m in the southern hemisphere, 18.6 m in the North Pacific, and about 17.3 m in the North Atlantic. Females are slightly longer than males. Calculated maximum body mass is about 30 tn (International Whaling Commission 1977; Gambell 1985). The head is 20–25 percent of the body length. The ventral grooves extend from the chin to a point anterior to the umbilicus. The grooves number from 36 to 62 and vary with geographic location. Body color is dark gray dorsally and paler beneath. The ventral surfaces of the flukes and flippers are dark. The body often has a number of circular, grayish-white scars, probably from the cookie-cutter shark, *Isistius*. The dorsal fin is located about two-thirds of the way from the rostrum to the tail and may be 60 cm tall. The baleen plates number from 296 to 402 per row and are uniformly ash-black in color with fine white fringes. The vertebral formula is: C7, T14, L13, Ca22–23, totaling 56–57 (Gambell 1985; Horwood 1987).

POPULATION SIZE AND TREND: The total world population prior to

Balaenoptera borealis

Sei whale, *Balaenoptera borealis*. (Photo by James G. Mead; courtesy of the Marine Mammal Program, Smithsonian Institution)

harvesting has been estimated at 100,000–300,000 (Gambell 1985; Horwood 1987; National Marine Fisheries Service 1991). The current world population is estimated at >25,000 with 9110 in the North Pacific, >4000 in the North Atlantic and >12,000 in the Southern Ocean (National Marine Fisheries Service 1991). There is no current information on population trends. The populations may be stable or increasing, however, since all stocks of sei whales are protected (Gambell 1985; National Marine Fisheries Service 1988; International Whaling Commission 1988).

DISTRIBUTION AND HISTORY OF DISTRIBUTION: Sei whales are distributed worldwide from warm-temperate to subarctic waters. They appear to avoid polar and tropical waters. They are not restricted to coastal areas. They undertake seasonal north-south migrations typical of balaenopterid whales (International Whaling Commission 1977; Leatherwood and Reeves 1983). The identity and distribution of the several stocks are not well known (Gambell 1985). In the western North Atlantic, two stocks have been identified: one in the Labrador Sea and one in the Nova Scotian shelf (Mitchell and Chapman 1977). The Nova Scotian stock may migrate as far south as northeastern Florida (Mitchell and Chapman 1977). There are no sei whale stranding records from Florida (Mead 1977; Schmidly 1981; Odell 1991). Gunter and Overstreet (1974) cite a newspaper article about a sei whale stranding near Tarpon Springs, Florida, on 30 May 1974. This animal was a Bryde's whale, *B. edeni*. Although there are two records of sei whale strandings from the northern Gulf of Mexico, they may be extralimital occurrences. Mead (1977) speculates that there may be a resident population of sei whales in the Caribbean/Gulf of Mexico, based on the season (summer) of a stranding from the Dominican Republic.

GEOGRAPHICAL STATUS: With the exception of some animals wintering off northeastern Florida, there are no known sei whale stocks in Florida waters. All stocks are protected by international agreement (International Whaling Commission 1988).

HABITAT REQUIREMENTS AND HABITAT TREND: Aside from food habits and distributional information obtained from whaling activities, little is known about the sei whale's habitat requirements. Sei whales are primarily skim feeders, similar to the right whale, and feed extensively on copepods. They also feed on euphausids, squid, and small schooling fish (International Whaling Commission 1977; Schmidly 1981; Leatherwood and Reeves 1983; Gambell 1985; Horwood 1987). Habitat trends are unknown.

VULNERABILITY OF SPECIES AND HABITAT: Sei whales are protected worldwide from commercial whaling (International Whaling Commission 1988) and there are no known fishery conflicts (Prescott et al. 1980). The primary area of vulnerability appears to be food resources.

CAUSES OF THREAT: Threats to food resources could come from ocean pollution and competition from commercial fisheries.

RESPONSES TO HABITAT MODIFICATION: Unknown. Some speculate that severe reduction of right whale populations provides an excess of food (copepods), which may have allowed the sei population to expand. This may be a factor in the apparent lack of recovery of the western North Atlantic right whales (Prescott et al. 1980). Sei whale reproductive parameters in the southern hemisphere may have been affected by changes in food abundance resulting from heavy exploitation of other species of whales (Horwood 1987).

DEMOGRAPHIC CHARACTERISTICS: Sei whales reach sexual maturity at 13.5 m (males) and 13.9 m (females) in the southern hemisphere and

12.8 m (males) and 13.3 m (females) in the North Pacific (Gambell 1985; Horwood 1987). Sei whales of the North Atlantic are probably similar to those of the North Pacific. Age at sexual maturity is 10–11 years for both sexes in the southern hemisphere, 10 years for both sexes in the North Pacific, and 12 years for males and 18 years for females in the North Atlantic (Gambell 1985; Horwood 1987). However, the sample size from the North Atlantic is small (International Whaling Commission 1977). In both the southern hemisphere and the North Pacific, age at sexual maturity was reduced by 4–5 years (both sexes) as a result of whaling. Calves are 4.4–4.5 m long at birth and weigh about 0.65 tn (Laws 1959; Gambell 1985). Gestation is about 10.5–12.5 months followed by a lactation period of at least 6 months (Gambell 1985; Horwood 1987). A 2-year breeding cycle is possible but not fully documented (Gambell 1985). Calves are weaned at a length of about 9 m. Pregnancy rates, where known from whaling, have ranged from 25 to 75 percent depending on the stock and extent of exploitation. Overall mortality rate of adults has been estimated at 0.06 (International Whaling Commission 1977).

KEY BEHAVIORS: Sei whale behavior is not well known. They are thought to be shallow divers and feed primarily by skimming plankton at the surface but can feed by gulping a single mouthful of water (Gambell 1985; Horwood 1987).

CONSERVATION MEASURES TAKEN: Sei whales are protected by the Marine Mammal Protection Act of 1972 (National Marine Fisheries Service 1988), the Endangered Species Act of 1973 (U.S. Fish and Wildlife Service 1989), the International Convention for the Regulation of Whaling, 1946 (International Whaling Commission 1988), and the Convention on International Trade in Endangered Species (CITES 1989). All stocks are protected from commercial whaling. The Iceland–Denmark Strait stock was still being exploited (383 animals) in 1985 (International Whaling Commission 1988). The sei whale is listed as endangered by the Florida Game and Fresh Water Fish Commission (1990).

CONSERVATION MEASURES PROPOSED: Factors that could affect the sei whale, including competition from commercial fisheries and ocean pollution, should be controlled wherever possible.

ACKNOWLEDGMENTS: This is Sea World of Florida Technical Contribution No. 8911-F.

Literature Cited

CITES. 1989. Convention on International Trade in Endangered Species of Wild Fauna and Flora. Regulations and appendices published by the Federal Wildlife Permit Office, U.S. Fish and Wildlife Service, Washington, D.C.

Florida Game and Fresh Water Fish Commission. 1990. Official lists of endangered and potentially endangered fauna and flora in Florida. Tallahassee. 23 pp.

Gambell, R. 1985. Sei whale—*Balaenoptera borealis*. Pp. 171–192 *in* S. H. Ridgway and R. Harrison, eds., Handbook of marine mammals. Vol. 3. The sirenians and baleen whales. Academic Press, London. 362 pp.

Gunter, G., and R. Overstreet. 1974. Cetacean notes. I. Sei and rorqual whales on the Mississippi coast, a correction. II. A dwarf sperm whale in Mississippi Sound and its helminth parasites. Gulf Research Reports 4:479–481.

Honacki, J. H., K. E. Kinman, and J. W. Koeppl, eds. 1982. Mammal species of the world. A taxonomic and geographic reference. Allen Press and Assoc. of Systematics Collections, Lawrence, Kansas. 694 pp.

Horwood, J. 1987. The sei whale: Population biology, ecology and management. Croom Helm, London. 375 pp.

International Whaling Commission. 1977. Report of the special meeting of the scientific committee on sei and Bryde's whales. Rep. Int. Whal. Commn. (Special Issue 1), International Whaling Commission, Cambridge, United Kingdom. 150 pp.

International Whaling Commission. 1988. International whaling statistics. XCV and XCVI. International Whaling Commission, Cambridge, United Kingdom. 68 pp.

International Whaling Commission. 1988. Schedule, international convention for the regulation of whaling, 1946. International Whaling Commission, Cambridge, United Kingdom. 27 pp.

Laws, R. M. 1959. The foetal growth rates of whales with special reference to the fin whale, *Balaenoptera physalus* Linn. Disc. Rep. 29:281–308.

Leatherwood, S., and R. R. Reeves. 1983. The Sierra Club handbook of whales and dolphins. Sierra Club Books. San Francisco. 302 pp.

Lesson, R. P. 1828. "Complements des Oeuvres de Buffon ou Historie Naturelle des Animaux Rares." Paris.

Mead, J. G. 1977. Records of sei and Bryde's whales from the Atlantic coast of the United States, the Gulf of Mexico and the Caribbean. Pp. 113–116 *in* Report of the special meeting of the scientific committee on sei and Bryde's whales. Rep. Int. Whal. Commn. (Special Issue 1), International Whaling Commission, Cambridge, United Kingdom. 150 pp.

Mitchell, E. D., and D. G. Chapman. 1977. Preliminary assessment of stocks of northwest Atlantic sei whales. Pp. 117–120 *in* Report of the special meeting of the scientific committee on sei and Bryde's whales. Rep. Int. Whal. Commn. (Special Issue 1), International Whaling Commission, Cambridge, United Kingdom. 150 pp.

National Marine Fisheries Service. 1988. Marine Mammal Protection Act of 1972. Annual Report 1987/88. USDOC/NOAA/NMFS, Washington, D.C. 68 pp.
National Marine Fisheries Service. 1991. Endangered Whales: Update. USDOC, NOAA, National Marine Fisheries Service, Silver Spring, Maryland. 52 pp.
Odell, D. K. 1991. A review of the southeastern United States marine mammal stranding network: 1978–1987. Pp. 19–23 *in* J. E. Reynolds III and D. K. Odell, eds., Marine Mammal Strandings in the United States. Proceedings of the Second Marine Mammal Stranding Workshop, Miami, Florida. December 3–5, 1987. NOAA Technical Report NMFS 98, USDOC, NOAA, National Marine Fisheries Service, Seattle, Washington. 157 pp.
Omura, H. 1966. Bryde's whale in the northwest Pacific. Pp. 70–78 *in* K. S. Norris, ed., Whales, dolphins and porpoises. University of California Press, Berkeley and Los Angeles. 789 pp.
Prescott, J. E., S. D. Kraus, and J. R. Gilbert, eds. 1980. East coast/Gulf coast cetacean and pinniped research workshop. 25–26 September 1979, Boston, Massachusetts. Final report, contract MM1533558-2, U.S. Marine Mammal Commission, Washington, D.C. NTIS PB80-160 104. 142 pp.
Schmidly, D. J. 1981. Marine mammals of the southeastern United States coast and the Gulf of Mexico. FWS/OBS-80/41. U.S. Fish and Wildlife Service, Washington, D.C. 165 pp.
U.S. Fish and Wildlife Service. 1989. Endangered and threatened wildlife and plants; animal notice of review. Federal Register 54:554–579.

Prepared by: Daniel K. Odell, Sea World of Florida, 7007 Sea World Drive, Orlando, FL 32821.

Endangered

Fin Whale
Balaenoptera physalus
FAMILY BALAENOPTERIDAE
Order Cetacea

TAXONOMY: *B. physalus* was described by Linnaeus (1758). The type locality is Norway, near Svalbard, Spitsbergen Sea (Honacki et al. 1982). Gambell (1985) reviews the taxonomy of the fin whale. As with the sei whale, fin whales of the northern hemisphere are slightly smaller than those in the southern hemisphere, and some subspecies have been designated: *B. p. physalus* for the northern hemisphere and *B. p. quoyi* for the southern hemisphere (Fischer 1829; Gambell 1985). Gambell (1985) lists a number of common names for the fin whale.

DESCRIPTION: The female fin whale reaches a length of about 24 m in the northern hemisphere and 27 m in the southern hemisphere. Males are about 2 m shorter. Maximum body mass is about 80 tn (Gambell 1985). The 50–100 ventral grooves extend from the chin to the umbilical region. The flippers are small. The dorsal fin may be up to 60 cm tall and is located about two-thirds of the way back from the tip of the rostrum. Dorsal color is dark gray, grading into white ventrally. The flukes and flippers are white ventrally. The color pattern on the head is characteristically asymmetrical. The right jaw is white and the left gray. There are 260-480 baleen plates in each row. The baleen is also asymmetrically colored, with the front fourth of the right row being white or yellowish and the remainder blue-gray with lighter streaks. The vertebral formula is: C7, T15–16, L13–16, Ca24–27, total usually 60–63. There are 14–16 ribs, with only the first pair joined to the sternum (Gambell 1985).

POPULATION SIZE AND TREND: The preexploitation population level has been estimated at >464,000, including >18,000 in the North Atlantic, <45,000 in the North Pacific and 400,000 in the Southern Ocean

Fin whale, *Balaenoptera physalus*. (Photo by John E. Heyning; by permission of the Natural History Museum of Los Angeles County)

(National Marine Fisheries Service 1991). Estimates for large parts of the North Atlantic are lacking. Current world population is estimated at about 119,000 animals, with 17,221 in the North Atlantic, 16,625 in the North Pacific, and 85,200 in the Southern Ocean (National Marine Fisheries Service 1991). All fin whale stocks are protected (Gambell 1985; International Whaling Commission 1988).

DISTRIBUTION AND HISTORY OF DISTRIBUTION: The fin whale is distributed in all oceans but is less common in tropical waters than in cooler waters (Leatherwood and Reeves 1983; Gambell 1985). Although all stocks have been severely depleted by commercial whaling, the current distribution is probably the same as the historic distribution. Fin whales show the typical balaenopterid migratory pattern: poleward in summer and into lower latitudes in winter. In the western North Atlantic, fin whales summer from the New England area to the Arctic and winter from the ice edge southward to the Gulf of Mexico and the Caribbean (Leatherwood and Reeves 1983; Gambell 1985; Scott 1985). Fin whales have stranded on both coasts of Florida.

GEOGRAPHICAL STATUS: All stocks of fin whales are protected. There is some evidence for two stocks of fin whale in the western North Atlantic and speculation about a resident stock in the Gulf of Mexico (Gambell 1985; Scott 1985).

HABITAT REQUIREMENTS AND HABITAT TREND: Little is known about habitat requirements of fin whales. They feed heavily on euphausids (krill), capelin (*Mallotus*), herring (*Clupea*), cod (*Gadus*), pollack, mackerel, and anchovies (*Engraulis*). Food habits vary geographically. Fin whales feed by gulping masses of water rather than by skimming (Prescott et al. 1980; Gambell 1985).

VULNERABILITY OF SPECIES AND HABITAT: All stocks of fin whales are protected except for limited aboriginal take (International Whaling Commission 1988). Fin whales are vulnerable to other human activities.

CAUSES OF THREAT: Fin whales are occasionally entangled in fishing gear (Prescott et al. 1980) and may be hit by ships. Fin whale behavior

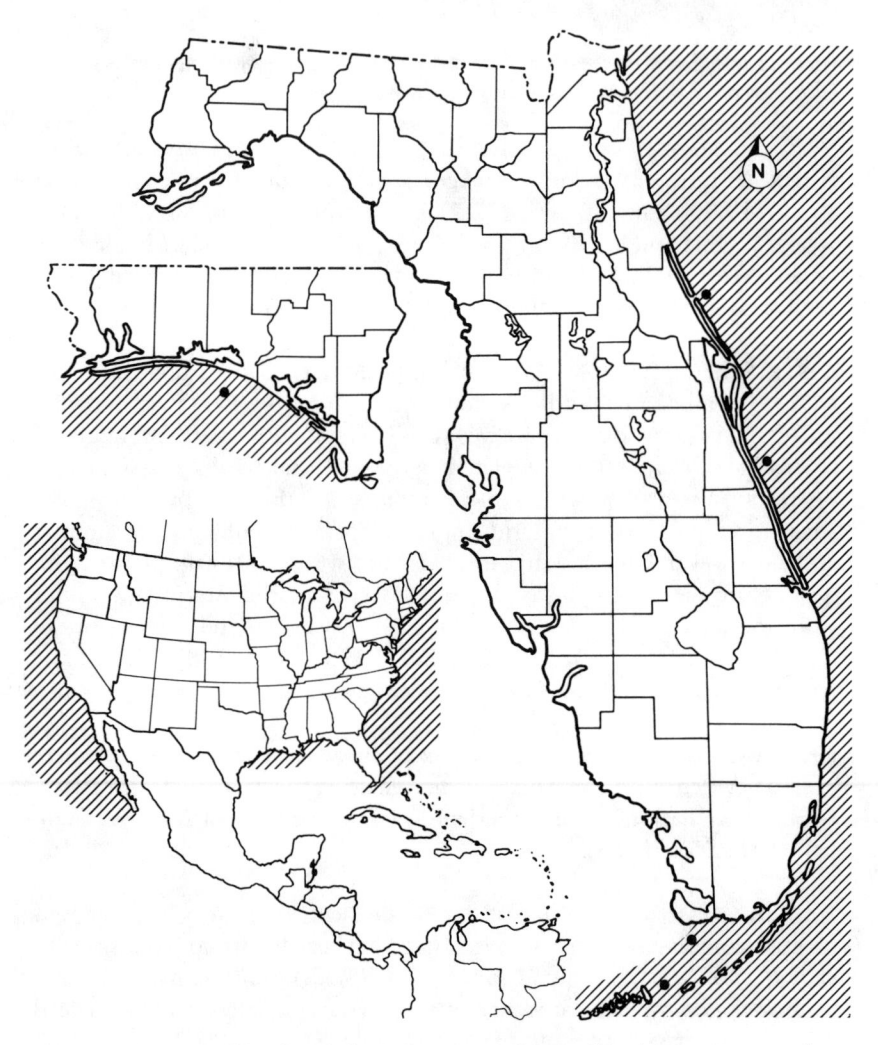

Distribution map of the fin whale, *Balaenoptera physalus*. Hatching represents the species; the dots represent sightings and strandings.

Balaenoptera physalus 157

could be affected by excessive whale-watching activities in New England (Prescott et al. 1980). They may be vulnerable to pollutants (heavy metals, polychlorinated hydrocarbons) passed through the food chain (Prescott et al. 1980). Some fisheries do compete with fin whales, but to an unknown extent.

RESPONSES TO HABITAT MODIFICATION: Unknown.

DEMOGRAPHIC CHARACTERISTICS: Fin whales may live 90–100 years (Roe 1967). Sexual maturity is reached at 10–13 years of age and a length of 18.5 m for males and 19.8 m for females. The average age at sexual maturity is lower in heavily exploited populations. Gestation is 11–12 months. The newborn is about 6.4 m long and weighs 1.9 tn (Laws 1959). Lactation is 6–7 months, and the calf is weaned at about 12 m in length (Leatherwood and Reeves 1983; Minasian et al. 1984; Gambell 1985).

KEY BEHAVIORS: Fin whales feed by engulfing masses of water. Variations include rushing, lunging, and circling at or near the surface (Prescott et al. 1980). The blow is 4–6 m tall. Fin whales are one of the fastest of the balaenopterids and may reach a speed of 20 knots in a short burst. Average speed is about 9 km/hr (Ray et al. 1978; Gambell 1985). Fin whales produce low frequency sounds from 20–100 Hz.

CONSERVATION MEASURES TAKEN: In the United States, fin whales are protected by the Marine Mammal Protection Act of 1972 (National Marine Fisheries Service 1988) and the Endangered Species Act of 1973 (U.S. Fish and Wildlife Service 1989). All stocks are protected from commercial whaling by the International Convention for the Regulation of Whaling, 1946 (International Whaling Commission 1988) and the Convention on International Trade in Endangered Species (CITES 1989). The Florida Game and Fresh Water Fish Commission (1990) lists the fin whale as endangered.

CONSERVATION MEASURES PROPOSED: Whale-watching activities should be monitored to prevent unnecessary disturbance of fin whales. Ocean pollution in general should be reduced to limit concentration of pollutants through the food chain.

ACKNOWLEDGMENTS: This is Sea World of Florida Technical Contribution No. 8910-F.

Literature Cited

CITES. 1989. Convention on International Trade in Endangered Species of Wild Fauna and Flora. Regulations and appendices published by the Federal Wildlife Permit Office, U.S. Fish and Wildlife Service, Washington, D.C.

Fischer. 1829. Synopsis Mammalium. J. G. Cottse, Stutgart.

Florida Game and Fresh Water Fish Commission. 1990. Official lists of endangered and potentially endangered fauna and flora in Florida. Tallahassee. 23 pp.

Gambell, R. 1985. Fin whale—*Balaenoptera physalus*. Pp. 171–192 *in* S. H. Ridgway and R. Harrison, eds., Handbook of marine mammals. Vol. 3. The sirenians and baleen whales. Academic Press, London. 362 pp.

Honacki, J. H., K. E. Kinman, and J. W. Koeppl, eds. 1982. Mammal species of the world. A taxonomic and geographic reference. Allen Press and Assoc. of Systematics Collections, Lawrence, Kansas. 694 pp.

International Whaling Commission. 1988. Schedule, International Convention for the Regulation of Whaling, 1946. International Whaling Commission, Cambridge, United Kingdom. 27 pp.

Laws, R. M. 1959. The foetal growth rates of whales with special reference to the fin whale, *Balaenoptera physalus* Linn. Disc. Rep. 29:281–308.

Leatherwood, S., and R. R. Reeves. 1983. The Sierra Club handbook of whales and dolphins. Sierra Club Books, San Francisco. 302 pp.

Linnaeus, C. 1758. Systema Naturae. 10th ed. Vol. 1. p. 75.

Minasian, S. M., K. C. Balcomb III, and L. Foster. 1984. The world's whales, the complete illustrated guide. Smithsonian Books, Washington, D.C. 244 pp.

Moore, J. C. 1953. Distribution of marine mammals to Florida waters. Amer. Midland Nat. 49:117–158.

National Marine Fisheries Service. 1988. Marine Mammal Protection Act of 1972. Annual Report 1987/88. USDOC/NOAA/NMFS, Washington, D.C. 68 pp.

National Marine Fisheries Service. 1991. Endangered Whales: Update. USDOC, NOAA, National Marine Fisheries Service, Silver Spring, Maryland. 52 pp.

Odell, D. K. 1991. A review of the southeastern United States marine mammal stranding network: 1978–1987. Pp. 19–23 *in* J. E. Reynolds III and D. K. Odell, eds., Marine Mammal Strandings in the United States. Proceedings of the Second Marine Mammal Stranding Workshop, Miami, Florida. December 3–5, 1987. NOAA Technical Report NMFS 98, USDOC, NOAA, National Marine Fisheries Service, Seattle, Washington. 157 pp.

Prescott, J. E., S. D. Kraus, and J. R. Gilbert, eds. 1980. East coast/Gulf coast cetacean and pinniped research workshop. 25–26 September 1979, Boston, Massachusetts. Final report, contract MM1533558-2, U.S. Marine Mammal Commission, Washington, D.C. NTIS PB80-160 104. 142 pp.

Ray, G. C., E. D. Mitchell, D. Wartzok, M. Kozicki, and R. Maiefski. 1978. Radio tracking of a fin whale (*Balaenoptera physalus*). Science 202:521–524.

Roe, H. S. J. 1967. Seasonal formation of laminae in the ear plug of the fin whale. Disc. Rep. 35:1–30.
Schmidly, D. J. 1981. Marine mammals of the southeastern United States coast and the Gulf of Mexico. FWS/OBS-80/41. U.S. Fish and Wildlife Service, Washington, D.C. 165 pp.
Scott, G. P., ed. 1985. Report of the working group on NEFC/SEFC marine mammal research. Results of the meeting held 8–9 January 1985. NOAA Tech. Memorandum NMFS-SEFC-168. National Marine Fisheries Service, Washington, D.C. 27 pp.
U.S. Fish and Wildlife Service. 1989. Endangered and threatened wildlife and plants; animal notice of review. Federal Register 54:554–579.

Prepared by: Daniel K. Odell, Sea World of Florida, 7007 Sea World Drive, Orlando, FL 32821.

Endangered

Humpback Whale
Megaptera novaeangliae
FAMILY BALAENOPTERIDAE
Order Cetacea

TAXONOMY: *Megaptera novaeangliae* was described by Borowski (1781). The type locality is United States of America, coast of New England (Honacki et al. 1982). While the humpback is included in the Balaenopteridae, its distinct anatomical characteristics separate it from other balaenopterids (Winn and Reichley 1985).

DESCRIPTION: The humpback whale is characterized by its large, white flippers about one-third of body length. Body length reaches 15 m in males and 16 m in females. Body mass has been estimated at 24–30 tn (Winn and Reichley 1985). Blowholes are paired and a small dorsal fin (about 30 cm) is present. The dorsal surface of the body is dark colored, but the ventral surface has variable patches of white. Knobs (tubercles) occur on the leading edge of the flippers and on the head in front of the blowholes (Reeves and Brownell 1982; Leatherwood and Reeves 1983; Swartz 1989). The humpback's body is more robust than that of the other balaenopterids. There are 12–36 pleats (grooves) running from the chin to the navel. There are 270–400 baleen plates in each row. Maximum plate length is about 70 cm. Baleen color is black, blackish brown, or gray. The bristles are black or olive-black. The vertebral formula is C7, T14, L11(10), Ca21, total 52–53 (Winn and Reichley 1985). The skull is about 3 m long.

POPULATION SIZE AND TREND: The preexploitation world population of humpbacks has been estimated at over 127,000, including 102,000 in the southern hemisphere, 15,000 in the North Pacific, and 10,000 in the North Atlantic (Winn and Reichley 1985). Current world population estimate is 9500–10,000, including 2500–3000 in the southern hemi-

Humpback whale, *Megaptera novaeangliae*. (Photo by James G. Mead; courtesy of the Marine Mammal Program, Smithsonian Institution)

sphere, 1200 in the North Pacific, and 5800 in the North Atlantic (National Marine Fisheries Service 1988, 1991). All populations are probably increasing. All stocks are protected from commercial whaling (International Whaling Commission 1988). There is a low level of subsistence take (Winn and Reichley 1985).

DISTRIBUTION AND HISTORY OF DISTRIBUTION: The humpback whale is distributed in all oceans and exhibits the seasonal north-south migration pattern typical of the rorquals (Swartz 1989). Humpbacks are much more coastal, especially in the feeding and calving/breeding grounds, than are other balaenopterids. In the western North Atlantic there is a single stock that summers from the coast of New England northward and winters in the Caribbean region. Humpback whales have been stranded and sighted on the east coast of Florida, but have only been sighted off the Gulf coast (Odell in press). Humpbacks seen in the Gulf of Mexico may have come from the Caribbean via the Yucatan channel and ridden the loop current, which ultimately takes them through the straits of Florida and to the summer feeding grounds via the Gulf Stream. The majority of the western North Atlantic stock migrates to and from the Caribbean via Bermuda (Reeves and Brownell 1982; Leatherwood and Reeves 1983; Winn and Reichley 1985; Stone et al. 1987). However, Schmidly (1981) cites two summer (June and July) sightings in the Gulf of Mexico as possible evidence for a breeding stock there.

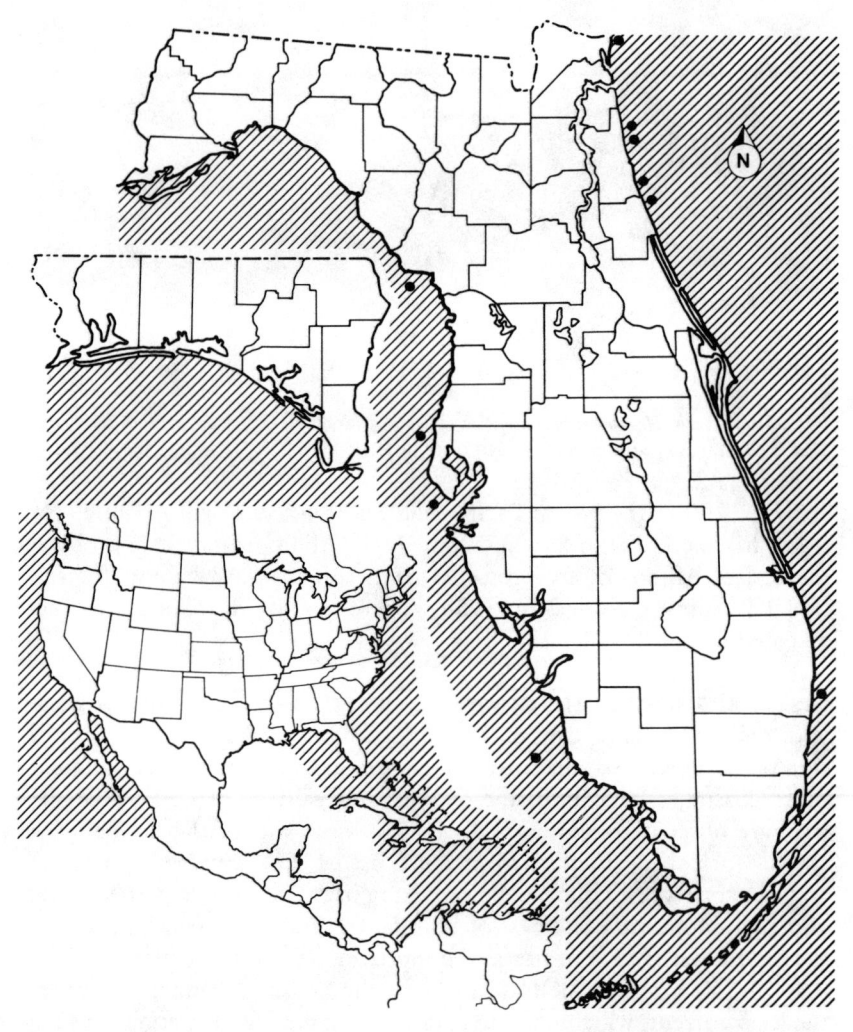

Distribution map of the humpback whale, *Megaptera novaeangliae*. Hatching represents the species; the dots represent sightings and strandings.

GEOGRAPHICAL STATUS: Present information suggests that the humpback whales seen in Florida waters belong to the western North Atlantic stock and that there are no resident stocks in Florida waters.

HABITAT REQUIREMENTS AND HABITAT TREND: As with other baleen whales, the habitat requirements and thus habitat trend are not well known. Humpbacks summer and winter in shallow coastal areas that must be considered important components of their habitat. They feed on euphausids (krill), mackerel, capelin, herring, sand lance, and pollack. In the western North Atlantic, small schooling fish are the main prey (Leatherwood and Reeves 1983; Winn and Reichley 1985; Katona 1986; Weinrich et al. 1986*a*, 1986*b*; Swartz 1989).

VULNERABILITY OF SPECIES AND HABITAT: Humpbacks are protected from all commercial whaling and only a few are taken each year by aboriginal whaling (Winn and Reichley 1985; International Whaling Commission 1988). Humpbacks are vulnerable to impact from other human activities.

CAUSES OF THREAT: Humpbacks have become entangled in fishing gear (Beamish 1979; Prescott et al. 1980; Scott 1985) and may collide with coastal vessels. Intense whale-watching activities on summer and winter grounds can affect whale behavior (National Marine Fisheries Service 1987, 1988). Whale watching in U.S. waters is subject to regulation under the Marine Mammal Protection Act. Pollutants concentrated through the food chain may affect humpbacks. Recent events also indicate that natural biotoxins (i.e., saxitoxin) may be responsible for some humpback mortality (Geraci 1989; Geraci et al. 1989; Swartz 1989). Offshore oil development activities could affect humpbacks and their habitat.

RESPONSES TO HABITAT MODIFICATION: Generally unknown. Intense, unregulated whale watching may cause humpbacks to move away from preferred habitat. Human activities that affect the distribution and abundance of prey could cause shifts in distribution.

DEMOGRAPHIC CHARACTERISTICS: Humpback whales reach sexual maturity at about 11–12 m in both sexes (Leatherwood and Reeves 1983). This corresponds with an age of 4–9 years (Winn and Reichley 1985; Scott 1985). The calving interval is usually 2 or 3 years with a gestation of 11–11.5 months. The newborn is 4–5 m long and nurses for 10–12 months (Leatherwood and Reeves 1983; Scott 1985). The nursing period is about 5 months long if there is a 2-year calving cycle (Winn and Reichley 1985). Weaned calves are 7.5–9 m long (Winn and Reichley 1985). These parameters may vary with population status relative to carrying capacity and degree of exploitation.

KEY BEHAVIORS: Humpbacks are well known for their spectacular aerial leaps, spy hopping, lob tailing (Winn and Reichley 1985), and acoustic behavior (songs; Payne and McVay 1971; Thompson et al. 1979; Payne 1983). Humpbacks feed by gulping masses of water containing prey. There are several variations of this behavior that the whales may do singly or in groups. These include bubble rings and lunging (Jurasz and Jurasz 1979; Winn and Reichley 1985; Swartz 1989).

CONSERVATION MEASURES TAKEN: Humpback whales are protected by the Marine Mammal Protection Act of 1972 (National Marine Fisheries Service 1988), the Endangered Species Act of 1973 (U.S. Fish and Wildlife Service 1989), the International Convention for the Regulation of Whaling, 1946 (International Whaling Commission 1988), and the Convention on International Trade in Endangered Species (CITES 1989). They are listed as endangered by the Florida Game and Fresh Water Fish Commission (1990).

CONSERVATION MEASURES PROPOSED: Because the humpback is primarily a coastal animal, all activities that may affect the coastal ecosystem should be carefully evaluated for possible effects on the whales (see Swartz 1989).

ACKNOWLEDGMENTS: This is Sea World of Florida Technical Contribution No. 8913-F.

Literature Cited

Beamish, P. 1979. Behavior and significance of entrapped baleen whales. Pp. 291–309 *in* H. E. Winn and B. L. Olla, eds., Behavior of marine animals. Current perspectives in research. Vol. 3. Cetaceans. Plenum Press, New York. 438 pp.
Borowski, G. H. 1781. Gemein. Naturgesch. Thier. 2:21.

CITES. 1989. Convention on International Trade in Endangered Species of Wild Fauna and Flora. Regulations and appendices published by the Federal Wildlife Permit Office, U.S. Fish and Wildlife Service, Washington, D.C.
Florida Game and Fresh Water Fish Commission. 1990. Official lists of endangered and potentially endangered fauna and flora in Florida. Tallahassee. 23 pp.
Geraci, J. R. 1989. Clinical investigation of the 1987–88 mass mortality of bottlenose dolphins along the U.S. central and south Atlantic coast. Unpubl. report, University of Guelph, Guelph, Ontario, Canada. 63 pp.
Geraci, J. R., D. M. Anderson, R. J. Timperi, D. J. St. Aubin, G. A. Early, J. H. Prescott, and C. A. Mayo. 1989. Humpback whales (*Megaptera novaeangliae*) fatally poisoned by dinoflagellate toxin. Canad. J. Fish. Aquat. Sci. 46: 1895–1898.
Honacki, J. H., K. E. Kinman, and J. W. Koeppl, eds. 1982. Mammal species of the world. A taxonomic and geographic reference. Allen Press and Assoc. of Systematics Collections, Lawrence, Kansas. 694 pp.
International Whaling Commission. 1988. International Convention for the Regulation of Whaling, 1946. Schedule. As amended by the Commission at the 40th Annual Meeting, May/June 1988, and replacing that dated November 1987. International Whaling Commission, Cambridge, United Kingdom. 27 pp.
Jurasz, C. M., and V. P. Jurasz. 1979. Feeding modes of the humpback whale, *Megaptera novaeangliae*, in southeast Alaska. Sci. Rep. Whale Res. Inst. 31: 69–83.
Katona, S. K. 1986. Biogeography of the humpback whale, *Megaptera novaeangliae*, in the North Atlantic. UNESCO Tech. Pap. Mar. Sci. 46:166–171.
Layne, J. N. 1965. Observations on marine mammals in Florida waters. Bull. Florida State Mus., Biol. Sci. 9:131–181.
Leatherwood, S., and R. R. Reeves. 1983. The Sierra Club handbook of whales and dolphins. Sierra Club Books, San Francisco. 302 pp.
Moore, J. C. 1953. Distribution of marine mammals to Florida waters. Amer. Midland Nat. 49:117–158.
Muller. 1776. Zool. Danicae Prodr., p. 7.
National Marine Fisheries Service. 1987. Marine Mammal Protection Act of 1972. Annual Report 1986/87. USDOC/NOAA/NMFS, Washington, D.C. 65 pp.
National Marine Fisheries Service. 1988. Marine Mammal Protection Act of 1972. Annual Report 1987/88. USDOC/NOAA/NMFS, Washington, D.C. 68 pp.
National Marine Fisheries Service. 1991. Endangered Whales: Update. USDOC, NOAA, National Marine Fisheries Service, Silver Spring, Maryland. 52 pp.
Odell, D. K. 1991. A review of the southeastern United States marine mammal stranding network: 1978–1987. Pp. 19–23 *in* J. E. Reynolds III and D. K. Odell, eds., Marine Mammal Strandings in the United States. Proceedings of the Second Marine Mammal Stranding Workshop, Miami, Florida, December

3–5, 1987. NOAA Technical Report NMFS 98, USDOC, NOAA, National Marine Fisheries Service, Seattle, Washington. 157 pp.

Payne, R., ed. 1983. Communication and behavior of whales. AAAS Selected Symposium 76. Amer. Assoc. for the Advancement of Science, Washington, D.C. 643 pp.

Payne, R., and S. McVay. 1971. Songs of humpback whales. Science 173:585–597.

Prescott, J. E., S. D. Kraus, and J. R. Gilbert, eds. 1980. East coast/Gulf coast cetacean and pinniped research workshop. 25–26 September 1979, Boston, Massachusetts. Final report, contract MM1533558-2, U.S. Marine Mammal Commission, Washington, D.C. NTIS PB80-160 104. 142 pp.

Reeves, R. R., and R. L. Brownell, Jr. 1982. Baleen whales (*Eubalaena glacialis* and allies). Pp. 415–444 *in* J. A. Chapman and G. A. Feldhammer, eds., Wild mammals of North America. Biology, management, and economics. Johns Hopkins University Press, Baltimore. 1147 pp.

Schmidly, D. J. 1981. Marine mammals of the southeastern United States coast and the Gulf of Mexico. FWS/OBS-80/41. U.S. Fish and Wildlife Service, Washington, D.C. 165 pp.

Scott, G. P., ed. 1985. Report of the working group on NEFC/SEFC marine mammal research. Results of the meeting held 8–9 January 1985. NOAA Tech. Memorandum NMFS-SEFC-168. National Marine Fisheries Service, Washington, D.C. 27 pp.

Stone, G. S., S. K. Katona, and E. B. Tucker. 1987. History, migration and present status of humpback whales *Megaptera novaeangliae* at Bermuda. Biol. Conserv. 42:133–145.

Swartz, S. L. 1989. The humpback whale. Pp. 386–403 *in* W. J. Chandler, ed., Audubon Wildlife Report 1989/1990. Academic Press, San Diego. 585 pp.

Thompson, T. J., H. E. Winn, and P. J. Perkins. 1979. Mysticete sounds. Pp. 403–431 *in* H. E. Winn and B. L. Olla, eds., Behavior of marine animals. Current perspectives in research. Vol. 3. Cetaceans. Plenum Press, New York. 438 pp.

U.S. Fish and Wildlife Service. 1989. Endangered and threatened wildlife and plants; animal notice of review. Federal Register 54:554–579.

Weinrich, M. T., C. R. Belt, H. J. Iken, and M. R. Schiling. 1986*a*. Individual feeding patterns of humpback whales (*Megaptera novaeangliae*) in New England. Pp. 28–40 *in* Humpback whales of the southern Gulf of Maine: Recent findings on habitat use, social behavior, and feeding patterns. Cetacean Research Unit, Gloucester, Massachusetts. 41 pp.

Weinrich, M. T., C. R. Belt, M. R. Schilling, and M. E. Cappellino. 1986*b*. Habitat use patterns as a function of age and reproductive status in humpback whales (*Megaptera novaeangliae*). Pp. 3–17 *in* Humpback whales of the southern Gulf of Maine: Recent findings on habitat use, social behavior, and feeding patterns. Cetacean Research Unit, Gloucester, Massachusetts. 41 pp.

Winn, H. E., and N. E. Reichley. 1985. Humpback whale—*Megaptera novaeangliae*. Pp. 241–274 *in* S. H. Ridgway and R. Harrison, eds., Handbook of marine mammals. Vol. 3. The sirenians and baleen whales. Academic Press, London. 362 pp.

Prepared by: Daniel K. Odell, Sea World of Florida, 7007 Sea World Drive, Orlando, FL 32821.

Endangered

Sperm Whale
Physeter macrocephalus
FAMILY PHYSETERIDAE
Order Cetacea

TAXONOMY: Other scientific names: *Physeter catodon*. Other common names: cachalot. Schevill (1986) explains the recent dilemma over the correct specific name for the sperm whale. Both names (*P. catodon* and *P. macrocephalus*) appear on the same page of the Systema Naturae, 10th edition, 1758, and are considered synonyms. *P. catodon* has line priority (Rice 1989). Boschma (1938) and Husson and Holthuis (1974) determined that *P. macrocephalus* had precedence based on the Principle of the First Reviser. Recent literature has tended to use *P. macrocephalus*. Schevill argues that the First Reviser argument is incorrect, in part because of the wording of the original description, and therefore, *P. catodon* is the proper scientific name for the sperm whale. Holthuis (1987) and Schevill (1987) continue the discussion. Rice (1989) reviews the entire controversy and agrees that *P. macrocephalus* is the correct name.

DESCRIPTION: The sperm whale is the largest and most sexually dimorphic of the toothed whales (Best 1979). Males rarely reach a length exceeding 18.5 m (Berzin 1972) and females 12.5 m (Rice 1989). While few sperm whales have been weighed, an 11 m female weighed 24 metric tn and a 18.1 m male weighed 57.1 metric tn (Rice 1989). The head is rectangular in profile and composes up to one-third of total body length. The single s-shaped blowhole is at the front of the head to the left of the centerline. The characteristic blow angles to the left and forward (Leatherwood et al. 1976; Leatherwood and Reeves 1982). The narrow, underslung jaw holds 17–29 pairs of functional teeth that do not erupt until puberty (about 8 years of age; Rice 1989). Vestigial teeth are occasionally found in the upper jaw. The color of the sperm whale has been variously described as uniform dark gray, bluish black, slate gray, and brownish

Physeter macrocephalus

Sperm whale, *Physeter macrocephalus*. (Photo by D. M. Burn)

(Schmidly 1981; Leatherwood and Reeves 1982; Rice 1989). The abdomen may be lighter in color than the dorsum. The edges of the mouth are pale to white and particularly noticeable in the calves. The dorsal "fin" is hump-like. The skin posterior to the head is wrinkled in appearance (Leatherwood and Reeves 1982). Rice (1989) presents a detailed review of internal and external anatomy of the sperm whale. The vertebral formula is: C7, T11, L8, and C21–25, total 47–51 (Rice 1989). Skull condylobasal length is about 25 percent of body length.

170   SPERM WHALE

POPULATION SIZE AND TREND: Rice (1989) and National Marine Fisheries Service (1991) estimate the worldwide population to be about 1.8 to 1.9 million whales with 190,000 in the North Atlantic, 930,000 in the North Pacific and 780,000 in the Southern Ocean. The National Marine Fisheries Service (1991) estimates the preexploitation population to be

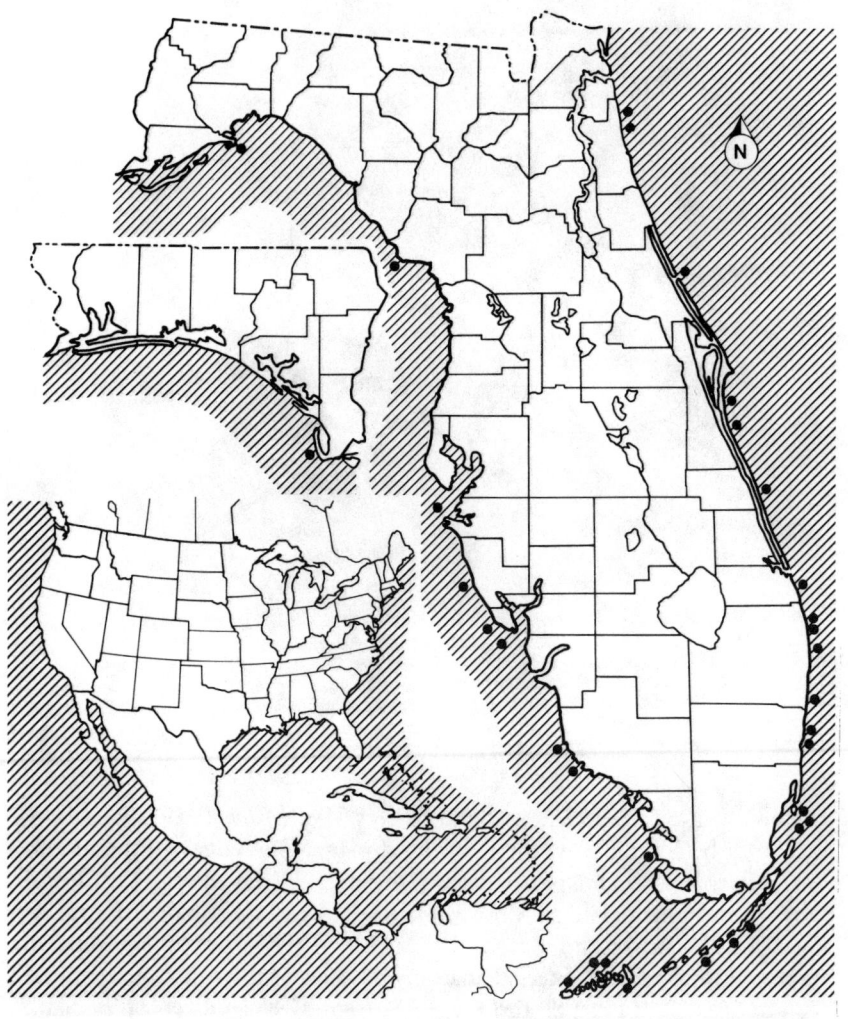

Distribution map of the sperm whale, *Physeter macrocephalus*. Hatching represents the species; the dots represent strandings.

about 2.8 million. Assessment of sperm whale population sizes and trends has relied almost exclusively on historical whaling records (Tillman and Donovan 1983) and catch-per-unit-effort data from modern whaling (Donovan 1980) to estimate initial stock sizes, current stock sizes, and trends.

DISTRIBUTION AND HISTORY OF DISTRIBUTION: The sperm whale is distributed worldwide from polar (70°N and 70°S latitude) to tropical waters, but most of the animals occur in temperate and tropical waters (40°N–40°S latitude). They inhabit deep waters occurring along the edge of the continental shelf. In the southeastern United States, most of the sperm whale sightings occur outside the 200-m depth curve (Schmidly 1981). Most sperm whales winter in equatorial regions and summer in higher latitudes (Best 1969). Only adult males occur outside the 50°N–50°S region. There is some evidence for a resident stock of sperm whales in the Gulf of Mexico (Fritts et al. 1983). See Rice (1989) for a detailed discussion of worldwide distribution.

Schmidly (1981, Fig. 10) summarized sperm whale stranding and sighting records for the southeastern United States through 1979. Reports of sperm whale stranding sites in Florida since 1979 are from the unpublished records of the Southeastern U.S. Marine Mammal Stranding Network through September 1987. These records are archived with the Smithsonian Institution's Marine Mammal Event Program.

GEOGRAPHIC STATUS: The single species of sperm whale is distributed throughout its historic range.

HABITAT REQUIREMENTS AND HABITAT TREND: As with most cetaceans, the habitat requirements and trend are basically unknown. Sperm whales live in deep (>200 m) waters beyond the edge of the continental shelf. They feed primarily on cephalopods and deep-water fish (Leatherwood and Reeves 1982; Rice 1989).

VULNERABILITY OF SPECIES AND HABITAT: The sperm whale is protected in all oceans and only limited shore-based whaling occurs (Leatherwood and Reeves 1982; International Whaling Commission 1988). The species is still vulnerable to certain human activities.

CAUSES OF THREAT: Sperm whales may be entangled in fishing gear and submarine cables. The effect of pollutants accumulated through the food chain is unknown (Prescott et al. 1980). Offshore oil development activities in certain areas (e.g., the Gulf of Mexico) where the ecology of the sperm whale is virtually unknown could affect the species.

RESPONSES TO HABITAT MODIFICATION: Unknown.

DEMOGRAPHIC CHARACTERISTICS: Females become sexually mature at about 7–13 years of age and males at 20–27 years (Best 1970; Leatherwood and Reeves 1982; Rice 1989). Gestation is 14–16 months (Best 1968, 1974). The calving interval may vary from 3 to 6 years and lactation may last 1–2 years (Scott 1985), or closer to 2 years in most populations (Rice 1989). The calving season is generally between May and September in the northern hemisphere (Rice 1989). Calves are about 4 m long at birth and weigh about 750 kg (Best 1968). Pregnancy rates range from 20 to 25 percent for unexploited and exploited stocks, respectively (International Whaling Commission 1980). Natural mortality rates for sperm whales in the North Pacific have been estimated at about 0.05 for juveniles and 0.05–0.33 for sexually mature individuals (International Whaling Commission 1980).

KEY BEHAVIORS: Sperm whales are deep divers and may go as deep as 3000 m (Leatherwood and Reeves 1982). The basic social unit appears to be a female-dominated school containing adult females, calves, and juveniles of both sexes (Best 1970, 1979; Leatherwood and Reeves 1982). Adult males associate with these schools during the breeding season (Rice 1989). Although distribution may shift seasonally, sperm whales do not migrate in the sense that most baleen whales do.

CONSERVATION MEASURES TAKEN: Sperm whales are protected by the Marine Mammal Protection Act of 1972 (National Marine Fisheries Service 1988), the Endangered Species Act of 1973 (U.S. Fish and Wildlife Service 1989), the International Convention for the Regulation of Whaling, 1946 (International Whaling Commission 1988), and the Convention on International Trade in Endangered Species (CITES 1989). They are listed as endangered by the Florida Game and Fresh Water Fish Commission (1990).

CONSERVATION MEASURES PROPOSED: Because there is some evidence for a stock of sperm whales in the Gulf of Mexico, offshore oil development activities should be evaluated for potential effects on the sperm whale.

ACKNOWLEDGMENTS: This is Sea World of Florida Technical Contribution No. 8909-F.

Literature Cited

Berzin, A. A. 1972. The sperm whale *in* A. V. Yablokov, ed., Pishchevaya Promyshlennost. Moscow. Transl. Israel Prog. Sci. Transl., Jerusalem, 1972. 394 pp.
Best, P. B. 1968. The sperm whale (*Physeter catodon*) off the west coast of South Africa. 2. Reproduction in the female. Republic of South Africa, Dept. Industries, Div. Sea Fish., Invest. Rep. 66:1–32.
Best, P. B. 1969. The sperm whale (*Physeter catodon*) off the west coast of South Africa. 4. Distribution and movements. Republic of South Africa, Dept. Industries, Div. Sea Fish., Invest. Rep. 78:1–12.
Best, P. B. 1970. The sperm whale (*Physeter catodon*) off the west coast of South Africa. 5. Age, growth and mortality. Republic of South Africa, Dept. Industries, Div. Sea Fish., Invest. Rep. 79:1–27.
Best, P. B. 1974. The biology of the sperm whale as it relates to stock management. Pp. 257–293 *in* W. E. Schevill, ed., The whale problem: A status report. Harvard University Press, Cambridge. 419 pp.
Best, P. B. 1979. Social organization in sperm whales, *Physeter macrocephalus*. Pp. 227–289 *in* H. E. Winn and B. L. Olla, eds., Behavior of marine animals: Current perspectives in research. Vol. 3. Cetacea. Plenum Press, New York and London. 438 pp.
Boschma, H. 1938. On the teeth and some other particulars of the sperm whale (*Physeter macrocephalus* L.). Temminckia 3:151–278.
CITES. 1989. Convention on International Trade in Endangered Species of Wild Fauna and Flora. Regulations and appendices published by the Federal Wildlife Permit Office, U.S. Fish and Wildlife Service, Washington, D.C.
Donovan, G. P., ed. 1980. Sperm whales: Special issue. Rep. Int. Whal. Commn. (Special Issue 2). 275 pp.
Florida Game and Fresh Water Fish Commission. 1990. Official lists of endangered and potentially endangered fauna and flora in Florida. Tallahassee. 23 pp.
Fritts, T. H., A. B. Irvine, R. D. Jennings, L. A. Collum, W. Hoffman, and M. A. McGhee. 1983. Turtles, birds and mammals in the northern Gulf of Mexico and nearby Atlantic waters. Minerals Management Service contract 14-16-0009-81-949. U.S. Dept. Interior, Fish and Wildlife Service, Div. Biological Services, Washington, D.C. FWS/OBS-82-65. 456 pp.
Holthuis, L. B. 1987. The scientific name of the sperm whale. Mar. Mammal Sci. 3:87–88.
Husson, A. M., and L. B. Holthuis. 1974. *Physeter macrocephalus* Linnaeus, 1758, the valid name for the sperm whale. Zool. Mededelingen, Leiden 48:205–217, 3 pls.

International Whaling Commission. 1980. Report of the Sub-Committee on sperm whales. Cambridge, June 1978. Pp. 65–74 *in* G. Donovan, ed., Rep. Int. Whal. Commn. (Special Issue 2). International Whaling Commission, Cambridge, United Kingdom. 275 pp.

International Whaling Commission. 1988. International Convention for the Regulation of Whaling, 1946. Schedule. As amended by the Commission at the 40th Annual Meeting, May/June 1988, and replacing that dated November 1987. International Whaling Commission, Cambridge, United Kingdom. 27 pp.

Layne, J. N. 1965. Observations on marine mammals in Florida waters. Bull. Fla. State Mus., Biol. Sci. 9:131–181.

Leatherwood, S., and R. R. Reeves. 1982. Bottlenose dolphin, *Tursiops truncatus*, and other toothed whales. Pp. 369–413 *in* J. A. Chapman and G. Feldhamer, eds., Wild mammals of North America. Johns Hopkins University Press, Baltimore, Maryland. 1147 pp.

Leatherwood, S., D. K. Caldwell, and H. E. Winn. 1976. Whales, dolphins, and porpoises of the western North Atlantic. A guide to their identification. NOAA Tech. Rept. NMFS Circ-396. USDOC, NOAA, NMFS, Washington, D.C. 176 pp.

Moore, J. C. 1953. Distribution of marine mammals in Florida waters. Amer. Midland Nat. 49:114–158.

National Marine Fisheries Service. 1988. Marine Mammal Protection Act of 1972. Annual Report 1987/88. USDOC/NOAA/NMFS, Washington, D.C. 68 pp.

National Marine Fisheries Service. 1991. Endangered Whales: Update. USDOC, NOAA, National Marine Fisheries Service, Silver Spring, Maryland. 52 pp.

Odell, D. K. 1991. A review of the southeastern United States marine mammal stranding network: 1978–1987. Pp. 19–23 *in* J. E. Reynolds III and D. K. Odell, eds., Marine Mammal Strandings in the United States. Proceedings of the Second Marine Mammal Stranding Workshop, December 3–5, 1987, Miami, Florida. NOAA Technical Report NMFS 98, USDOC, NOAA, National Marine Fisheries Service, Seattle, Washington. 157 pp.

Prescott, J. H., S. D. Kraus, and J. R. Gilbert. 1980. East coast/Gulf coast cetacean and pinniped workshop. Final report for U.S. Marine Mammal Commission contract MM1533558-2. Nat. Tech. Info. Serv., Springfield, Virginia. NTIS PB80-160 104. 142 pp.

Rice, D. W. 1989. Sperm whale *Physeter macrocephalus* Linnaeus, 1758. Pp. 177–233 *in* S. H. Ridgway and R. Harrison, eds., Handbook of marine mammals. Vol. 4. River dolphins and the larger toothed whales. Academic Press, London. 442 pp.

Schevill, W. E. 1986. The international code of zoological nomenclature and a paradigm: The name *Physeter catodon* Linnaeus 1758. Mar. Mammal Sci. 2:153–157.

Schevill, W. E. 1987. The scientific name of the sperm whale. Mar. Mammal Sci. 3:89–90.

Schmidly, D. J. 1981. Marine mammals of the southeastern United States coast and the Gulf of Mexico. U.S. Fish and Wildlife Service, Office of Biological Services, Washington, D.C. FWS/OBS-80/41. 163 pp.
Scott, G. P., ed. 1985. Report of the working group on NEFC/SEFC marine mammal research. Results of the meeting held 8–9 January, 1985. NOAA Tech. Memorandum NMFS-SEFC-168. USDOC/NOAA/NMFS/SEFC, Miami, Florida. 27 pp.
Tillman, M. F., and G. P. Donovan, eds. 1983. Special issue on historical whaling records. Rep. Int. Whal. Commn. (Special Issue 5). 269 pp.
U.S. Fish and Wildlife Service. 1989. Endangered and threatened wildlife and plants; animal notice of review. Federal Register 54:554–579.

Prepared by: Daniel K. Odell, Sea World of Florida, 7007 Sea World Drive, Orlando, FL 32821.

Endangered

Florida Panther
Felis concolor coryi
FAMILY FELIDAE
Order Carnivora

TAXONOMY: Other names: puma, cougar, mountain lion, painter, catamount. A treatment of the subspecies of *F. concolor* can be found in Young and Goldman (1946). The most recent morphologic description of a Florida panther is by Belden and Forrester (1980). Preliminary analyses of genetic profiles of Florida panthers captured in the wild indicate the Florida race is genetically different from western pumas and exhibits a slightly reduced variability in chromosomal material (Roelke et al. 1986). This may be an inherent trait of the panther or the result of isolation and reduced outbreeding opportunities.

DESCRIPTION: The Florida panther is a large, slender cat, tawny above and whitish below. Lengths and weights vary from 188 cm and 32–45 kg for females to 220 cm and 50–72 kg for males. Kittens exhibit distinct black spots on a buffy background until 9–12 months old. Front pad widths of adult panthers range from 4 to 5 cm and 5 to 6 cm for females and males, respectively. Kittens older than 6 months may have feet as large or larger than adult females and always larger than bobcats (*Felis rufus*). A typical panther track exhibits 4 unclawed toes arranged asymmetrically around the 3-lobed heel pad. Front foot pads are slightly wider than rear pads. Characteristics believed to be typical of the subspecies include a dorsal hair whorl (cowlick), a crook at the end of the tail, and white flecking around the neck and shoulders (Belden and Forrester 1980). These flecks may be environmentally induced (perhaps by ticks). Panthers from southeastern Florida usually lack the crooked tail and cowlick.

Panthers are distinctively different from any other large mammal in Florida, yet mistaken identifications are frequently made. Other animals

Felis concolor coryi

Florida panther, *Felis concolor coryi*. (Photo by David S. Maehr)

or their sign that have been confused with panthers include domestic dog (*Canis familiaris*), black bear (*Ursus americanus*), and bobcat. There is no evidence that black panthers ever existed in Florida.

POPULATION SIZE AND TREND: The actual number of Florida panthers remaining in the wild is probably between 30 and 50. Williams' (1978) comment probably is still appropriate: "The exact number is anybody's guess." If a population trend were measurable, it likely would be downward in response to development and concomitant habitat loss in southern Florida. Populations of this subspecies from other states probably are extirpated. Although the panther may not be on the verge of extinction, unchecked habitat destruction and human disturbance ultimately will have the same result as overharvest.

DISTRIBUTION AND HISTORY OF DISTRIBUTION: The puma (*F. con-*

color) has the widest distribution of any native mammal in the western hemisphere. There are 27 subspecies of puma recognized in North, Central, and South America (Anderson 1983). *F. c. coryi* is the last known extant subspecies of mountain lion in the eastern United States. The

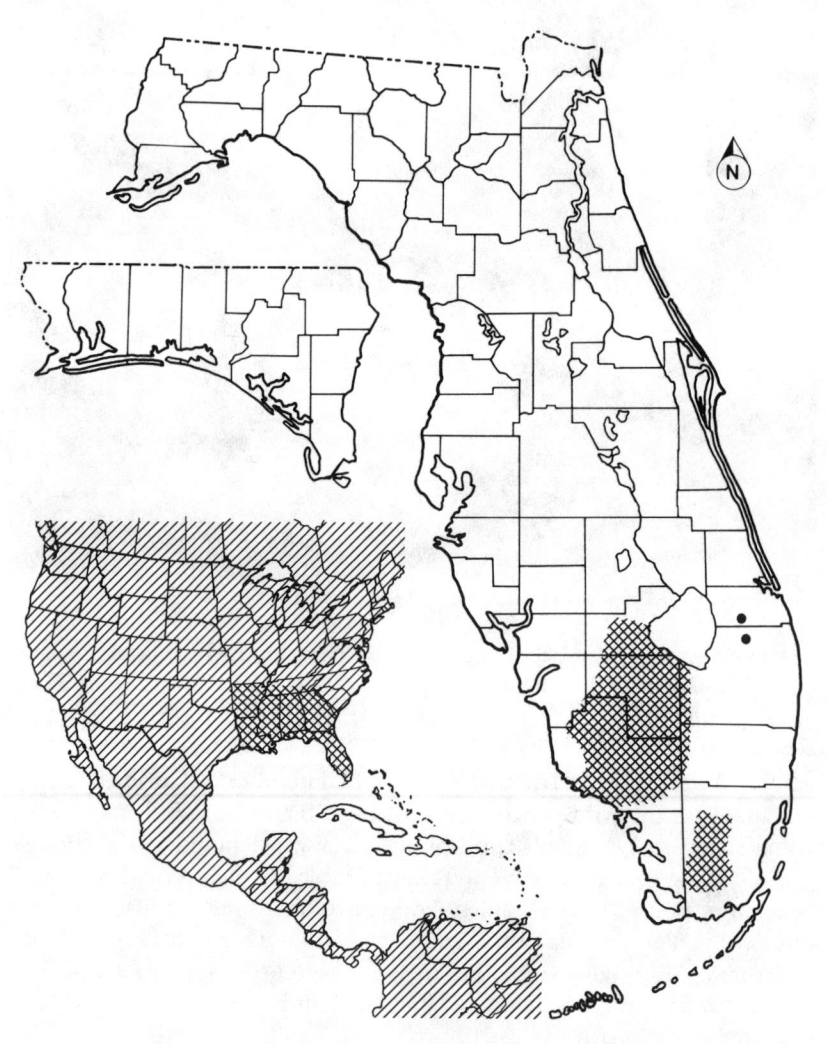

Distribution map of the Florida panther, *Felis concolor coryi*. Hatching represents the species; crosshatching and dots represent the subspecies. The main map shows current range, whereas the inset shows the original range.

southeastern subspecies was presumed to have been distributed throughout the southeastern coastal plain from Arkansas and southwestern Tennessee, east to South Carolina (Hall 1981). Although a known population exists only in Florida, some speculate that individuals persist in Arkansas (Lewis 1969; Sealander and Gipson 1973). However, no solid evidence of occupation has been presented (Neal 1987). Even in Florida, the distribution of *F. c. coryi* has been conjectural (Stevenson 1976:515; Williams 1978). Reports of panthers continue to emanate from all parts of Florida, but consistent, verified sign is found only in southern Florida.

Accumulated evidence on road kills and verified sign (tracks, kills, etc.) since the early 1970s and radiotelemetry of Florida panthers (Belden et al. 1988; Maehr 1986, 1987) have documented a reproducing population only in southern Florida. Everglades National Park, Big Cypress National Preserve, Fakahatchee Strand State Preserve, Big Cypress Seminole Indian Reservation, and the Florida Panther National Wildlife Refuge are the primary public tracts supporting this remnant population. Private ranches and preserves in parts of Collier, Hendry, Lee, and Glades counties also are inhabited. In addition to this established distribution, a number of recent, verified reports or specimens have come from Highlands (Layne and Wassmer 1988), Palm Beach, Broward, Martin, Osceola, Volusia, and St. Johns counties. However, no reproduction has been recorded in these areas. Restricted access to potentially occupied lands in private ownership has hindered efforts to survey and understand distribution patterns of panthers in large parts of central and southern Florida.

Historical records indicate that panthers were encountered frequently in southern Florida and the St. Johns River drainage. Outside of this area, records are less abundant. *F. concolor* apparently was widespread in northern Florida during the Pleistocene (Webb 1974).

GEOGRAPHICAL STATUS: In North America the Florida panther represents the only known population of *F. concolor* remaining in the eastern United States; in Florida it exists as a remnant in the southern end of the peninsula. Its continued presence there is the result of large interconnected blocks of woodlands remaining in public and private ownership.

HABITAT REQUIREMENTS AND HABITAT TREND: The panther is considered a wilderness species, yet it has shown considerable flexibility adapting to human intrusions and habitat alterations. Historically, a prerequisite for panthers apparently was an abundant deer population. The introduction of domestic livestock such as cattle (*Bos taurus*) and swine

(*Sus scrofa*), colonization by armadillos (*Dasypus novemcinctus*), and recent intensive agricultural practices have considerably changed environmental conditions and prey resources for this native lion.

Panthers in southern Florida have been found in most types of vegetation, including tropical hammocks, pine flatwoods, cabbage palm (*Sabal palmetto*) forests, mixed swamp, cypress (*Taxodium distichum*) swamp, live oak (*Quercus virginiana*) hammocks, sawgrass (*Cladium jamaicense*) marshes, and Brazilian pepper (*Schinus terebinthefolius*) thickets. Most telemetry data are from cats using the Big Cypress National Preserve, Fakahatchee Strand State Preserve, and surrounding private lands, where cypress and mixed hardwood forests are interspersed with herbaceous marsh, old field, and pinelands. Open agricultural lands are common around most publicly owned land in southern Florida and receive some use by panthers if cover nearby is adequate. Day-use sites typically are dense patches of saw palmetto (*Serenoa repens*) surrounded by swamp, pine flatwoods, or hammock (Maehr et al., unpubl. data).

Considerable variation in the diet has been noted even within the subspecies' restricted range south of Lake Okeechobee. This variation, however, probably reflects habitat quality more than actual food preferences. Whereas white-tailed deer (*Odocoileus virginianus*) appear to be used throughout this area, feral pigs seem to be important in private ranch lands north of Big Cypress National Preserve, and smaller mammals such as raccoons (*Procyon lotor*) and armadillos are important in the southern part of the range (Maehr 1989).

Studies of deer in southwestern Florida suggest that the Bear Island herd in the Big Cypress National Preserve exists at a higher density and nutritional level than deer in the eastern Monument Unit (Raccoon Point) or Fakahatchee Strand State Preserve (McCown 1988). These differences probably result from poorer habitat conditions at Raccoon Point and the Fakahatchee Strand State Preserve. There panthers appear to exist at relatively low densities and with relatively little reproduction; two females produced two known kittens in five years. In Bear Island, two females produced at least eight kittens during the same period.

VULNERABILITY OF SPECIES AND HABITAT: Because the Florida panther is found only in one small part of its original range and in one geographic locale, continued habitat loss and fragmentation would cause extinction of the subspecies. Once a population becomes isolated it may be only a matter of time before it fluctuates out of existence (Diamond 1978). Continued fragmentation of the population and shrinking numbers increase the likelihood that a natural disaster (e.g., disease, hurricane) will push the panther beyond the brink of recovery. In addition, genetic problems are more likely in small, isolated populations with no outbreeding opportunities (Frankel and Soule 1981).

CAUSES OF THREAT: Although individual panthers are susceptible to mortality caused by illegal shootings and highway collisions, these losses can be compensated by reproduction and recruitment. Permanent losses in numbers result primarily from habitat lost to expanding urbanization and agriculture. The recent southward retreat of the state's freeze line has placed new pressures on southwestern Florida's wildlands to produce citrus. This, in addition to other intensifying agricultural practices, will undoubtedly reduce the ability of the region to support panthers. A moderate level of human disturbance seems to be tolerated as long as cover and food requirements are met.

RESPONSES TO HABITAT MODIFICATION: The wholesale conversion of forested landscapes to agricultural monocultures has eliminated the panther throughout much of its range. Where improved pasture or vegetable crops exist in a mosaic of forest cover, however, panthers may persist. This interspersion of forested and early successional habitats seems to benefit the panther's primary prey, deer and feral swine. In some areas panthers subsist primarily on swine, the result of introductions by Europeans centuries ago. This was perhaps the most significant alteration caused by man because it provided a large-sized, abundant alternative to its primary native prey.

Where wild pigs are not available, habitat management favorable to white-tailed deer benefits panthers. Frequent burning in suitable vegetation over large areas maintains many plant species in a successional stage favorable to deer. Artificial feeding, including food plots, is a common practice on private lands to increase deer numbers. Although the effect of supplemental feeding in southern Florida is not known, it may serve, at least, to concentrate deer. Direct supplemental feeding of panthers does not appear to be a viable management option at this time (Maehr et al. 1989a).

Development related to oil exploration has been extensive in occupied panther habitat. The construction of roads, pads, and associated petroleum production activities have changed localized areas, but effects on panthers are difficult if not impossible to measure. The most conspicuous habitat modification was the construction of major highways through once-contiguous habitat. Collisions with vehicles cause the greatest known mortality in panthers; however, road kills also are the easiest form of death to document.

DEMOGRAPHIC CHARACTERISTICS: Although details of Florida panther reproduction and population dynamics are based on small samples, preliminary information suggests they are similar to western subspecies of *F. concolor*. Mountain lions produce from 1 to 6 young approximately every 2 years (Anderson 1983:33). Prenatal litter size is 3.4, and litters 6–12 months old contain an average of 2.2 kittens. Observed litter sizes in Florida panthers range from 1 to 4 (Maehr et al., unpubl. data), while prenatal litters range from 3 to 4. Gestation lasts 90–96 days. Mountain lions in the western United States probably are sexually mature at about 2 years but do not reproduce until established on a territory (Hornocker 1970). The earliest published instance of first reproduction in the Florida panther was an 18–19-month-old female panther that raised 4 kittens in her mother's home range (Maehr et al. 1989b). Kittens have been observed to remain with their mothers for 12 to 18 months ($n = 4$).

Males are polygamous (Seidensticker et al. 1973:53) and breed with several females residing in their home ranges. Peaks in breeding activities in Florida seem to occur during fall and winter, although reproduction probably can occur year-round. Anderson (1983:35) calculated expected longevity of male and female mountain lions as 11.2 and 10.5 years, respectively.

Florida panthers are susceptible to a number of diseases and parasites that range from benign infections to potential epizootics. Parasites include one protozoan, two trematodes, three cestodes, eight nematodes, and seven arthropods (Forrester et al. 1985; Roelke et al. 1986). The hookworm (*Ancylostoma pluridentatus*) is of particular concern because it can cause severe anemia and weight loss and may contribute significantly to kitten mortality. Little is known about the lifecycle of this nematode. Two viruses found in panthers include feline panleukopenia (feline parvovirus) and feline calicivirus. Most panthers examined have shown indications of previous exposure to parvovirus and about half have been exposed to calicivirus. Although no deaths have been attributed to either virus, parvovirus is potentially devastating. Variation in parvovirus prevalence rates among bobcats in southern Florida suggests that environmental conditions and population density may influence the dynamics of this disease. Little is known about the prevalence and status of pseudorabies in wild felids but, because it also is potentially devastating and is carried commonly by feral hogs, the disease deserves close attention (Roelke et al. 1986). Rabies caused the death of a radio-collared subadult male in Hendry County in 1989.

Perhaps the most publicized mortality factor in Florida panthers is collisions with highway vehicles. Since 1979, 12 panthers have been killed on roadways in southern Florida. The regularity of these deaths (41 percent of known mortality, averaging 1.2 per year) may keep the local social structure of panthers in a state of flux, depressing reproduction. The highway deaths seem to be centered around the Fakahatchee Strand, so not all panthers are equally vulnerable to this mortality agent.

The Florida panther represents a classic example of insularization of a population at a distributional extreme (Simpson 1964; Mayr 1982). Partial isolation of the panther resulting from Florida's peninsular geography has been exacerbated by recent trends in land use by humans. Although the biological significance of isolation and small size of this population are only speculative, they may provide a good case for examining the consequences and developing remedies.

KEY BEHAVIORS: Until the advent of radiotelemetry, the Florida panther remained a mysterious inhabitant of the state's wilderness. Its secretive nature and tendency to avoid humans led to many unfounded fears, myths, and misconceptions. Unlike western mountain lions, Florida panthers are not regular livestock killers, and attacks on humans are unknown.

Except for mating pairs, the female-kitten group, and occasional brief encounters among individuals, panthers are solitary animals. Most mating activities appear to occur between November and March, when copulating pairs spend up to a week together. The kittens are born in a densely vegetated thicket, usually weighing about 500 g each. A collared female chose the same large palmetto thicket surrounded by hammock and freshwater marsh for her natal den in 1986 and 1988 (Maehr et al. 1989c). Kittens remain at the den for 2 months and then begin accompanying their mother during hunting forays. Young become independent at about 18 months, when dispersal occurs. A male subadult (18–20 months old) traveled 15 km from his natal area in about a week before being killed by a resident adult male (Maehr et al., unpubl. data). Another subadult male (24–30 months old) used an area interspersed by human residences and roads between Lake Trafford (Collier Co.) and Ft. Myers (Lee Co.), before traveling to southeastern Hendry County via Lake Hicpochee.

Home ranges average about 400 km^2 and 200 km^2 for males and females, respectively. Considerable geographic overlap among females and males is facilitated by temporal isolation. Habitat quality and prey density appear to determine female home-range size and possibly litter size (Maehr et al. 1989*c*). Females probably select areas with relatively high prey densities. Activities associated with kitten rearing greatly reduce movements and area used. Male home ranges may reflect the density and distribution of adult females.

Annual movement patterns vary little. However, in winter panthers appear more active during the day than in warmer months. Diel activities appear to peak around sunset and sunrise.

CONSERVATION MEASURES TAKEN: Over the last century and a half, the Florida panther has been subject to a variety of public attitudes, ranging from government-sanctioned persecution to strong government and public support for its protection and recovery (see chronology following). The progression of increasing concern parallels the economic development and human population growth in the state. Intensive hunting once was normal; for example, Florida sportsman Dave Newell killed eight panthers in five weeks, circa 1935 (Tinsely 1970). As direct persecution decreased, habitat loss and modification increasingly threatened the panther's existence.

Since the use of radiotelemetry studies in 1981, movement data have been important in developing conservation measures. Reduced night speed limits have been posted on S.R. 29 and 84 and U.S. Highway 41 in areas of frequent crossings. Telemetry and road mortality data also have been used to design wildlife underpasses for the conversion of S.R. 84 (Alligator Alley) to Interstate 75. These efforts may improve crossing conditions for panthers in Collier County. As of March 1990, three underpasses west of S.R. 29 on I-75 have been used by male panthers.

Efforts by several government agencies have resulted in an increase in acreage set aside in public trust for panther conservation. Specifically, additions to Fakahatchee Strand State Preserve and acquisition of the Florida Panther National Wildlife Refuge have increased areas that can be managed for panthers.

In light of the booming human population and concomitant development in Florida, it is remarkable that a wilderness species like the panther has remained. In 1982 it became the state mammal by popular vote of Florida's school children. Its popularity, the political controversies surrounding it, and the obstacles involved in studying it make conservation of the Florida panther one of the most challenging wildlife problems in America.

Important Dates in Florida Panther Conservation

1832	Bounty paid for dead panthers.
1930s	Tick fever/deer eradiction program (deer herd decline).
1950	Panther designated a game animal (restricted hunting).
1950s	Screw worm fly eliminated (deer herd increase).
1954	National Audubon Society's Corkscrew Swamp Sanctuary established.
1958	Declared an endangered species by Game and Fresh Water Fish Commission (complete protection).
1967	Listed as endangered by U.S. Department of Interior.
1972	Field searches for panthers began.
1973	Endangered Species Act passed.
1974	Fakahatchee Strand State Preserve established.
1976	Florida Panther Conference held.
1976	Florida Panther Record Clearinghouse established.
1976	Florida Panther Recovery Team appointed.
1978	Big Cypress National Preserve (BCNP) established.
1981	Florida Game and Fresh Water Fish Commission radiotelemetry studies began.
1981	Florida Panther Recovery Plan submitted to U.S. Fish and Wildlife Service.
1982	Panther made state mammal.
1982	Oil Exploration and Recovery Plan developed for BCNP.
1983	State legislature passed Florida Panther Bill establishing Panther Trust Fund and Advisory Council.
1983	Save Our Everglades program started.
1984	Reduced night-time speed limits on S.R. 84 and 29 and U.S. 41.
1985	S.R. 29 widened and realigned north of S.R. 84.
1986	Conversion of S.R. 84 to I-75 began.
1986	Florida Panther National Wildlife Refuge acquisition began.
1986	Acquisition of Big Cypress National Preserve addition initiated.
1986	Florida Panther Interagency Committee formed.
1986	National Park Service initiated telemetry studies of panthers in Everglades National Park.
1987	Florida Game and Fresh Water Fish Commission initiated telemetry studies of panthers on private land.
1989	Florida Panther National Wildlife Refuge acquired.
1989	Panthers documented using underpasses to cross I-75.

CONSERVATION MEASURES PROPOSED: A sense of urgency should surround activities associated with Florida panther conservation. Intensification of agriculture on private lands must be stemmed to protect panthers outside areas in public ownership; little has been done to maintain panther habitat on private land (Maehr 1990). First, however, land-use trends need to be documented, current land uses mapped, future patterns forecast, and a strategy for preserving key parcels developed. Outright purchases, wildlife easements, the encouragement of wildlife uses on these lands (e.g., hunting leases, commercial ventures, preserves), and other economic incentives to keep large areas undeveloped could combine to stabilize habitat losses in southern Florida. Panther-conservation problems on private lands and possible solutions were discussed by Maehr (1990). Studies must continue on public lands to determine the effects of various human activities on panthers and their prey, and to ensure that management decisions are made in the best interest of the panther. In addition, potential competition for food with other carnivores (bobcats and black bears) should be investigated. Radiotelemetry studies should emphasize kitten survival and mortality, dispersal patterns and home-range establishment by juveniles, and social interactions among adults. This information can provide important insights into population status and trends. Radiotelemetry data also should be used to evaluate the effects of I-75 through Collier County. Surveys of large tracts of private lands (see Roof and Maehr 1988), especially in southern and central Florida, should be conducted to identify additional population centers of the subspecies before development pressures render these areas uninhabitable by panthers.

Examination of the characteristics and dynamics of prey populations should continue. The relationships of deer and hogs with their environment (including harvest and predation) need to be evaluated to find ways to maximize prey opportunities for panthers.

Investigations should continue to outline genetic characteristics of Florida panthers and to make recommendations for genetically based population management. Dynamics of feline diseases and lifecycles of pathogenic parasites need continuing study to develop treatments or management strategies for individuals or populations when appropriate.

Captive breeding should concentrate on securing potential release sites for captive-reared animals. Site suitability should be evaluated according to the success of released radio-instrumented western cougars.

Public education is important in generating and maintaining support. Consistent and frequent dissemination of advances and accomplishments in panther conservation is needed. Similarly, setbacks in efforts can emphasize the problems facing the Florida panther and the need for public support.

ACKNOWLEDGMENTS: The original group of individuals concerned and involved in panther research in Florida was small but dedicated. The most prominent of this group include R. Nowak, R. McBride, L. Williams, and R. Belden. They set the stage for the work continuing today. Others involved in field work and administration include T. Logan, B. Frankenberger, T. Hines, T. Schwikert, J. Harris, M. Ramsey, J. Schortemeyer, O. Bass, D. Jansen, and J. Roboski. Those involved in current activities have a combined presence in the field spanning many years of dedication and include J. Roof, J. McCown, D. Land, R. McBride, M. Roelke, T. Ruth, L. Wilkins, J. Kappes, and R. Bell. Numerous Game and Fresh Water Fish Commission law enforcement officers are owed a debt of gratitude for their dogged patrols of highways in Collier County and assistance with field work. From all of these professionals and others too numerous to mention have come the current knowledge and advancements in conservation of the Florida panther. S. Humphrey and T. Logan made helpful suggestions on the manuscript.

Literature Cited

Anderson, A. E. 1983. A critical review of literature on puma (*Felis concolor*). Colorado Div. Wildl. Special Rep. 54. 91 pp.
Belden, R. C., and D. J. Forrester. 1980. A specimen of *Felis concolor coryi* from Florida. J. Mammal. 61:160–161.
Belden, R. C., W. B. Frankenberger, R. T. McBride, and S. T. Schwikert. 1988. Panther habitat use in southern Florida. J. Wildl. Manage. 52:660–663.
Diamond, J. M. 1978. Critical areas for maintaining viable populations of species. Pp. 27–40 *in* M. W. Holdgate and M. J. Woodman, eds., The breakdown and restoration of ecosystems. Plenum Press, New York.
Forrester, D. J., J. A. Conti, and R. C. Belden. 1985. Parasites of the Florida panther (*Felis concolor coryi*). Proc. Helminthol. Soc. Wash. 52:95–97.
Frankel, O. H., and M. E. Soule. 1981. Conservation and evolution. Cambridge University Press, Cambridge, Massachusetts. 327 pp.
Hall, E. R. 1981. The mammals of North America. 2d ed. John Wiley and Sons, New York. 1081 + 90 pp.
Hornocker, M. G. 1970. An analysis of mountain lion predation upon mule deer and elk in the Idaho Primitive area. Wildl. Monogr. 21:1–39.
Layne, J. N., and D. A. Wassmer. 1988. Records of the panther in Highlands County, Florida. Florida Field Nat. 16:70–72.

Lewis, J.C. 1969. Evidence of mountain lions in the Ozarks and adjacent areas, 1948–1968. J. Mammal. 50:371–372.
Maehr, D.S. 1986. Florida panther habitat utilization. E-1-10 annual performance report. Florida Game and Fresh Water Fish Commission, Tallahassee.
Maehr, D.S. 1987. Florida panther movements, social organization, and habitat utilization. E-1-11 annual performance report. Florida Game and Fresh Water Fish Commission, Tallahassee.
Maehr, D.S. 1989. Florida panther food habits. E-1-13 annual performance report. Florida Game and Fresh Water Fish Commission, Tallahassee.
Maehr, D.S. 1990. The Florida panther and private lands. Conservation Biology 4:167–170.
Maehr, D.S., J.C. Roof, E.D. Land, J.W. McCown, R.C. Belden, and W.B. Frankenberger. 1989*a*. Fates of wild hogs released into occupied Florida panther home ranges. Florida Field Nat. 17:42–43.
Maehr, D.S., J.C. Roof, E.D. Land, and J.W. McCown. 1989*b*. First reproduction of a panther (*Felis concolor coryi*) in southwestern Florida, U.S.A. Mammalia 53:129–131.
Maehr, D.S., E.D. Land, J.C. Roof, and J.W. McCown. 1989*c*. Early maternal behavior in the Florida panther. Amer. Midland Nat. 122:34–43.
Mayr, E. 1982. The growth of biological thought. Harvard University Press, Cambridge, Massachusetts. 974 pp.
McCown, J.W. 1988. Big Cypress watershed deer/hog/panther relationships. E-1-12 annual performance report. Florida Game and Fresh Water Fish Commission, Tallahassee.
Neal, W.A. 1987. Report on mountain lions in Arkansas. Report to John Christian, U.S. Fish and Wildl. Serv., Endangered Species Office, Atlanta, Georgia. 15 pp. mimeo.
Roelke, M.E., E.R. Jacobson, G.V. Kollias, and D.J. Forrester. 1986. Florida panther health and reproduction. E-1-10 annual performance report. Florida Game and Fresh Water Fish Commission, Tallahassee.
Roof, J.C., and D.S. Maehr. 1988. Sign surveys for Florida panthers in peripheral areas of their known range. Florida Field Nat. 16:81–85.
Sealander, J.A., and P.S. Gipson. 1973. Status of the mountain lion in Arkansas. Proc. Arkansas Acad. Sci. 27:38–41.
Seidensticker, J.C., IV, M.G. Hornocker, W.V. Wiles, and J.P. Messick. 1973. Mountain lion social organization in the Idaho Primitive area. Wildl. Monogr. 35:1–60.
Simpson, G.G. 1964. Species density of North American recent mammals. Syst. Zool. 13:57–73.
Stevenson, H.M. 1976. Vertebrates of Florida. University Presses of Florida, Gainesville. 607 pp.
Tinsley, J.B. 1970. The Florida panther. Great Outdoors Publishing Co., St. Petersburg, Florida. 60 pp.
Webb, S.D., ed. 1974. Pleistocene mammals of Florida. University Presses of Florida, Gainesville. 270 pp.

Williams, L. E. 1978. Florida panther. Pp. 13–15 *in* J. N. Layne, ed., Rare and endangered biota of Florida. Vol. 1. Mammals. University Presses of Florida, Gainesville. xx + 52 pp.

Young, S. P., and E. A. Goldman. 1946. The puma, mysterious American cat. The Amer. Wildl. Inst., Washington, D.C. 358 pp.

Prepared by: David S. Maehr, Florida Game and Fresh Water Fish Commission, 566 Commercial Boulevard, Naples, FL 33942.

Endangered

Florida Manatee
Trichechus manatus latirostris
FAMILY TRICHECHIDAE
Order Sirenia

TAXONOMY: Also called sea cow. Earlier subspecific designation of the Florida manatee (*T. m. latirostris*) has recently been supported by a modern analysis of cranial morphometrics (Domning and Hayek 1986). Except where noted, this account pertains to the Florida subspecies. Domning and Hayek (1986) recommend the usage of the common names West Indian manatee for the species *Trichechus manatus*, Florida manatee for *T. m. latirostris*, and Antillean manatee for the subspecies *T. m. manatus*.

DESCRIPTION: The West Indian manatee is a robust, gray to gray-brown, nearly hairless aquatic mammal. Its body is fusiform. The skin is thick and may bear barnacles and algae. The head is indistinct from the body; forelimbs are paddle-like and have nails. Hind limbs are lacking, and the tail is spatulate. The upper lip is cleft, set with bristles, and capable of grasping. Sexes are similar, the male having a more anterior genital aperture. The typical adult ranges from 400 to 900 kg and from 2.8 to 3.5 m total length. Newborn calves are 1.0–1.5 m in length and weigh about 20–30 kg. The maximum recorded weight was 1620 kg for a 3.75-m female.

POPULATION SIZE AND TREND: No estimates are available for total numbers of West Indian manatees throughout the species' range, because of a lack of appropriate census methods. Populations have been reduced or extirpated in many regions of former occurrence (Thornback and Jenkins 1982; Lefebvre et al. 1989). Except perhaps for a few extralimital stragglers, Florida harbors the entire U.S. population in winter. Movements into neighboring states occur in summer. The population is divided about equally on the Gulf and Atlantic coasts (Hartman 1974; Irvine and Campbell 1978).

Florida manatee, *Trichechus manatus latirostris*. (Photo by Robert M. Rattner, © Robert M. Rattner, 1981)

Variability in manatee density and sightability has prevented statistically reliable estimation of population size or trend (Packard 1985*a*). At least 1200 Florida manatees were accounted for at wintering areas in January 1985 (O'Shea 1988; Reynolds and Wilcox 1986), with no estimate of additional numbers not seen. A more comprehensive aerial survey conducted in February 1991 included both aggregation sites and intervening areas. This survey resulted in a new minimum count of 1465 (Florida Department of Natural Resources, unpubl. data). Historical estimates of Florida manatee population size are inconclusive, although a long history of hunting and incidental take as food may have suppressed abundance until recent decades, when increases may have occurred (O'Shea 1988). The best conditions for reliable counts exist at Crystal River in Citrus County and Blue Spring in Volusia County, which have shown substantial increases in manatees over the past two decades through recruitment and immigration. Over 60 manatees now winter at Blue Spring, compared with 11 in 1971. Nearly 300 have been counted in the Crystal River area, in comparison with about 45 in 1968 and a general absence during the 20th century prior to the 1950s (Moore 1951; Hartman 1979; O'Shea 1988). Relatively high genetic diversity with no regional differentiation is characteristic of the Florida population (McClenaghan and O'Shea 1988).

DISTRIBUTION AND HISTORY OF DISTRIBUTION: The West Indian manatee occupies coastal and major inland waterways of Florida, the Ca-

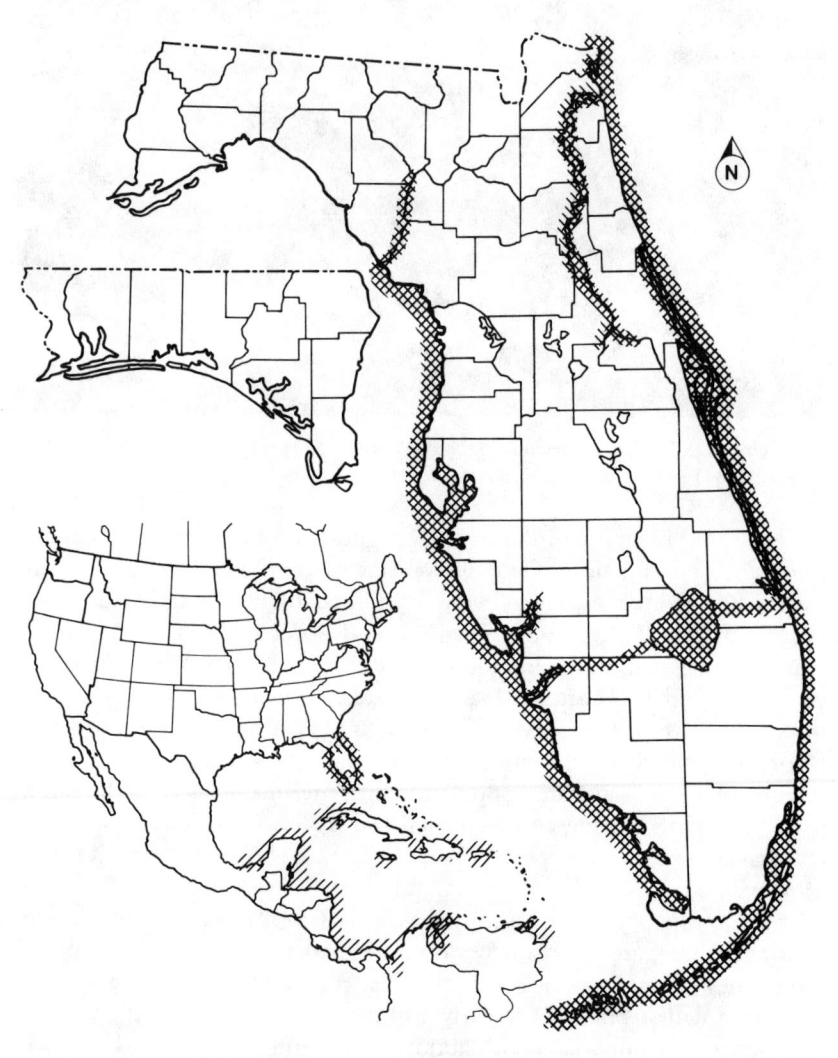

Distribution map of the Florida manatee, *Trichechus manatus latirostris*. Hatching represents the species; crosshatching represents the subspecies.

ribbean islands, Mexico, Central America, and northern South America. The northern and southern limits of the species' range are dictated by energetic constraints, coinciding with the 24°C isotherm (Whitehead 1977); 24°C is also the species' lower limit of thermoneutrality (Irvine 1983). The range limits remain similar to those known historically, but the distribution is fragmented due to areas of local extirpation (Thornback and Jenkins 1982; Lefebvre et al. 1989).

The range of the Florida manatee varies seasonally. During summer manatees are widely dispersed and frequently encountered at the lower reaches of rivers and canals, and in estuaries, lagoons, and bays from the Suwannee River south on the Gulf coast, and from coastal Georgia to Biscayne Bay on the Atlantic coast. Gulf coast records north of the Suwannee, along the panhandle, and between the Chassahowitzka River and Tampa Bay are infrequent. On the Atlantic coast manatees commonly occur in the St. Johns River, the Indian River Lagoon system, and various other waterways. In winter manatees may move southward, and they aggregate at natural and industrial warm-water sources. The most well-used wintering areas are at Crystal River, Homosassa River, Tampa Bay, Ft. Myers, Port Everglades, Riviera Beach, near Titusville, and Blue Spring. Detailed, county-by-county reviews of distribution information are available (Hartman 1974; Beeler and O'Shea 1988).

Records from the late 1800s through 1950 suggest that the northern winter range-limits in Florida were Charlotte Harbor on the Gulf coast and the Sebastian River on the Atlantic coast (Moore 1951). The northern aggregation sites therefore represent recent expansion, although former use of wintering areas on the Gulf coast north of Charlotte Harbor prior to the late 1800s has been documented (O'Shea 1988).

GEOGRAPHICAL STATUS: The only year-round manatee population in the United States occurs in Florida. The Florida manatee has differentiated to the subspecies level as a result of isolation by the Straits of Florida and the northern Gulf of Mexico (Domning and Hayek 1986). Historical records for southern Texas are probably representatives from the northern fringe of the range of the Antillean manatee, which extends northward through eastern coastal Mexico (Gunter 1941; Powell and Rathbun 1984; Domning and Hayek 1986).

HABITAT REQUIREMENTS AND HABITAT TREND: Florida manatees occupy coastal, estuarine, and some riverine habitats. Primary habitat requisites are access to vascular aquatic plants, freshwater sources, proximity to channels 1–2 m deep, and access to natural springs or man-made warm-water refugia during winter (Hartman 1978). Sheltered bays, coves, and canals are important for resting, feeding, and reproductive activity (Bengtson 1981; Powell and Rathbun 1984). Florida manatees forage on a wide variety of aquatic plants including seagrasses, bank grasses, overhanging mangrove, and submerged, rooted, or floating species (Hartman 1979; Best 1981). In summer individuals may range widely, and seasonal migrations of 850 km to wintering areas have been documented (U.S. Fish and Wildlife Service, unpubl. data). These wide-ranging movements require access to travel corridors that are unobstructed by dams, shallows, or congested boat traffic.

Manatee habitat in Florida has been and continues to be greatly altered by residential and commercial development of coastal land (Packard and Wetterqvist 1986). Dredge-and-fill activities may destroy areas of aquatic vegetation, whereas new channels and inlets may allow access to additional habitat. Tampa Bay, for example, has experienced an 81 percent decrease in seagrass acreage in the last century due to adjacent urbanization (Lewis et al. 1985). Water pollution poses a threat to aquatic plants as a food base. Manatees are not known to accumulate significant residues of most persistent environmental contaminants, because of their low position in the food chain. Exposure to copper-containing aquatic herbicides, however, has potential harmful effects (O'Shea et al. 1984). The number of boats using manatee habitat has increased rapidly (O'Shea 1988), creating substantial disturbance as well as greater potential for injury and death. An increase in artificial warm-water sources in the 1950s and 1960s and a proliferation of exotic aquatic vegetation have proven of short-term benefit to manatee populations (Powell and Rathbun 1984; Shane 1984), but some industrial sources of warm water will soon reach the end of their designed operating life.

VULNERABILITY OF SPECIES AND HABITAT: Man is the only significant predator on West Indian manatees (Rathbun 1984). Manatees are docile and react to danger only by flight. They are vulnerable to being struck by boats and barges, crushed in locks and flood control structures, and entangled in fishing nets and crab trap lines. Other sources of mortality include illegal intentional killing, red tides, and cold stress (O'Shea et al. 1985). Deaths due to parasites and disease have been documented, but their incidence is very low (Beck and Forrester 1988; Buergelt et al. 1984). A low maximum potential rate of population increase renders Florida manatees extremely vulnerable to increases in mortality factors, and survival of adults and subadults is especially critical to preventing population declines (Packard 1985*b*). In recent years, a disproportionate number of adults killed by boats in certain regions and subadults killed by record cold temperatures give cause for concern (Kinnaird 1985; O'Shea et al. 1985). Manatees are particularly vulnerable to catastrophic losses when gathered in large numbers at winter aggregation sites.

Although several areas used by manatees are protected as parks, wildlife refuges, or sanctuaries, most of the present range includes large areas of existing or probable future human activity and habitat alteration.

CAUSES OF THREAT: The greatest chronic threat to Florida manatees is an increase in accidental mortality due to collisions with boats. The number killed by boats each year has doubled over the past decade (O'Shea 1988), with a record total of 51 killed in collisions during 1989 (Florida Department of Natural Resources, unpubl. data). Crushing in gates of navigation locks and flood control dams, entanglement in lines, entrapment in culverts, and poaching are additional human-related threats (Odell and Reynolds 1979; O'Shea et al. 1985). Habitat modification resulting in increased disturbance, declines of aquatic vegetation, or blocked access to potential use areas are other long-term threats.

RESPONSES TO HABITAT MODIFICATION: Manatees are opportunistic and flexible and can readily adapt to a variety of habitat modifications. They use sanctuaries set aside in their behalf, aggregate at artificial warmwater sources, use man-made freshwater discharges for drinking, rest in quiet marinas, travel through dredged canals, and thrive on introduced exotic vegetation. However, they suffer disturbance and may try to avoid areas of high boat-traffic, must cope with loss of feeding areas through weed-control operations or deteriorating water quality, and may have access to traditional travel corridors blocked by water-control structures and other factors.

DEMOGRAPHIC CHARACTERISTICS: Aspects of demographic traits of Florida manatees detailed below are based almost entirely on long-term longitudinal studies of known individuals in the field initiated by Hartman (1979) and continued by the U.S. Fish and Wildlife Service (unpubl. data). Females have a minimum age of first reproduction of 4–5 years, although the modal individual may not successfully first rear young until 6 years or older. Gestation is about 12–14 months. One young is born, but twinning occurs rarely. Age to weaning is variable, ranging from 1 to 2 years. The minimum birth interval is about 2 years, but not all females maintain this rapid a schedule. Longevity and age at reproductive senescence are unknown. Some individuals continue to thrive after long periods (30–40 years) in captivity. The maximum potential rate of population increase has been estimated at 2 to 7 percent and is most sensitive to changes in adult survival and, secondarily, subadult survival (Packard 1985b). Preliminary population modeling does not rule out the possibility of an ongoing decline, but the ability to detect a population decline in Florida manatees is hampered by a lack of good population estimation techniques (Packard 1985b).

KEY BEHAVIORS: Manatees have reasonably good vision and well-developed tactile and hearing senses. They communicate with a variety of sounds in the human audible range, best described as squeaks, chirps, grunts, and groans (Hartman 1979). They are most commonly solitary, but frequently join in groups of two or more to embrace, nudge, roll, and play. Ranges of individuals show much overlap, and manatees may move widely. The nonwinter behavior and movements of manatees are best known for the upper St. Johns River (Bengtson 1981). In the St. Johns River, females tend to be more sedentary than males in the warm seasons. Males travel more often and over wider, overlapping circuits, and may range 200 km or more (Bengtson 1981). Overlapping ranges of several females are encompassed within those circuits. Groups of 2 to 12 or more manatees of mixed sex and age form during these wanderings, but these are highly labile in composition. Mating herds of up to 17 males will build up and persist for 7 to 30 days around a focal estrous female, who may copulate with more than one male (Hartman 1979; Bengtson 1981).

Florida manatees migrate seasonally. Long-distance moves of 200–300 km are made in eastern Florida at rates of travel up to 49 km per day without substantial stopovers (Reid and O'Shea 1989). Some individuals sighted in autumn in the upper St. Johns River have been resighted 850 km away in southeastern Florida in winter (U.S. Fish and Wildlife Service, unpubl. data). In winter, manatees adopt a refuging strategy, aggregating at warm-water sources following drops in ambient temperature below about 20°C (Hartman 1979; Powell and Waldron 1981). Numbers at aggregation sites are inversely correlated with water temperatures, with over 300 counted in the largest aggregations (Reynolds and Wilcox 1986). Diel activity patterns at warm-water sites are influenced by daily ambient water temperature cycles, with manatees tending to leave during warmer times of day (Bengtson 1981; Kochman et al. 1985). During cold spells manatees may linger at these sites to conserve energy, and they may not venture out to feed for a week or more (Bengtson 1981). During warm seasons manatees lack a distinct diel cycle (Hartman 1979; Bengtson

1981). Nonwinter activity budgets determined for manatees in the St. Johns River were categorized as 30 to 40 percent moving, 25 percent resting, 25 percent feeding, and 10 to 15 percent cavorting (Bengtson 1981).

CONSERVATION MEASURES TAKEN: The West Indian manatee is classified as endangered under the U.S. Endangered Species Act of 1973 and is protected under the U.S. Marine Mammal Protection Act of 1972. It also receives protection by the Florida Endangered and Threatened Species Act of 1977 and Manatee Sanctuary Act of 1978. A new Florida manatee recovery plan was completed in 1989 by a large recovery team composed of individuals from government agencies, non-government organizations, and private industry.

A number of protective measures have been implemented in the last decade. These include designation of sanctuaries such as those at Crystal River and Blue Spring, establishment and enforcement of boat speed zones, posting of cautionary signs, review and regulation of coastal development projects in sensitive areas, and the development of information and education programs. The Save the Manatee Club, an outgrowth of Florida Audubon, is a focal point for growing public sentiment in favor of manatee conservation and has active information and education programs. Substantial contributions to manatee research and public awareness have been made by Florida Power and Light Company, Sea World of Florida, and other private institutions. To obtain additional biological data needed to protect manatees, radiotelemetry studies, aerial surveys, habitat studies, and carcass salvage programs are being carried out in Florida by a variety of cooperating groups and agencies.

CONSERVATION MEASURES PROPOSED: As development and boat-traffic levels continue to increase in areas used by manatees, more stringent regulation and enforcement will be required to prevent further increases in manatee deaths. Additional sanctuaries will be required in the near future to protect critical habitats from development and heavy watercraft activity. Manatee protection elements will be incorporated in planning documents being prepared for key areas under endangered species requirements of the Florida Local Government Comprehensive Planning and Land Development Regulation Act of 1985. The Florida Manatee Recovery Plan (U.S. Fish and Wildlife Service 1989) details a lengthy series of multifaceted tasks that must be implemented in order to reach an interim goal of downlisting from endangered to threatened.

In October 1989 the governor and cabinet unanimously approved development of manatee protection plans (slow-speed zoning and boat-facility expansion policies) for 13 key counties, and they directed the Department of Natural Resources to recommend priority acquisition of critical manatee areas under the Conservation and Recreation Lands program and to strengthen Aquatic Preserve management plans to ensure protection of seagrass beds. The governor and cabinet also approved the department to proceed with legislative proposals for maximum boat speed-limits, vessel-operator licensing and education, amendments to the Florida Manatee Sanctuary Act for strengthened habitat protection, and other measures with direct and indirect effects on manatee conservation.

Literature Cited

Beck, C. A., and D. J. Forrester. 1988. Helminths of the Florida manatee, *Trichechus manatus latirostris*, with a discussion and summary of the parasites of sirenians. J. Parasitol. 74:628–637.

Beeler, I. E., and T. J. O'Shea. 1988. Distribution and mortality of the West Indian manatee (*Trichechus manatus*) in the southeastern United States: A compilation and review of recent information. Natl. Tech. Inf. Ser., PB88-207980/AS, Springfield, Virginia. 2 vols. 613 pp.

Bengtson, J. L. 1981. Ecology of manatees (*Trichechus manatus*) in the St. Johns River, Florida. Unpubl. Ph.D. diss., University of Minnesota, Minneapolis. 126 pp.

Best, R. C. 1981. Foods and feeding habits of wild and captive Sirenia. Mammal. Rev. 11:3–29.

Buergelt, C. D., R. K. Bonde, C. A. Beck, and T. J. O'Shea. 1984. Pathologic findings in manatees in Florida. J. Amer. Vet. Med. Assoc. 185:1331–1334.

Domning, D. P., and L. C. Hayek. 1986. Interspecific and intraspecific morphological variation in manatees (Sirenia: *Trichechus*). Marine Mammal Science 2:87–144.

Gunter, G. 1941. Occurrence of the manatee in the United States, with records from Texas. J. Mammal. 22:60–64.

Hartman, D. S. 1974. Distribution, status, and conservation of the manatee in the United States. Natl. Tech. Inf. Ser., PB81-140725, Springfield, Virginia. 246 pp.

Hartman, D. S. 1978. West Indian manatee. Pp. 27–39 *in* J. N. Layne, ed., Rare and endangered biota of Florida. Vol. 1. Mammals. University Presses of Florida, Gainesville. xx + 52 pp.

Hartman, D. S. 1979. Ecology and behavior of the manatee (*Trichechus manatus*) in Florida. Amer. Soc. Mammal., Spec. Publ. 5:1–153.

Irvine, A. B. 1983. Manatee metabolism and its influence on distribution in Florida. Biol. Conserv. 25:315–334.

Irvine, A. B., and H. W. Campbell. 1978. Aerial census of the West Indian manatee, *Trichechus manatus*, in the southeastern United States. J. Mammal. 59: 613–617.

Kinnaird, M. F. 1985. Aerial census of manatees in northeastern Florida. Biol. Conserv. 32:59–79.

Kochman, H. I., G. B. Rathbun, and J. A. Powell. 1985. Temporal and spatial distribution of manatees in Kings Bay, Crystal River, Florida. J. Wildl. Manage. 49:921–924.

Lefebvre, L. W., T. J. O'Shea, G. B. Rathbun, and R. C. Best. 1989. Distribution, status, and biogeography of the West Indian manatee. Pp. 567–610 *in* C. A. Woods, ed., The biogeography of the West Indies: Past, present, and future. Sandhill Crane Press, Gainesville, Florida. 878 pp.

Lewis, R. R., III, M. J. Durako, M. D. Moffler, and R. C. Phillips. 1985. Seagrass meadows of Tampa Bay—a review. Pp. 210–246 *in* S. F. Treat, J. L. Simon, R. R. Lewis, and R. L. Whitman, eds., Proceedings Tampa Bay area scientific information symposium [May 1982]. Florida Sea Grant Publ. 65. Burgess Publ. Co., Minneapolis. 663 pp.

McClenaghan, L. R., Jr., and T. J. O'Shea. 1988. Genetic variability in the Florida manatee (*Trichechus manatus*). J. Mammal. 69:481–488.

Moore, J. C. 1951. The range of the Florida manatee. Quart. J. Florida Acad. Sci. 14:1–19.

Odell, D. K., and J. E. Reynolds. 1979. Observations on manatee mortality in south Florida. J. Wildl. Manage. 43:572–577.

O'Shea, T. J. 1988. The past, present, and future of manatees in the southeastern United States: Realities, misunderstandings, and enigmas. Pp. 184–204 *in* R. R. Odom, K. A. Riddleberger, and J. C. Ozier, eds., Proceedings third southeastern nongame and endangered wildlife symposium. Georgia Dept. Nat. Resour., Game and Fish Division, Social Circle, Georgia. 253 pp.

O'Shea, T. J., J. F. Moore, and H. I. Kochman. 1984. Contaminant concentrations in manatees in Florida. J. Wildl. Manage. 48:741–748.

O'Shea, T. J., C. A. Beck, R. K. Bonde, H. I. Kochman, and D. K. Odell. 1985. An analysis of manatee mortality patterns in Florida, 1976–81. J. Wildl. Manage. 49:1–11.

Packard, J. M. 1985*a*. Development of manatee aerial survey techniques. Manatee Pop. Res. Rep. No. 7. Tech. Rep. No. 8-7. Florida Coop. Fish Wildl. Res. Unit, University of Florida, Gainesville. 68 pp.

Packard, J. M. 1985*b*. Preliminary assessment of uncertainty involved in modeling manatee populations. Manatee Pop. Res. Rep. No. 9. Tech. Rep. No. 8-9. Florida Coop. Fish Wildl. Res. Unit, University of Florida, Gainesville. 19 pp.

Packard, J. M., and O. F. Wetterqvist. 1986. Evaluation of manatee habitat systems on the northwestern Florida coast. Coastal Zone Manage. J. 14:279–310.

Powell, J. A., and G. B. Rathbun. 1984. Distribution and abundance of manatees along the northern coast of the Gulf of Mexico. Northeast Gulf Sci. 7:1–28.

Powell, J. A., and J. C. Waldron. 1981. The manatee population in Blue Spring,

Volusia County, Florida. Pp. 41–51 *in* R. L. Brownell, Jr., and K. Ralls, eds., The West Indian manatee in Florida. Florida Dept. Nat. Resour., Tallahassee. 157 pp.

Rathbun, G. B. 1984. Sirenians. Pp. 537–547 *in* S. Anderson and J. K. Jones, Jr., eds., Orders and families of recent mammals of the world. John Wiley and Sons, New York. 686 pp.

Reid, J. P., and T. J. O'Shea. 1989. Three years operational use of satellite transmitters on Florida manatees: Tag improvements based on challenges from the field. Pp. 217–232 *in* Proceedings of the 1989 North American Argos users conference and exhibit. Service Argos, Inc., Landover, Maryland. 361 pp.

Reynolds, J. E., III, and J. R. Wilcox. 1986. Distribution and abundance of the West Indian manatee, *Trichechus manatus*, around selected Florida power plants following winter cold fronts: 1984–85. Biol. Conserv. 38:103–113.

Shane, S. H. 1984. Manatee use of power plant effluents in Brevard County, Florida. Florida Sci. 47:180–187.

Thornback, J., and M. Jenkins. 1982. The IUCN mammal red data book, part 1: Threatened mammalian taxa of the Americas and the Australasian zoogeographic region (excluding Cetacea). IUCN, Gland, Switzerland.

U.S. Fish and Wildlife Service. 1989. Florida manatee (*Trichechus manatus latirostris*) recovery plan. Atlanta, Georgia. 98 pp.

Whitehead, P. J. P. 1977. The former southern distribution of New World manatees (*Trichechus* spp.). Biol. J. Linn. Soc. 9:165–189.

Prepared by: Thomas J. O'Shea and Mark E. Ludlow, U.S. Fish and Wildlife Service, National Ecology Research Center, Sirenia Project, 412 NE 16th Avenue, Gainesville, FL 32601 (TJO); and Florida Department of Natural Resources, Division of Recreation and Parks, District V Administration, Clermont, FL 34711 (MEL).

Endangered

Key Deer
Odocoileus virginianus clavium
FAMILY CERVIDAE
Order Artiodactyla

TAXONOMY: Although Hall and Kelson (1959) cited the law of priority to employ *Dama* for all New World deer, Opinion 581 of the International Commission on Zoological Nomenclature (1960) rejected *Dama*/Zimmerman 1780, making *Odocoileus*/Rafinesque 1832 available. Because there is uncertainty as to distribution of *O. v. osceola* and *O. v. seminolus* (Layne 1974), the precise relationship of *O. v. clavium* to these southern Florida races is uncertain, although Key deer undoubtedly originated from mainland stock. Barbour and Allen (1922) considered *clavium* a distinct race based on length of row of molariform teeth, whereas Klimstra et al. (1980) suggested Key deer show distinction beyond race level. Analyses of skull dimensions suggest several that distinguish Key deer (Maffei et al. 1988). Current study of skulls and genetics of Florida deer and of genetics of Key deer may resolve these questions.

DESCRIPTION: As documented by Barbour and Allen (1922) and substantiated by Dickson (1955) and Klimstra et al. (1974), the Key deer is on the average smaller than other eastern races of whitetails. Data recorded since 1967 suggest much variation in weight, with the average approximately 36 kg for adult males and 29 kg for females. Maxima for both sexes are about 58 kg and 50 kg, respectively, suggesting overlap with races of the Florida mainland. At birth, spotted fawns average 1.5 kg. Maximum size of females occurs at 4–5 years, whereas growth of males tends to continue at least through 8 years. This deer appears stockier than other races because of proportionately shorter legs; shoulder height of adults ranges from 64 to 76 cm. Readily apparent is a wider skull, especially obvious in males; ratios that compare interorbital and palate widths versus length of skull distinguish *clavium* from other races (Maffei et al. 1988).

Key deer, *Odocoileus virginianus clavium*. (Photo by Willard D. Klimstra)

Coat color varies from a melanistic reddish-brown to a grizzled near-gray; animals with the latter color often seem to be older. The rather obvious blackish facial mask in most fawns becomes more intensified in adults. Data accumulated from 1968 through 1973 suggest typical antler development is spike at 2 years, fork at 3 years, 6 points at 4 years, and 8 points at 5 years. More recent documentation suggests much variation to this pattern, however, with the appearance of several with small 6-point racks at 2 and 3 years of age. This may be associated with dynamics of a population seemingly stabilized at 350–400 animals during the 1970s (Klimstra et al. 1978, 1980, 1982) and then declining to 250–300 animals during the 1980s (Hardin et al. 1984; Klimstra 1985; U. S. Fish and Wildlife Service 1985).

POPULATION SIZE AND TREND: According to Dickson (1955) there may have been as few as 25–80 animals in 1951; the cause of its near demise is believed to be hunting (DePourtales 1877). Until the early 1970s, recruitment appeared to exceed mortalities as the population grew.

Odocoileus virginianus clavium

By about 1973–74 the herd seemed to be stable at 350–400 (Klimstra et al. 1974, 1978, 1982), of which 65–70 percent were on Big Pine Key (Klimstra et al. 1974). Although population levels of 600–1000 have been indicated (Leposky 1973; Layne 1974; Woodard 1980; Weiner 1985), verification is not possible. By the early 1980s the numbers had

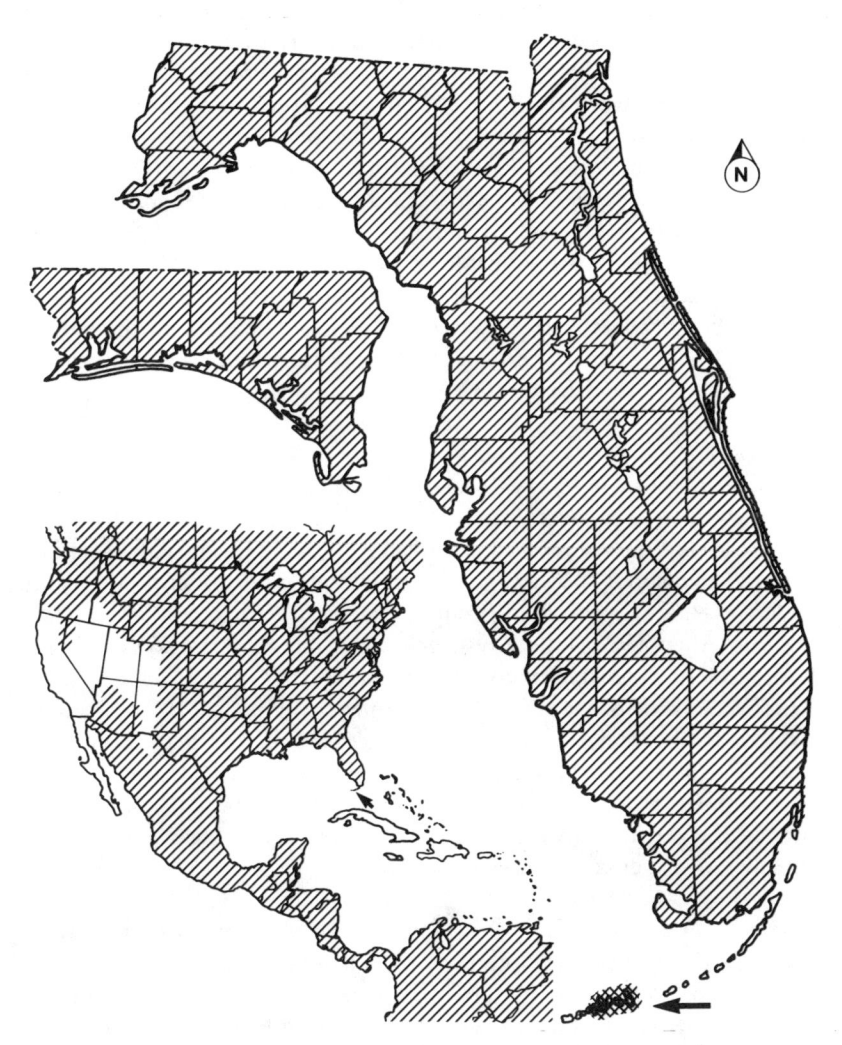

Distribution map of the Key deer, *Odocoileus virginianus clavium*. Hatching represents the species; crosshatching represents the subspecies.

declined by 100 to the current estimated population of 250-300 (Hardin et al. 1984; Klimstra 1985).

DISTRIBUTION AND HISTORY OF DISTRIBUTION: Key deer, restricted to the Lower Keys, once ranged from Key West to Duck Key (Barbour and Allen 1922); considering the 9.8 km of open water between Key Vaca and keys to the west there may be question about Boot, Grassy, Duck, and Vaca keys. During 1951-52, Dickson (1955) recorded sign on only 11 keys. Klimstra et al. (1974) established that Key deer probably use most keys from Boca Chica to the Johnsons. Sign of use, however, indicates only transient or seasonal occurrence for many islands, except for Big Johnson, Little Pine, No Name, Big Pine, Big Torch, Middle Torch, Little Torch, Cudjoe, Howe, Sugarloaf, Knockemdown, and Summerland, where there is reasonable access to fresh water. During the dry season the range is limited to locations with acceptable water; hence, deer populations of keys with fresh water may show an increase during winter (Jacobson 1974). The majority of the deer occur within the boundary of the National Key Deer Refuge.

GEOGRAPHICAL STATUS: The Key deer became endemic to the lower Florida Keys following a rise in ocean level with the retreat of the Wisconsin glacier (Hoffmeister and Muller 1968). The inundation of the land mass contiguous with the mainland created Florida Bay, isolating terrestrial fauna and subjecting it to constraints and influences of island conditions. Once believed an occupant of Duck Key to Key West (Barbour and Allen 1922), the Key deer now is recorded on 26 islands located between Spanish Harbor Bridge and Boca Chica (Klimstra et al. 1974; W. D. Klimstra, unpubl. data). Permanent residence is restricted by access to fresh water and diverse habitat, especially the pine-palm community. Regular seasonal use and occasional transient use of selected keys is accomplished by swimming across open water. The 9.8 km of open water to the south and west of Key Vaca appears a viable barrier except for truly unusual circumstances.

HABITAT REQUIREMENTS AND HABITAT TREND: Based on the character of Big Pine Key, which supports 65-70 percent of all the Key deer, diversity and extent of habitat are important, but top priority is available fresh water. According to Silvy (1975), pinelands and hardwoods are used most often; open-developed and mangrove areas show somewhat lesser use. Although hammock and buttonwood communities are of some importance, their use is related to special needs of a daily or seasonal nature. Sites burned, recently cleared, and sparsely developed are especially attractive (Klimstra et. al 1974; Klimstra 1986), providing new, more succulent early-successional vegetation as well as escape from mosquitoes due to more pronounced wind. Some habitat manipulations affect patterns of deer movement, enhance food supply, and increase area of use.

That fresh water is essential is shown by deer movements to keys with permanent supplies during the low rainfall normally occurring from October through April (Jacobson 1974; Klimstra et al. 1974). As a result, many keys have only temporary or transient deer use even though other habitat qualities may be minimally adequate. Although these temporary uses seem to be associated with periods of higher rainfall, dispersal or accelerated movement occurs in conjunction with breeding and fawning seasons (Hardin 1974; Silvy 1975).

The importance of pinelands is obvious from deer occupancy (Silvy 1975); however, reasons are not so clear. Because the pine-palm habitat depends on fire (Dickson 1955; Alexander and Dickson 1972; Klukas 1973), a greater diversity and availability of quality deer food and cover are presumed a consequence. Where there has not been fire for an extended period, the density of the understory, largely palms, precludes adequate food and desirable cover (Klimstra 1986). Although such keys afford diversity and fresh water, deer use is low; hence, habitat manipulations recommended by Klimstra et al. (1974, 1980) are appropriate.

Based on the range of activities recorded for sex and age classes (Silvy 1975), adequate space is an obvious need, but quality habitat and its diversity and water may be the most important. Keys with elevations less than 3–4 ft above sea level contribute only seasonally; hence, use by deer is temporary or transitory.

Although the diet of the Key deer is diverse as reflected by 164 plant foods identified in 129 samples from road mortalities, 28 plant species yielded 75 percent of the total volume (Dooley 1975). Woody plant browse and fruits yielded 67 percent; palm fruits, flowers, and spathes 14 percent; and forbs 13 percent. Use of browse was greatest during December–March and fruits and flowers during April–November, reflecting seasonal availability. In energy value, plant foods available to deer ranged from 1985 to 5940 cal/g; for those known to be prominent in their diet, the mean was 4534 cal/g (Morthland 1972). Analysis of 216 available plant foods (Widowski 1977) generally exhibited high ash and calcium, high Ca:P ratios, and low phosphorus in comparison to needs of white-tailed deer; the latter may contribute to low herd productivity.

Thatch palm (*Thrinax microcarpa*) ranked fifth in dietary importance value (Dooley 1975) and contained 5174 cal/g, but yielded the lowest predicted *in vivo* digestibility of the deer's staple foods (Donvito 1979).

VULNERABILITY OF SPECIES AND HABITAT: The Key deer is especially vulnerable to habitat loss to commercial land use that serves tourists and local residents, and to subdivision development (Klimstra 1985). Although the current rate and extent of habitat loss to development on Big Pine Key are somewhat less than the predicted 46 ha per year during 1969–78, most, if not all, private holdings could be developed early in the 21st century. This would result in a resident population of 6800 plus 4000 seasonal residents (Klimstra et al. 1974; Klimstra 1985; Sedway Cooke Associates 1989). Big Pine Key is especially significant as it supports 65–70 percent of the deer population because of its size, extent of public-owned lands, and quality of habitat. Although other keys subject to development have less high-quality habitat, have fewer deer, and are of smaller size, human impact can reduce deer use at least seasonally.

A major factor in suitability of habitats is not only quality, as reflected in food and cover needs, but especially fresh water. Because rainfall is the principal source, its availability is dependent upon annual occurrence in depressions that capture surface waters in the oolitic limestone. Further, its persistence is dependent upon elevation, intrusion of salt/brackish waters through the subsurface limestone (either from normal rise and fall of tides or from canals dredged into the interiors of some keys), and the presence and volume of freshwater lenses (i.e., Big Pine Key). Hence, developments and wells have an obvious effect on occurrence of fresh water and, in turn, time and level of deer use of otherwise suitable habitats. As a result, total usable area increases in the rainy season (May–October) and decreases in the dry season (November–April).

The Key deer is vulnerable to a variety of mortalities from human agents (Klimstra 1985). Heavy traffic on state, county, and subdivision roads results in approximately 80 percent of all deer losses (an average per year for 1968–88 of about 44), largely on U.S. Highway 1 across the southern portion of Big Pine Key (Drummond 1989). Hardin (1974) documented an 18 percent loss of marked fawns in 160 km of mosquito ditches. Although hunting has been banned since 1939, poaching continues; especially vulnerable are those animals in close association with people, readily enticed by feeding or baiting. With increasing human populations is a marked rise in the "guard" dog population, of which many are free-running. Documented deer mortality has increased both from direct dog attacks and from pursuit onto roads or into water where auto accidents or drowning occurs. There has been little documentation to suggest that parasites and diseases are a problem in Key deer. Schulte et al. (1976) found a variety of intestinal parasites but recorded no evidence of pathogenic effects. Of the most prevalent and potentially dangerous deer diseases in the southern United States (Newson 1984), Key deer are known to carry antibodies for epizootic hemorrhagic disease and blue tongue, indicating exposure. The potential for "exotic" diseases and parasites rises with the increasing number and variety of domestic animals associated with increased human residency and activity. Also, the attraction of maintained rights-of-way and subdivisions increases the perception that Key deer are a nuisance because of auto accidents and damage to landscape plantings. Many residents have resorted to fencing, which in turn reduces habitat and space available.

The Key deer has a low reproductive potential. Twins are uncommon, precocial breeding is clearly exceptional, and contribution by yearlings is infrequently documented (Hardin 1974; Klimstra et al. 1974; Hardin et al. 1984). This low level of reproductive output may be an adaptation of an island population with a rather stable lifestyle, especially in the absence of large terrestrial predators. Verme (1969) believed male-biased sex ratios such as documented during 1968–73 for Key deer was evidence of a reduced rate of population increase.

CAUSES OF THREAT: Although lack of protection and hunting reduced the population to 25–80 animals (Dickson 1955), the recent and current threat is a combination of extensive deer mortality and habitat loss resulting from development to serve an ever-increasing number of residents and tourists. Direct and indirect losses of deer occur from auto accidents, poaching, free-running dogs, mosquito ditches, and feeding. Known losses to auto accidents from 1980 through 1988 total 388 deer; an additional 95 animals were lost to other factors. For a herd estimated at 250–300 and with naturally low annual productivity, annual losses have approached or exceeded a given year's recruitment—hence the population decline of $\geq$100 animals since 1979.

RESPONSES TO HABITAT MODIFICATION: The response of Key deer to habitat modification is well documented (Klimstra et al. 1974, 1980, 1982; Hardin et al. 1984; Klimstra 1986). Disturbance such as subdivision development, road building, mosquito ditching, and clearing created openings and edges used extensively by deer at the subsequent early-successional stages of revegetation. The response to burning, whether planned or unplanned, was immediate, before all smoke or smoldering had ceased (Silvy 1975; Klimstra 1986). Especially attractive are frequently mowed subdivisions and rights-of-way that maintain early successional and palatable grasses and forbs. Cleared strips, whether firelane, roadway, or mosquito ditch, also serve as travel lanes through heavy cover.

The pine-palm communities considered important Key deer habitat are a consequence of fire and can be perpetuated in high-quality condition only through periodic burning (Alexander and Dickson 1972; Klukas 1973; Wade et al. 1980; Schomer and Drew 1982; Klimstra 1986; Carlson 1989). The advancement to hardwood hammock conditions in openings (i.e., grassy areas) can be effectively retarded or reduced by fire (Alexander and Dickson 1970). The food supply for deer is dramatically enhanced in these communities as a consequence, resulting in greatly increased deer use.

Deer depredation on landscape plants in subdivisions is a serious, long-standing problem. Hence, there is extensive fencing or individual plant isolation by wire cylinders. Some residents consider the deer a nuisance and an expense to be endured.

Developments to enhance available fresh water have included removing accumulated debris from selected water holes, enlargement and deepening of natural depressions, and use of a plastic "guzzler." Some deer response on Big Munson, Howe, Sugarloaf, Water, and Johnson keys has been documented. Subdivision and road construction have changed terrain on Big Pine Key, resulting in fresh water being surface-pocketed in a large area formerly influenced by tidal or subsurface intrusion of saline water. Also, mosquito ditches in former freshwater sites (i.e., solution basins) of hardwood hammock communities on Big Pine Key, now blocked off, greatly enhance permanency and distribution of available fresh water. During the lengthy drought of 1970 an obvious response of deer to these man-created water supplies was documented during routine censuses (Jacobson 1974), suggesting significant movement of deer from other parts of Big Pine and probably adjacent keys.

DEMOGRAPHIC CHARACTERISTICS: Breeding begins in September, peaks in October, and declines through November and December; occurrence of a spotted fawn in any month suggests occasional exceptions (Hardin 1974; Hardin et al. 1984). Parturition is primarily from mid-March through mid-May, following a 204-day gestation. Females 2 years and older contribute a major portion of annual recruitment; precocial and yearling breeding appear minimal at best. Twins are infrequent and rarely recorded among surviving newborn. Males >3 years of age seldom have full rack development and rarely participate in breeding; exceptions may occur late in the breeding season when dominant adult males are less aggressive.

There is a minimum of 20 percent mortality through the first 6 months of life (Hardin 1974). A 50 percent survival of males through 1.5 years has been recorded; none have been documented after 8 years. In contrast, 50 percent of females reach 2.5 years; most do not survive beyond 9 years. However, a female marked in March 1968 was known to have survived 20 years and another at least 19 years; each was believed to have produced a fawn until its year of death. During 1968–73, the sex ratio of fetuses was 1.75 males to 1.00 females, whereas for fawns it was 2.0 to 1.0. Such a high ratio of males to females (Verme 1969) is associated with relatively low recruitment and self-imposed constraint on reproduction.

KEY BEHAVIORS: Loose matriarchal groups consisting of an adult female with one or two generations of offspring is the norm (Hardin 1974; Hardin et al. 1976), but generally they are rather solitary animals, beginning with the wandering of fawns at 3–4 months in age. Bucks occur in groups during the nonbreeding season, especially following antler loss and until loss of velvet; occasionally adult males in velvet are in a gang of several females. Since 1973, mixed groups, usually more than one adult female and offspring, have been observed feeding, bedding, and escaping insects in subdivisions at almost any time (Klimstra 1985). This is in response to water and domestic foods presented by individual residents.

A given animal or group has well-defined patterns of activity and location (Klimstra et al. 1974). Well-established trails and areas of bedding are evident in some areas. Repeated use of certain sites to cross roads results in "hot spots" for automobile accidents.

Key deer readily swim from island to island; this occurs most when sources of fresh water are limited and as dispersal associated with the breeding season. Patterns of home-range use and dispersal are very similar to those of other whitetails (Hardin 1974; Silvy 1975).

CONSERVATION MEASURES TAKEN: As a result of local concern and the 1934 cartoon by "Ding" Darling, the Florida legislature banned hunting of Key deer in 1939 (Ryden 1978). The studies of Dickson (1955) during 1951–52; the hiring of Jack C. Watson, Sr., as warden with funds from the Boone and Crockett Club in 1951; the contributions and donations of private individuals; and the unrelenting efforts of representatives of members of the U.S. Congress, the Boone and Crockett Club, Wildlife Management Institute, National Audubon Society, North American Wildlife Federation, and U.S. Fish and Wildlife Service provided the initiative for establishment of the National Key Deer Refuge in 1957 (Ryden 1978). As a result of donation and purchase, the refuge now includes about 3000 ha. Beginning in 1964, U.S. Fish and Wildlife Service personnel initiated study of deer habitat use and attempted to monitor population levels (Stieglitz 1964, 1967; Yaw 1965, 1966). The Key deer was placed on the federal endangered species list in 1967, with full protection under the Endangered Species Act of 1973, and was documented among endangered biota of Florida in 1978 (Klimstra et al. 1978). A six-year research effort began in December 1967 (Klimstra et al. 1974), resulting in data on the biology of the deer, its habitat use, current and anticipated problems, and management recommendations. Subsequently a recovery plan was produced (Klimstra et al. 1980; U.S. Fish and Wildlife Service 1985). Key issues included habitat acquisition, protection, and maintenance; deer protection; monitoring of deer condition, population, and habitat; experimenting with habitat manipulation; educating the public; and conducting additional studies of natural history and population dynamics.

Because automobile accidents contribute 75–80 percent of known mortality, an effort was made to reduce such loss. Initially this was through use of deer crossing signs and reflectors. Subsequently a maximum speed of 45 mph was posted on U.S. 1; speed limits on all other roads were reduced from 45 to 30 mph. From 1957 to 1967, annual road losses ranged from 14 to 16, and during 1968–88 it varied little, averaging around 44. This suggests these techniques have not been adequate.

There have been several efforts to control dog harassment of deer, through dog capture and public education. A dog pound on Big Pine Key and appropriate personnel have been established through extensive collaboration of private, state, and federal interests.

Efforts to acquire land through gift or purchase have been under way since the establishment of the refuge in 1957. Only in the last few years has the effort focused on those habitats deemed essential to Key deer needs. There has been continuous monitoring of the population (numbers, location, sex, and age) since 1968 (Klimstra et al. 1974; Silvy 1975; Humphrey and Bell 1986) emphasizing road censuses (largely on Big Pine Key). Additionally, condition of deer has been evaluated during examination of mortalities, including analysis of selected tissue samples by the Southeast Deer Disease Study Group.

Beginning in 1969, controlled burning began on Big Pine Key. This effort was expanded in 1978 and most recently has been extended to other keys. Burning has perpetuated high-quality deer habitat (Alexander and Dickson 1970, 1972; Klimstra 1986; Carlson 1989). Additionally, disturbed areas resulting from development such as roadways, subdivisions, and mosquito ditching increased diversity valuable to deer. Water supply has been enhanced by removing accumulated sediment in selected sites on several keys and establishing plastic guzzlers on two.

Filling selected segments of mosquito ditches to reduce fawn loss has been approved and implemented. Through additional staff, the refuge has increased patrols to reduce harassment by dogs and possible poaching. Public education has been stepped up regarding people-related problems, especially feeding. Although the Florida legislature made feeding the deer a misdemeanor and provided for a $500 fine, deer feeding continues because few convictions have been obtained.

The recent emphasis by various organizations and agencies regarding land use in the Lower Keys has great importance to Key deer. The development of Monroe County's Comprehensive Land Use plan and its implementation will have far-reaching effects; it is to be hoped that its ecological integrity will not be diluted by political processes.

CONSERVATION MEASURES PROPOSED: Most important is the full implementation of the recovery plan (Klimstra et al. 1980; U.S. Fish and Wildlife Service 1985). Priority should be given to recommended acquisitions not yet in public ownership, with immediate attention to selected areas on Big Pine and Big Torch keys and all of No Name Key. No Name Key is of highest quality, reflecting outstanding habitat diversity and permanent fresh water. Its complete acquisition would maximize public control and restrict public access. In all acquisitions, every effort should be made to reduce private inholdings that hinder effective application of habitat management, contribute to undesirable people-deer interactions, and constrain effective monitoring of deer. Major emphasis should be on preservation of wetlands, especially those on Big Pine Key south of Watson Boulevard. Also, effective growth management and land-use planning for the Lower Florida Keys must be expedited.

Habitat management must be aggressively pursued with a workable controlled-burning plan that will enhance food supply, diversity, and edge (Klimstra 1986; Carlson 1989). Pine-palm communities and open grasslands must be maintained. Fresh water supplies must be enhanced by protection of wetlands, insuring permanence of strategically located sites, installation of guzzlers, and possibly replenishing water on selected keys during extreme drought.

The dog control plan being finalized must be effectively implemented.

Auto speed constraints must be established on U.S. 1 across all of Big Pine Key and there must be greater enforcement on all state and feeder roads. Rights-of-way must be assessed for ways to reduce their attractiveness to deer and enhance deer-driver visibility. The serious problem of deer interacting with residents and tourists must be significantly reduced. Aggressive public education must be mounted and maintained to enlighten the permanent and temporary residents and tourists regarding the Key deer and the negative impacts of treating them as pets.

Monitoring and research must be continued to understand the future status of the deer. This includes road censusing, age and sex-ratio analyses, documentation of mortalities, habitat evaluation and deer use, health of the deer based on mortalities, analysis of deer use of public and private lands, and evaluation of the various deer- and people-management practices implemented.

ACKNOWLEDGMENTS: The contributions of Jack C. Watson, Sr. (deceased), the first manager of the National Key Deer Refuge, were invaluable in establishing and conducting research during 1967–73. Many past and current staff of the refuge gave important assistance: refuge managers Donald Kosin and Deborah Holle; wildlife biologists Steven Klett, Bonnie Bell, and Thomas Wilmers; and Chester Eldard, Jack Watson, Jr., Lois Robinson, Valeen Silvy, and William Swicker. The baseline data contributed by John D. Dickson, III, Taylor R. Alexander, Edward D. Yaw, Stephen C. Johnson, and Walter O. Stieglitz provided an important foundation for the Key deer studies. Financial support was contributed by the U.S. Fish and Wildlife Service, National Geographic Society, North American Wildlife Foundation, National Wildlife Federation, Cooperative Wildlife Research Laboratory of Southern Illinois University, and W. D. Klimstra Distinguished Professor Fund. Scores of graduate students at Southern Illinois University participated in the field research, especially James W. Hardin, Nova J. Silvy, Bruce N. Jacobson, Todd R. Eberhardt, and Allan L. Dooley. Others studied materials collected: Jimmie R. McCain, Virginia A. Terpening, Kathy Jacobson, Douglas Morthland, John Schulte, James A. Ruchy, Martha K. Hunt, Larry M. David, Mark T. Donvito, and Stephen Widowski.

Literature Cited

Alexander, T. R., and J. D. Dickson, III. 1970. Vegetational changes in the National Key Deer Refuge. Q. J. Florida Acad. Sci. 32:81–89.

Alexander, T. R., and J. D. Dickson, III. 1972. Vegetational changes in the National Key Deer Refuge—II. Q. J. Florida Acad. Sci. 35:85–96.
Barbour, R., and G. M. Allen. 1922. The white-tailed deer of eastern United States. J. Mammal. 3:65–78.
Carlson, P. C. 1989. Effects of burning in the rockland pine community of the Key Deer National Wildlife Refuge, Florida Keys. Unpubl. M.S. thesis, University of Florida. 69 pp.
DePourtales, L. F. 1877. Hints on the origin of flora and fauna of the Florida Keys. Amer. Nat. 11:137–144.
Dickson, J. D., III. 1955. An ecological study of the Key deer. Florida Game and Fresh Water Fish. Comm. Tech. Bull. 3:1–104.
Donvito, M. T. 1979. An *in vitro* digestibility study of selected Florida Key deer foods. Unpubl. M.A. thesis, Southern Illinois University, Carbondale. 42 pp.
Dooley, A. L. 1975. Foods of the Florida Key deer. Unpubl. M.A. thesis, Southern Illinois University, Carbondale. 80 pp.
Drummond, F. 1989. Factors affecting road mortality of Key deer. Unpubl. M.S. thesis, Southern Illinois University, Carbondale. 85 pp.
Hall, E. R., and K. R. Kelson. 1959. The mammals of North America. Ronald Press, New York. 1083 pp.
Hardin, J. W. 1974. Behavior, socio-biology, and reproductive life history of the Florida Key deer, *Odocoileus virginianus clavium*. Unpubl. Ph.D. diss., Southern Illinois University, Carbondale. 226 pp.
Hardin, J. W., N. J. Silvy, and W. D. Klimstra. 1976. Group size and composition of the Florida Key deer. J. Wildl. Manage. 40:454–463.
Hardin, J. W., W. D. Klimstra, and N. J. Silvy. 1984. White-tailed population and habitats: Florida Keys. Pp. 381–390 *in* L. K. Halls, ed., White-tailed deer: Ecology and management. Stackpole Books, Harrisburg, Pennsylvania.
Hoffmeister, J. E., and H. G. Muller. 1968. Geology and origin of the Florida Keys. Geol. Soc. Amer. Bull. 79:1487–1502.
Humphrey, S. R., and B. Bell. 1986. The Key deer population is declining. Wildl. Soc. Bull. 14:261–265.
International Commission on Zoological Nomenclature. 1960. Opinion 581. Determination of the generic names for the fallow deer of Europe and the virginia deer of America (Class Mammalia). Bull. Zoological Nomenclature 17:267–275.
Jacobson, B. N. 1974. Effects of drinking water on habitat utilization by Key deer. Unpubl. M.S. research paper, Southern Illinois University, Carbondale. 43 pp.
Klimstra, W. D. 1985. The Key deer. Florida Naturalist 58(4):2–5.
Klimstra, W. D. 1986. Controlled burning in habitat management: Some observations, National Key Deer Refuge. Cooperative Wildlife Research Laboratory, Southern Illinois University, Carbondale. Unpubl. final report. 35 pp.
Klimstra, W. D., J. W. Hardin, and N. J. Silvy. 1978. Endangered Key deer. Pp. 15–17 *in* J. N. Layne, ed., Rare and endangered biota of Florida. Vol. 1. Mammals. University Presses of Florida, Gainesville. xx + 52 pp.

Klimstra, W. D., J. W. Hardin, and N. J. Silvy. 1978. Population ecology of Key deer (*Odocoileus virginianus clavium*). Pp. 313–321 *in* P. H. Achser and J. S. Lea, eds., National Geographic Research Reports: 1969 Projects.

Klimstra, W. D., J. W. Hardin, N. J. Silvy, B. N. Jacobson, and V. A. Terpening. 1974. Key deer investigations final report. Period of study: December 1967–June 1973. Cooperative Wildlife Research Laboratory, Southern Illinois University, Carbondale. Mimeo. 184 pp.

Klimstra, W. D., J. W. Hardin, M. P. Carpenter, and S. Jenkusky. 1980. Key deer recovery plan. U.S. Fish and Wildlife Service. 52 pp.

Klimstra, W. D., N. J. Silvy, and J. W. Hardin. 1982. The Key deer: Its status and prospects for the future. Proceedings of the Nongame and Endangered Wildlife Symposium. Georgia Dept. Nat. Resour. Tech. Bull. WL5:137–141.

Klukas, R. W. 1973. Control burn activities in Everglades National Park. Proc. Tall Timbers Fire Ecol. Conf. 12:397–425.

Layne, J. N. 1974. The land mammals of south Florida. Pp. 386–413 *in* P. J. Gleason, ed., Environments of south Florida: Present and past. Miami Geol. Soc. Memoir 2:1–452.

Leposky, G. 1973. Comeback of the Key deer. Florida Wildlife 26:12–14.

Maffei, M. D., W. D. Klimstra, and T. J. Wilmers. 1988. Cranial and mandibular characteristics of Key deer (*Odocoileus virginianus clavium*). J. Mammal. 69(2):403–407.

Morthland, D. 1972. Energy values of selected vegetation from Big Pine Key, Florida. Unpubl. M.S. research paper, Southern Illinois University, Carbondale. 32 pp.

Ryden, H. 1978. Saga of the toy deer. Audubon 80:92–103.

Schomer, N. A., and R. D. Drew. 1982. An ecological characterization of the Lower Everglades, Florida Bay and the Florida Keys. FWS/OBS-82/58.1. 246 pp.

Schulte, J. W., W. D. Klimstra, and W. G. Dyer. 1976. Protozoan helminth parasites of Key deer. J. Wildl. Manage. 40:479–581.

Sedway Cooke Associates. 1989. Synopsis of selected data: Growth patterns Big Pine Key community plan. Mimeo. 11 pp.

Silvy, N. J. 1975. Population density, movements, and habitat utilization of Key deer, *Odocoileus virginianus clavium*. Unpubl. Ph.D. diss., Southern Illinois University, Carbondale. 152 pp.

Stieglitz, W. O. 1964. Wildlife management study outline: Key deer investigations and amendments. U.S. Fish and Wildlife Service, Branch of Wildlife Refuges, Region 4, Project-Key Deer 1. 3 pp.

Stieglitz, W. O. 1967. Progress report: Key deer investigations (wildlife management study—Key deer #1). U.S. Fish and Wildlife Service, Branch of Refuges, Region 4. 27 pp.

U.S. Fish and Wildlife Service. 1985. Key deer recovery plan. U.S. Fish and Wildlife Service. Atlanta, Georgia. 46 pp.

Verme, L. J. 1969. Reproductive patterns of white-tailed deer related to nutritional plane. J. Wildl. Manage. 33:881–887.

Wade, D., J. Ewel, and R. Hofstetter. 1980. Fire in south Florida ecosystems. U.S. Dept. Agric. For. Serv., Gen. Tech. Rept. SE-17. 125 pp.

Weiner, A. H. 1978-86. The hardwood hammocks of the Florida Keys: An ecological study. The Florida Keys Land Trust, Inc., Key West. 42 pp. + appendices and atlas.

Widowski, S. 1977. Calcium and phosphorus in selected vegetation from Big Pine Key, Florida. Unpubl. M.S. thesis, Southern Illinois University, Carbondale. 80 pp.

Woodard, D.W. 1980. Selected vertebrate endangered species of the southeast coast of the United States—the Key deer. FWS/OBS-80/01.48. 5 pp.

Yaw, E. No date. Progress report: Key deer (*Odocoileus virginianus clavium*) investigations (June 15, 1965-June 15, 1966). Key Deer National Wildlife Refuge, U.S. Fish and Wildlife Service, Branch of Refuges, Region 4. 13 pp.

Yaw, E. 1965. Progress report: Key deer (*Odocoileus virginianus clavium*) investigations (winter 1964-65). Key Deer National Wildlife Refuge, U.S. Fish and Wildlife Services, Branch of Refuges, Region 4. 5 pp. + 5 tables and 1 figure.

Yaw, E. 1966. Key deer notes #1: Saltwater damage to deer habitat. 1 p.; #2: first six weeks of deer trapping. 1 p.; and, #3: ageing of deer pellets on Big Pine Key. 1 p. Key Deer National Wildlife Refuge, U.S. Fish and Wildlife Service, Branch of Refuges, Region 4.

Prepared by: Willard D. Klimstra, Cooperative Wildlife Research Laboratory, Southern Illinois University, Carbondale, IL 62901.

Threatened

Florida Mastiff Bat
Eumops glaucinus floridanus
FAMILY MOLOSSIDAE
Order Chiroptera

TAXONOMY: Other names: Wagner's mastiff bat, mastiff bat. Eger (1977) revised the genus and recognized *E. glaucinus floridanus* (Allen 1932) in Florida and *E. g. glaucinus* in Cuba, Jamaica, Central America, and South America. The subspecies was described by Allen (1932).

DESCRIPTION: This is the largest bat in Florida. Adults collected from the Miami area measure from 134 to 165 mm in total length (Owre 1978). Three specimens in the Florida Museum of Natural History, from the western coast of Florida, measured 126–162 mm. Other external measurements (combined for individuals from both coasts) are: forearm 57.9–69.2 mm, hind foot 10.8–15.0 mm, and ear (from notch) 19.9–31.0 mm. Mass ranges from 30.2 to 46.6 g. A pregnant female, far from full term, weighed 55.4 g (Belwood 1981). Robson et al. (1989) captured a pregnant female in August 1988 that weighed 39 g.

The lower half of the tail projects beyond the interfemoral membrane, and the ears are joined at the midline. The latter, and the large size of this animal, distinguish it from the smaller *Tadarida brasiliensis*, the only other molossid to occur in the state. In addition, the anterior tragus is broad and truncated distally. Color varies from black or brownish-gray to cinnamon-brown. Underparts are lighter and grayish. The fur is short and glossy. Hairs are bicolored and lighter at their bases.

POPULATION SIZE AND TREND: The number of *E. glaucinus* in Florida is unknown. Once believed to be common on Florida's eastern coast, in Miami and Coral Gables, these bats have been reported there only once since 1967 (see below). Prior to that, Gordon Hubbell (personal communication) routinely obtained several individuals per year that were col-

Florida mastiff bat, *Eumops glaucinus floridanus*. (Photo by Roger W. Barbour)

lected during the winter months in people's houses. He also reports having regularly heard the loud, distinctive calls of these bats at night as they flew and foraged high above buildings.

Bats in southern Florida, including *E. g. floridanus*, appear to have declined drastically in numbers in recent years. In 1982, a survey of more than 100 pest contol companies on the southeastern coast showed that requests to remove "nuisance" bats in this area have all but ceased in the last 20 years (J.J. Belwood, unpubl. data). Few bats other than *Tadarida brasiliensis* (G.T. Hubbell, personal communication) are removed from buildings in this part of the state.

Only one sighting of *E. g. floridanus* has been reported on the western coast of Florida (Belwood 1981). This occurred in 1979 and consisted of a small colony of eight individuals, including a pregnant female. A sixweek field trip to both the eastern and western coasts of southern Florida in 1980 to locate more *E. g. floridanus* was unsuccessful (J.J. Belwood and L. Wilkins, unpubl. data), leading to the belief that these bats were probably extinct in Florida (Belwood 1981).

The recent capture of a pregnant female in Coral Gables (Robson et al. 1989) in August 1988 is important because it indicates the presence of breeding animals in southern Florida. This individual was found on the seventh-floor balcony of an office building.

E. glaucinus in Cuba is rare (Silva-Taboada 1979), and it does not roost in large groups.

DISTRIBUTION AND HISTORY OF DISTRIBUTION: *E. g. floridanus* has the most restricted distribution of any subspecies of bat in the United

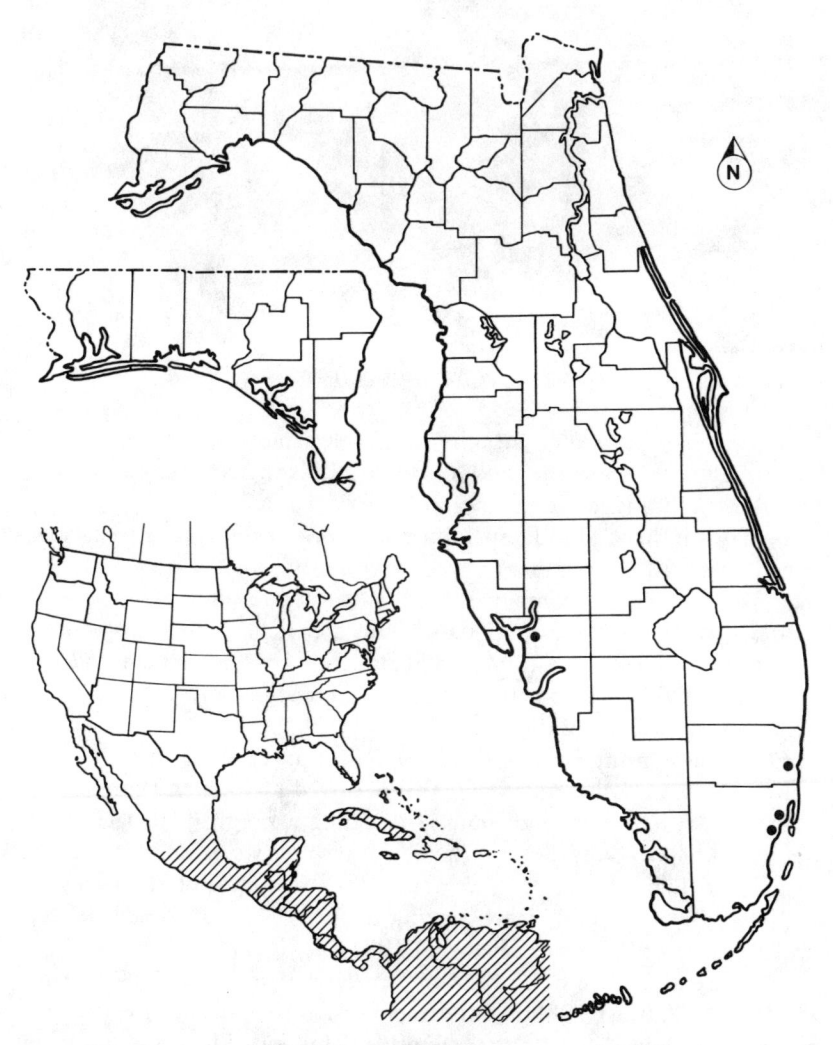

Distribution map of the Florida mastiff bat, *Eumops glaucinus floridanus*.
Hatching represents the species; crosshatching and dots represent the subspecies.

States. Most sightings of this bat have been from the Miami and Coral Gables areas, although G. T. Hubbell (personal communication) reported a female with young from Fort Lauderdale. All his sightings of mastiff bats were near human dwellings. Belwood's (1981) western coast colony was found in a tree, near Punta Gorda, Charlotte County, where the bats appeared to have been permanent residents. Pleistocene remains are known from Melbourne, Brevard County, 250 km NNW of Miami (Allen 1932), from Monkey Jungle Hammock, Dade County, 25 km SSW of Miami (Martin 1977), and from Indian River County (Morgan 1985).

Temperature tolerances of this species are unknown, but its presence in southern Florida, the Caribbean, Central America, and South America indicates that it is essentially a tropical bat whose distribution may be restricted by temperature. Tamsitt and Valdivieso (1963), however, reported one specimen from Bogotá, Colombia, at an elevation of 2750 m, where the mean annual temperature is 60–70° F. It is noteworthy that the Melbourne fossil site is late Wisconsinan, during which time climatic conditions are thought to have been drier and cooler than they are at present (Morgan 1985).

GEOGRAPHICAL STATUS: The subspecies is endemic to southern Florida. Extent of former geographical range is unknown.

HABITAT REQUIREMENTS AND HABITAT TREND: Although these bats are known to occur in cities as well as in forested areas, precise foraging and roosting habits and long-term requirements of *E. g. floridanus* are unknown. Past sightings in Miami indicate that their favorite diurnal roosts may be the shingles under Spanish tile roofs, although some specimens in Coral Gables have also been found in shafts of royal palm (*Roystonea regia*) leaves (G. T. Hubbell, personal communication). Belwood's (1981) small western coast colony was found in a longleaf pine, *Pinus palustris*, in a flatwoods community adjacent to the 65,000-acre Webb Wildlife Management Area. The bats were in a cavity 4.6 m high, which had been excavated by a red-cockaded woodpecker, *Picoides borealis*, and enlarged by a pileated woodpecker, *Dryocopus pileatus*.

Silva-Taboada (1979) reports similar habitat associations for *E. g. glaucinus* from 33 collecting localities in Cuba. Ten known roosts were in buildings and nine in tree hollows. The latter include woodpecker holes and cavities in royal palms, "dagame" trees (*Callycophyllum candidissimum*), and mastic trees (*Bursera simaruba*). He also found one bat roosting in the foliage of the palm *Copernicia vespertilionum*. Approximately 80 specimens of *E. glaucinus* from Venezuela are in the U.S. National Museum.

Most of these animals also were collected in tree cavities in heavily forested areas.

For most bats, the availability of suitable roosts is an important limiting factor (Humphrey 1975). Human population growth in southern Florida is thriving, which would appear to benefit the urban *E. glaucinus*, because Spanish tile roofs on dwellings are as common today as in the past. But forested areas are becoming rare as a result of human encroachment. This will severely affect the forest populations of this bat. For example, the Punta Gorda roost discussed here (Belwood 1981) was destroyed as part of a highway construction project. Nonurban localities such as Cecil Webb and Corbett wildlife management areas, Everglades National Park, and the Big Cypress/Fakahatchee areas are probably the only areas in which *E. g. floridanus* are likely to be found in the future, provided they are allowed to retain their older trees with hollows and cavities. Tree cavities are rare in southern Florida and competition for them is fierce. Other animals found inhabiting such holes include flying squirrels, corn snakes, red-headed woodpeckers, and various hymenopterous insects (J.J. Belwood and L. Wilkins, unpubl. data).

In the past, *Eumops* have been observed foraging high in the skies over Miami, where they fed on insects. In Punta Gorda, Belwood (1981) found that Coleoptera (55%), Diptera (15%), Hemiptera (10%), and other assorted insects made up the bulk of the bats' diets. Silva-Taboada (1979) identified Coleoptera, Lepidoptera, and Orthoptera as the main dietary items of the Cuban subspecies. All these results are anecdotal.

VULNERABILITY OF SPECIES AND HABITAT: *E. glaucinus* are vulnerable to habitat loss (in urban and forested areas), habitat alteration (removal of older trees with roosting cavities in favor of younger stands of trees for commercial use in forested areas), and heavy pesticide spraying for mosquitoes in urban areas. Despite an abundance of roosting sites, G. T. Hubbell (personal communication) believes the latter to be the main reason for the decline of these bats in the Miami area.

CAUSES OF THREAT: Unknown.

RESPONSES TO HABITAT MODIFICATION: Beginning in the early to mid-1960s, destruction of bat colonies and roosts increased as a result of rapidly expanding human populations in southern Florida, particularly in southern Dade County. The immediate results of this, in terms of responses of the bats, are not known.

DEMOGRAPHIC CHARACTERISTICS: In Cuba, members of this species appear to be polyestrous and give birth to only one young (Silva-Taboada 1979). Polyestry also appears to be the case in Florida; G. T. Hubbell (personal communication) reports young in June, and Belwood (1981) and M. S. Robson (personal communication) found pregnant females in September and August, respectively. The single embryo in Belwood's study had a crown-to-rump length of 23 mm, that of Robson et al. 38 mm.

KEY BEHAVIORS: G. T. Hubbell (personal communication) believes that *Eumops* in Miami roost singly. In contrast, Belwood's (1981) colony consisted of seven females and one male, all adult. Five of the females were postlactating, the sixth was pregnant, and the male had enlarged testicles. The sex ratio of this colony, and the reproductive condition of the male, suggest that this group may have consisted of a male and his harem. This social organization has been recorded in some bats including *Tadarida midas*, another molossid (Bradbury 1977). Such social groupings may be facilitated by roosting in tree cavities, which can be defended from other males. These bats have a pungent musky odor (J. J. Belwood, unpubl. data). The male *E. glaucinus* has a scent gland on its chest, the exact function of which is unknown. It may be used to mark females or a group's roosting site. In Cuba, males and females have been found roosting together throughout the year in tree hollows in groups of up to 32 individuals. Tyson (1964) also found a group of 10 *E. g. glaucinus* roosting together in the attic of a house in Panama.

G. T. Hubbell (personal communication) reports that these bats leave their roosts to forage after dark, seldom occur below 10 m in the air, and produce loud, audible calls that are easily recognized as they fly. Unlike most molossids, which need to drop 8–10 m from a roost before they can fly, *E. glaucinus* can take flight from horizontal surfaces.

E. glaucinus have a high wing-aspect ratio, which results in relatively efficient flight (Norberg and Rayner 1987). They are capable of long, straight, and sustained flight, which should allow them to travel far. They are "fast hawking" bats (Simmons et al. 1979) that rely on speed and agility to catch insects in the absence of clutter. These bats are confined to foraging in open spaces and use echolocation to detect prey at relatively long range (3–5 m).

CONSERVATION MEASURES TAKEN: *E. g. floridanus* is under review for listing by the U.S. Fish and Wildlife Service. Studies to determine the status of the bat in southern Florida also are under way by the Florida Game and Fresh Water Fish Commission (M.S. Robson, personal communication).

CONSERVATION MEASURES PROPOSED: Very little is known about the biology, ecology, requirements, or status of these animals. A thorough survey of forested areas, and to a lesser extent cities, in southern Florida should be undertaken to determine the approximate distribution of this bat. This could be accomplished most efficiently using acoustic censusing techniques. If concentrations are found, they should be studied in detail, and known roosts and the surrounding areas should be preserved. Pesticides are believed to have played a role in the demise of *E. g. floridanus* populations in Miami (G. T. Hubbell, personal communication). Sympatric populations of the closely related *Tadarida brasiliensis* still occur, however, in some areas of southern Florida (J. J. Belwood, personal observation), suggesting that some bats may tolerate pesticide exposure. The effects of pesticides on bat populations in Florida are not known but should be investigated. Last, because bats are feared by most people, some effort should be made to educate the general public as to their importance. Bat education campaigns have recently been implemented in England, where all bats are considered endangered. The response of the general public to this effort has been favorable.

ACKNOWLEDGMENTS: This was written during a Smithsonian postdoctoral fellowship. Oscar T. Owre deserves credit for having written the original species account for this bat (Owre 1978), and special thanks go to Gordon T. Hubbell for providing a great deal of unpublished information. Both were also kind enough to provide an update (in August 1987) on the status of *Eumops* in the Miami area. Mark S. Robson made available his unpublished manuscript on the 1988 sighting. In 1980, Laurie Wilkins helped look for *Eumops* in the field in an expedition partly funded by the Florida Museum of Natural History. At that time, Larry Campbell provided accommodations at the Webb Wildlife Management Area and the late Bruce Bowman provided facilities at the Corbett Wildlife Management Area. Charles O. Handley, Jr. (U.S. National Museum), allowed access to unpublished material on *Eumops* from Venezuela and Roger W. Barbour kindly supplied the photograph. Laurie Wilkins, Mark S. Robson, and Jeffrey A. Gore read the manuscript.

Literature Cited

Allen, G. M. 1932. A Pleistocene bat from Florida. J. Mammal. 13:256–259.
Belwood, J. J. 1981. Wagner's mastiff bat, *Eumops glaucinus floridanus* (Molossidae) in southwestern Florida. J. Mammal. 62:411–413.
Bradbury, J. W. 1977. Social organization and communication. Pp. 1–72 *in* W. A. Wimsatt, ed., Biology of bats. Vol. 3. Academic Press, New York. 651 pp.
Eger, J. L. 1977. Systematics of the genus *Eumops* (Chiroptera: Molossidae). Life Sci. Contrib., Royal Ont. Museum 110:1–69.
Humphrey, S. R. 1975. Nursery roosts and community diversity of Nearctic bats. J. Mammal. 56:321–346.
Martin, R. A. 1977. Late Pleistocene *Eumops* from Florida. Bull. New Jersey Acad. Sci. 22:18–19.
Morgan, G. S. 1985. Fossil bats (Mammalia: Chiroptera) from the late Pleistocene and Holocene Vero Fauna, Indian River County, Florida. Brimleyana 11:97–117.
Norberg, U. M., and J. M. V. Rayner. 1987. Ecological morphology and flight in bats (Mammalia: Chiroptera): Wing adaptations, flight performance, foraging strategy and echolocation. Phil. Trans. R. Soc. London B. 316:335–427.
Owre, O. T. 1978. The Florida mastiff bat, *Eumops glaucinus floridanus*. Pp. 43–44 *in* J. N. Layne, ed., Rare and endangered biota of Florida. Vol. 1. Mammals. University Presses of Florida, Gainesville. xx + 52 pp.
Robson, M. S., F. J. Mazzotti, and T. Parrott. 1989. Recent evidence of the mastiff bat in southern Florida. Florida Field Nat. 17:81–82.
Silva-Taboada, G. 1979. Los murciélagos de Cuba. Editorial Academia, Havana. 421 pp.
Simmons, J. A., M. B. Fenton, and M. J. O'Farrell. 1979. Echolocation and pursuit of prey by bats. Science 203:16–21.
Tamsitt, J. R., and D. Valdivieso. 1963. Records and observations on Colombian bats. J. Mammal. 44:168–180.
Tyson, E. L. 1964. Two new records of bats (Molossidae) from Panama. J. Mammal. 45:495–496.

Prepared by: Jacqueline J. Belwood, Bat Conservation International, P.O. Box 162603, Austin, TX 78716.

Threatened

Big Cypress Fox Squirrel
Sciurus niger avicennia
FAMILY SCIURIDAE
Order Rodentia

TAXONOMY: *S. n. avicennia* was described from a single specimen by Howell (1919), and the taxon was reevaluated by Moore (1956) based on 56 specimens. Moore confirmed the distinctiveness of *avicennia* by careful quantitative analysis, and he stated that fox squirrels from the Miami Ridge were intergrades of *avicennia* and *shermani*. The subspecific name and often-used common name of mangrove fox squirrel incorrectly imply restriction to one habitat. To reflect its approximate distribution, the common name Big Cypress fox squirrel is preferable. D.A. Turner and J. Laerm (personal communication) also have confirmed the distinctiveness of *avicennia* by rigorous statistical analysis of cranial characters.

DESCRIPTION: The Big Cypress fox squirrel is distinctly smaller than Sherman's fox squirrel (Moore 1956). Diagnostic skull measurements are a condylobasal length less than 62 mm and a mastoid breadth less than 27 mm. The head plus body averages 278 mm long versus 315 mm for *S. n. shermani*, and the hind foot is 75 mm long instead of 85 mm. Like other subspecies of the fox squirrel in the southeastern United States, the Big Cypress fox squirrel is highly variable in color, ranging from buff to black. The most frequent color phase of this subspecies is buff, with a buff venter and buff basal bands on the tail hairs, white toes, lips, nose, and eartips, a black or blackish crown, and an agouti or blackish-agouti back. Second in frequency is black or a black-dominated intermediate between black and buff. In the black phase, all the pelage is black except for white lips, nose, and eartips, white or blackish toes, and blackish-agouti feet. The least frequent color phase is tan, with a tan venter and tan basal bands of the tail hairs, white lips, nose, and eartips, white to tan limbs, and agouti cheeks, nape, back, and sides.

Sciurus niger avicennia

Big Cypress fox squirrel, *Sciurus niger avicennia*. (Photo by Patrick G. R. Jodice)

POPULATION SIZE AND TREND: In a 1978 field survey based on interviews of local residents, Williams and Humphrey (1979) reported approximately 340 individuals. Because most observations were made along roads, presumably many more were present but not detected. Williams and Humphrey recorded crude density estimates at two sites: 4 on 4338 ha (1/1084 ha) at Corkscrew Swamp Sanctuary and 30 on 1554 (1/52 ha) on a ranch in Hendry County. Both values are much too low, because they included large areas of unoccupied habitat (wetlands and pastures). Field surveys conducted in the Big Cypress National Preserve during 1989-90, along with questionnaire responses and interviews with Big Cypress users, also indicate low densities of fox squirrels (Jodice 1990). Thus densities appear to be quite low, and on this basis the subspecies can be considered inherently rare. Based on a questionnaire survey of wildlife officials, Brady (1977) reported populations to be stable in Monroe, Collier, and Hendry counties, but found no consensus of opinion

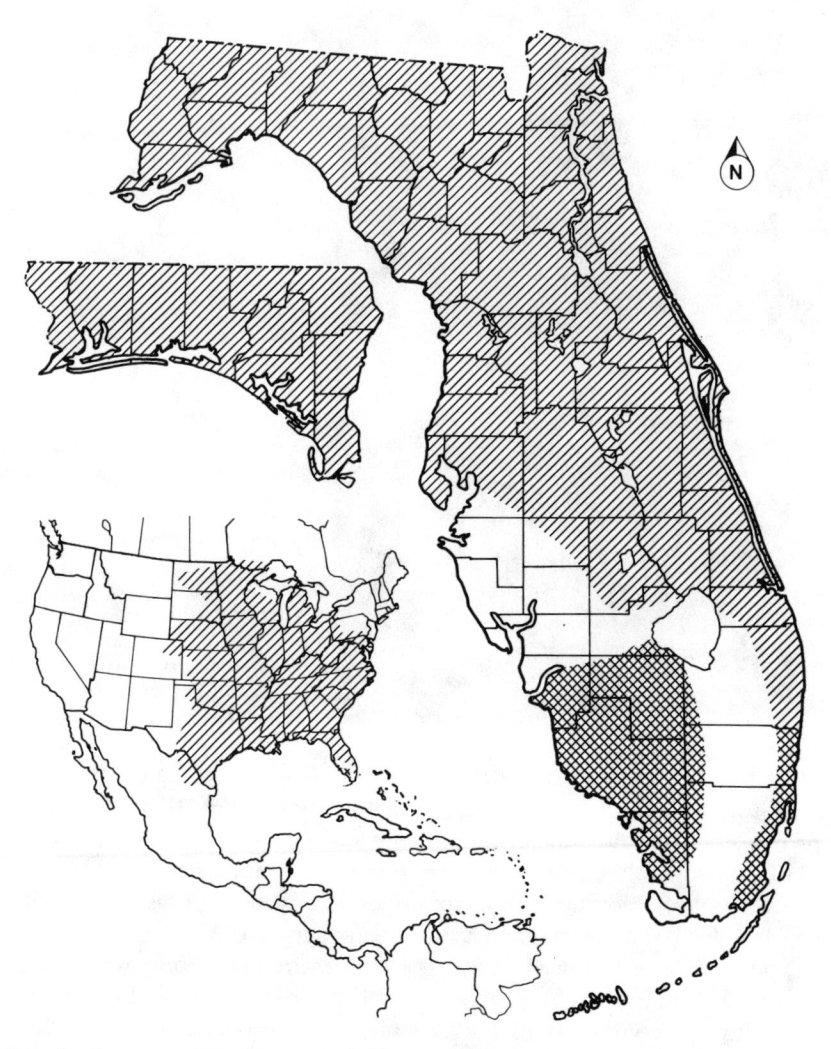

Distribution map of the Big Cypress fox squirrel, *Sciurus niger avicennia*. Hatching represents the species; crosshatching represents the subspecies. Crosshatching east of the Everglades represents intergrades with another subspecies.

for Lee County. Subsequently, ranchers in Hendry County reported a population increase following prohibition of hunting (Williams and Humphrey 1979). Since the Big Cypress National Preserve was established in 1974, preserve staff have recorded progressively fewer fox squirrels, concluding that the population is not prospering there.

DISTRIBUTION AND HISTORY OF DISTRIBUTION: This subspecies occurs in the Immokalee Rise, Big Cypress Swamp, and Devil's Garden areas of Lee, Hendry, Collier, Monroe, and extreme western Dade counties in southwestern Florida. This distribution is south of the Caloosahatchee River and west of the Everglades, and disjunct from the range of other subspecies of fox squirrels. Former populations in eastern and southern Dade County on pinelands from Miami to Paradise Key were intergrades (Howell 1919; Moore 1956) between *S. n. avicennia* and *S. n. shermani*; these populations apparently are extirpated, as the last published report was from 1951 (Moore 1956). The recent distribution (Williams and Humphrey 1979) of *avicennia* proper, however, is essentially unchanged from that described by Moore (1956). Some areas in this range have become vacated (e.g., Corkscrew Swamp Sanctuary, Everglades City), while many other suitable areas are being altered or becoming isolated through development.

GEOGRAPHICAL STATUS: This subspecies is endemic to southwestern Florida, but the species is widespread in eastern and central North America.

HABITAT REQUIREMENTS AND HABITAT TREND: Habitat use by the Big Cypress fox squirrel is complex and poorly understood. The subspecies uses most types of forest occurring in its range. Only the dense interiors of mixed cypress-hardwood strands seem to be avoided—these are occupied by dense populations of gray squirrels (*Sciurus carolinensis*). *S. n. avicennia* has been reported in cypress swamp, pine flatwoods, tropical hardwood forest, live oak woods, mangrove forest, and suburban habitats including golf courses, city parks, and residential areas in native vegetation (Williams and Humphrey 1979). Moore (1956) judged that mangrove habitat would not provide suitable food and water year-round. J. Beever (personal communication) stated that fox squirrels use black mangrove habitat with open understory during autumn seed set, and fringing red mangrove less often.

Few data on diet are available. Slash pine (*Pinus elliotii* var. *densa*) seems to be a preferred food source for Big Cypress fox squirrels. Individuals have been observed feeding on microstrobili (male cones) in winter, while seeds from megastrobili (female cones) form the bulk of the diet during the summer rainy season. Mega- and microstrobili from cypress (*Taxodium distichum*) are consumed in autumn and early winter, respectively. Cabbage palm (*Sabal palmetto*) fruits, bromeliad buds, and acorns also have been recorded as food items. Fox squirrels on golf courses in Naples have been observed eating all of the above, plus, on a regular basis, queen palm (*Cocos pamosa*) fruits, fig fruits (*Ficus* spp.), and fungi. A variety of exotic plant species are also consumed in small quantities. Ground foraging is also common. Caching food has been observed, with palm fruits being cached frequently. Caching of pine seeds has not been observed.

Platform nests in pines and hardwoods and moss and stick nests in cypress, tops of cabbage palms, and large clumps of bromeliads have been observed. Cabbage palms and bromeliads seem especially important, because they allow fox squirrels to range over large areas without the need to construct nests on a daily basis. Fox squirrels often strip bark from cypress trees for incorporation into the nest (Jodice 1990). This is an excellent indication of the presence of fox or gray squirrels.

Many questions about habitat use remain unanswered. For example, overprotection of the Big Cypress from fire may be responsible for the apparent decline of fox squirrels there.

VULNERABILITY OF SPECIES AND HABITAT: Hunter survey records of the Florida Game and Fresh Water Fish Commission estimated the 3-year harvest during 1952–55 at 258 individuals in 5 management areas of the Big Cypress Swamp. Like other kinds of fox squirrels (Baumgartner 1943; Nixon et al. 1974), this subspecies is vulnerable to overhunting. Habitat of this species is vulnerable; although it is mostly forested wetlands and hence relatively expensive to convert to other uses, still that can be done.

CAUSES OF THREAT: Early reports (Brown 1973, 1978) considered habitat destruction to endanger the Big Cypress fox squirrel. While public acquisition of large parcels of land in the subspecies' range has reduced the maximum potential of this threat, large-scale commercial and residential development of pinelands west of the Big Cypress National Preserve (specifically the Naples area), conversion of forested wetlands to citrus north of the Big Cypress National Preserve, and expansion of Interstate Highway 75 (along with increased local traffic on State Road 29) still pose serious threats to habitat quality and quantity. This loss of habitat will make the distribution smaller and will further isolate remaining populations of squirrels. Then the major problem will be whether each fragment of the population can persist locally. At Corkscrew Swamp Sanctuary, for example, where the former population was recently extirpated by unknown forces, recolonization may be prevented by the site's isolation from current fox squirrel range by agricultural development. Some of the populations at golf courses in Naples now are completely isolated from source and dispersal areas by urban growth that has claimed native habitats formerly connecting the golf courses. Future development is expected to isolate other golf course populations that still are near undeveloped pinelands.

RESPONSES TO HABITAT MODIFICATION: Ground fires apparently are valuable to the habitats of Big Cypress fox squirrels (Robertson 1953; Brown 1978) because they retard plant succession, but this relationship has not been studied. Suppression of wild fires has been the historical norm in the Big Cypress Swamp. Consequently, many pinelands in the Big Cypress National Preserve currently have a dense understory of palmetto, which is considered an undesirable habitat for fox squirrels. In the 1980s, fires typically burned only about 40,000 ac a year in the Big Cypress National Preserve (20,000 ac prescribed, plus 20,000 ac wildfire and arson usually suppressed). Future fire management plans call for an increase in prescribed burning to 50,000 ac a year. Pinelands are expected to be burned on a 5-7 year rotation, providing a more open understory and improved seed bed for slash pine seedlings. This management plan may improve habitat quality and quantity for fox squirrels. Similarly, light grazing maintains the sparse ground cover favored by fox squirrels (Baumgartner 1938; Allen 1943); it probably contributes to the value of ranchland oak habitat (Williams and Humphrey 1979). Large populations occurring at golf courses show that populations can exist in park-like habitat protected from poaching and dogs. Recently attention has been focused on fox squirrel populations in residential Naples. Fox squirrels are commonly reported in the Golden Gate area, a pinewoods undergoing development but temporarily retaining much undeveloped land. In a recent mail survey of 30 golf courses in Naples, 10 reported regular observations of fox squirrels (Jodice 1990). Golf courses may represent an important residential habitat and may provide suitable resources. It is too early to judge exactly how golf courses affect fox squirrel populations. Golf course design, vegetation type, housing density, and, possibly most important, access to other squirrel populations in adjoining golf courses or undeveloped habitats, all may play critical roles in determining long-term viability of urban fox squirrel populations. If golf-course design can be influenced to provide habitat components necessary for fox squirrels, golf courses may provide acceptable habitats for fox squirrels. Much more information is needed on responses of this subspecies to various kinds of habitat modification.

DEMOGRAPHIC CHARACTERISTICS: Few data are available on demographic characteristics. Mating has been observed on golf courses in June, July, and August. Young fox squirrels have been observed out of the nest in September and March (Jodice 1990).

KEY BEHAVIORS: Like other fox squirrels, this subspecies spends much time on the ground, accounting for its vulnerability to hunting and the value of fire and grazing in helping squirrels see and escape from predators. During the winter dry season, Williams and Humphrey (1979) noted that individuals retreated to cypress habitat at midday, after foraging in pine flatwoods all morning. During the summer rainy season, staff at Corkscrew Swamp Sanctuary reported the squirrels to "disappear," perhaps representing a shift in habitat use. Seasonal fluctuations in water levels and shifting food resources may affect foraging behavior, seasonal habitat use, and travel.

Observations of *avicennia* on golf courses in Naples have provided new information on behavior. Fox squirrels spend slightly more time active than inactive, although this varies with season, food availability, and weather. Foraging is the most frequent activity. Foraging is terrestrial and arboreal.

Inactive time varies seasonally, peaking in the humid summer months. Diurnal inactivity may last from a few minutes between foraging bouts to several hours during extreme weather. During warmer weather fox squirrels often lie draped over a branch or palm frond, with legs dangling. Animals resting this way often remain motionless even when approached by humans. During cooler weather, fox squirrels perch on branches, covering their backs with their tails.

Mating chases are a large proportion of social interactions and are observed most frequently during May to August. Mating chases resemble those of other squirrel species, with two or more males chasing a female. Vocalizing is common during this time, but takes place infrequently outside of mating season. Cheek rubbing, a form of scent marking (Benson 1980), is often observed during this time but also occurs throughout the year.

Nonmating social interactions on golf courses are infrequent; those that do occur are typically observed near a food source. One individual may approach or chase another. Individuals often do not leave an area when chased, but continue to feed after a brief interaction. Chases may occur repetitively with similar outcomes. Some individuals appear to be more aggressive than others in initiating chases. It is not uncommon, however, to observe three or four fox squirrels feeding on a golf course within a 10-m area with no aggression.

Fox squirrels on golf courses spend about 10 percent of their time traveling, often between feeding sites or to a nest, usually on the ground. Before traveling across the ground, fox squirrels will often position themselves on a tree trunk 2–3 m above ground in an alert posture. It is common to observe golf course fox squirrels traveling 200–300 m, frequently pausing to take up a bipedal alert posture. Lack of understory and closed canopy on golf courses inhibit extensive travel through the trees.

Fox squirrels in the Big Cypress National Preserve seem to travel extensively through the trees as well as along the ground. During the wet season, radio-collared fox squirrels were observed traveling long distances in cypress strands through the canopy, usually by crossing at overlapping branches (although they would also make leaps of over 2 m). Ground travel becomes more common when water is low. Telemetry data from translocated fox squirrels indicate that traveling >2 km/day over dry ground may not be uncommon (Jodice 1990). Individuals have been observed walking through shallow water but seem to prefer staying dry when possible.

Fox squirrels react in a variety of ways to disturbances. The most common reaction seems to be for the squirrel to climb a tree and sit motionless and silent until the disturbance passes. Fox squirrels rarely vocalize when disturbed. Lack of vocalization and movement when disturbed may add to the difficulty in observing fox squirrels in the wild.

CONSERVATION MEASURES TAKEN: This subspecies is listed as threatened by the Florida Game and Fresh Water Fish Commission (1990), and it is under review for possible listing by the U.S. Fish and Wildlife Service (1989). Although the Big Cypress fox squirrel is protected from hunting by state law, Williams and Humphrey (1979) reported that poaching was frequent. Recent questionnaire responses and hunter-exit interviews indicate that this is still true (P. G. R. Jodice, unpubl. data). Numerous illegal hunting camps are being removed from the backcountry of the Big Cypress National Preserve. This may benefit fox squirrels by reducing the level of poaching, lessening human effects on hardwood hammocks where many of the camps are located, and allowing a more liberal approach to fire suppression. Very substantial portions of the subspecies' range have been purchased for conservation purposes, including Corkscrew Swamp Sanctuary, Everglades National Park, Collier-Seminole State Park, Big Cypress National Preserve, Fakahatchee Strand State Preserve, and Florida Panther National Wildlife Refuge.

CONSERVATION MEASURES PROPOSED: Existing lands dedicated to conservation should be enlarged and consolidated, including the planned acquisitions of Big Cypress National Preserve additions, Fakahatchee Strand State Preserve completion and additions, and Bird Rookery Swamp-Flintpen Strand. Strong protection from poaching should be provided. A study of habitat use and diet would help management agencies protect the populations under their care. The relationship of burning to habitat use by fox squirrels should be documented in the course of accelerated fire management of the Big Cypress National Preserve. Further study of the populations on golf courses and on adjacent undeveloped lands would help to clarify fox squirrel use of available habitat. Maintaining these suburban populations may be critical for the subspecies' viability.

ACKNOWLEDGMENTS: The careful work of Joseph C. Moore on the nature of this taxon and its natural history has been a strong basis for understanding the Big Cypress fox squirrel. Subsequent field surveys by Kathleen S. Williams and Lance Gunderson have provided the monitoring necessary to judge its survival status.

Literature Cited

Allen, D. L. 1943. Michigan fox squirrel management. Mich. Dept. Cons., Game Division Publ. 100:1–404.
Baumgartner, L. L. 1938. Population studies of the fox squirrel in Ohio. Trans. N. Amer. Wildl. Conf. 3:685–689.
Baumgartner, L. L. 1943. Population studies of the fox squirrel in Ohio. Trans. Third North Amer. Wildlife Conf.
Benson, B. N. 1980. Dominance relationships, mating behaviour and scent marking in fox squirrels (*Sciurus niger*). Mammalia 44:143–160.
Brady, J. R. 1977. Status of the fox squirrel in Florida. Pittman-Robertson Federal Aid Project W-41-24, Job XV-A-1. Progress report, Florida Game and Fresh Water Fish Commission, 4005 S. Main Street, Gainesville.
Brown, L. N. 1973. The Everglades fox squirrel (*Sciurus niger avicennia*). Pp. 222–223 *in* Threatened wildlife of the United States. U.S. Fish and Wildlife Serv., Resource Publ. 114.

Brown, L. N. 1978. Mangrove fox squirrel. Pp. 5–6 *in* J. N. Layne, ed., Rare and endangered biota of Florida. Vol. 1. Mammals. University Presses of Florida, Gainesville. xx + 52 pp.
Florida Game and Fresh Water Fish Commission. 1990. Official lists of endangered and potentially endangered fauna and flora in Florida. Tallahassee, Florida. 23 pp.
Howell, A. H. 1919. Notes on the fox squirrels of the southeastern United States, with description of a new form from Florida. J. Mammal. 1:36–38.
Jodice, P. G. R. 1990. Ecology and translocation of urban populations of Big Cypress fox squirrels (*Sciurus niger avicennia*). Unpubl. M.S. thesis, University of Florida, Gainesville. 89 pp.
Moore, J. C. 1956. Variation in the fox squirrel in Florida. Amer. Midland Nat. 55:41–65.
Nixon, C. M., R. W. Donohoe, and T. Nash. 1974. Overharvest of fox squirrels from two woodlots in western Ohio. J. Wildl. Manage. 38:67–80.
Robertson, W. B., Jr. 1953. A survey of the effects of fire in Everglades National Park. U.S. Dept. Interior, Nat. Park Serv.
U.S. Fish and Wildlife Service. 1989. Endangered and threatened wildlife and plants; animal notice of review. Federal Register 54:554–579.
Williams, K. S., and S. R. Humphrey. 1979. Distribution and status of the endangered big cypress fox squirrel (*Sciurus niger avicennia*) in Florida. Florida Sci. 42:201–205.

Prepared by: Stephen R. Humphrey, Florida Museum of Natural History, University of Florida, Gainesville, FL 32611; and Patrick G. R. Jodice, Florida Game and Fresh Water Fish Commission, Route 7, Box 440, Lake City, FL 32055.

Threatened

Sherman's Fox Squirrel
Sciurus niger shermani
FAMILY SCIURIDAE
Order Rodentia

TAXONOMY: No other names. Sherman's is one of three subspecies of fox squirrel in Florida. Others are *S. n. niger*, found in the panhandle west of the Aucilla River, and the Big Cypress fox squirrel (*S. n. avicennia*), in southwestern Florida (Moore 1956). Both are smaller than *S. n. shermani*, and the venter of *S. n. niger* is white or whitish rather than tan. The distinctness of *shermani* as a subspecies has been confirmed by rigorous statistical evaluation of cranial characters (D. A. Turner and J. Laerm, personal communication).

DESCRIPTION: The fox squirrels of the southeastern coastal plain are the largest and most variably colored tree squirrels in the western hemisphere (Nowak and Paradiso 1983). They are quite different from midwestern fox squirrels in size, appearance, and ecology. Sherman's is the largest of the southeastern fox squirrels. Adults weigh 900–1200 g and measure 600–700 mm in total length. The top of the head is typically black, with nose and ears white. The remainder of the pelage is quite variable, falling within one of three basic color morphs: all dark (agouti to black) in 1/16 of fox squirrels examined, all tan (light agouti to tan) in 9/16, and evenly divided between dark over tan and tan over dark in 6/16.

POPULATION SIZE AND TREND: Total population size of Sherman's fox squirrel is unknown. Populations have declined and continue to do so in proportion to loss of the squirrel's primary habitat: the mature, fire-maintained longleaf pine (*Pinus palustris*)-turkey oak (*Quercus laevis*) sandhill and flatwood communities. Based on the amount of known habitat loss, fox squirrel populations have undoubtedly declined at least 85 per-

Sherman's fox squirrel, *Sciurus niger shermani*.
(Photo by John B. Wooding)

cent from presettlement levels. A questionnaire survey conducted by Brady (1977) indicated that population levels of fox squirrels were only "low to medium" and were declining over much of their range. Densities are relatively low, reported as $38/km^2$ (Moore 1957) and $8-15/km^2$ (Humphrey et al. 1985; Kantola 1986).

DISTRIBUTION AND HISTORY OF DISTRIBUTION: Moore (1956) described the range of this squirrel as most of peninsular Florida, extending north into at least Nassau County, possibly into southeastern Georgia, west into Gilchrist and Levy counties, and south to a line extending from Tampa Bay through Lake Okeechobee. D. A. Turner and J. Laerm (personal communication) have extended the range of this subspecies north into the piedmont of Georgia.

GEOGRAPHICAL STATUS: The subspecies is endemic to northern and central Florida (Allen 1871; Maynard 1872; Bangs 1896). Historically

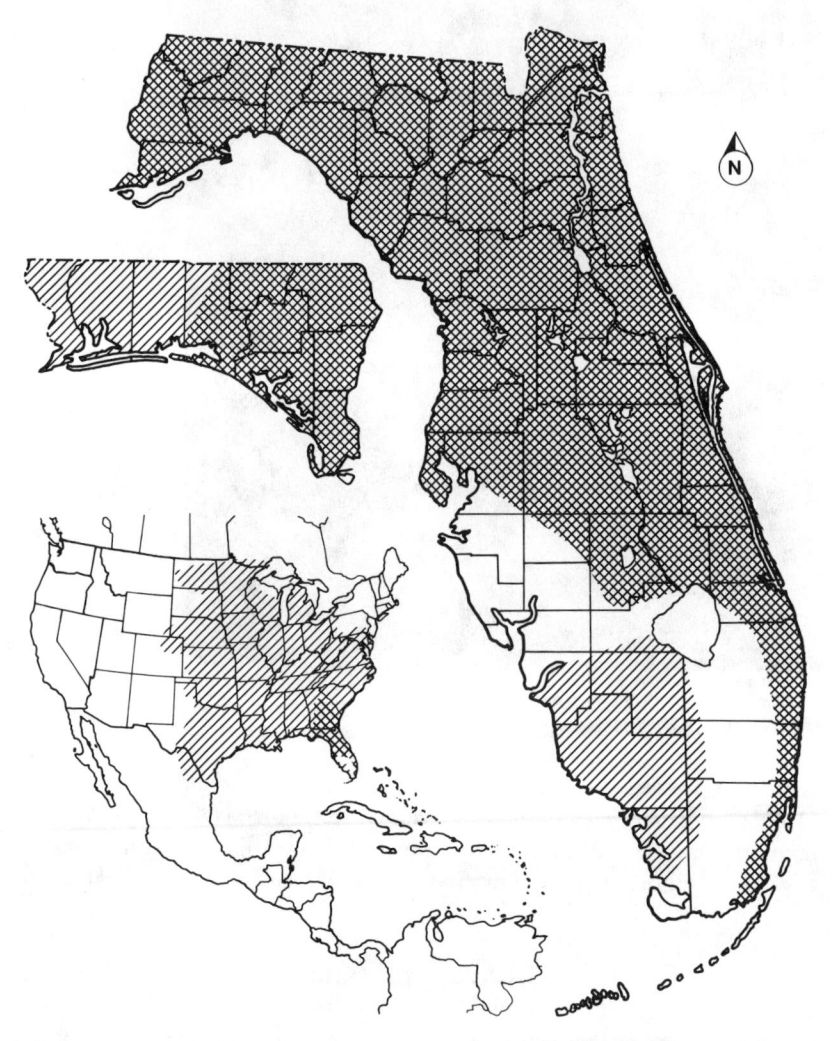

Distribution map of Sherman's fox squirrel, *Sciurus niger shermani*. Hatching represents the species; crosshatching represents the subspecies. Crosshatching east of the Everglades represents intergrades with another subspecies.

Sherman's fox squirrel was common throughout the extensive sandhills and pine flatwoods.

HABITAT REQUIREMENTS AND HABITAT TREND: The mature, fire-maintained longleaf pine-turkey oak sandhills and flatwoods are the optimal habitat for Sherman's fox squirrels. Only 10–20 percent of the original habitat is still intact, however, having been greatly altered through extensive logging; conversion to pasture, single-stand short-rotation forestry, agricultural, commercial, and residential development; and by lack of fire (Bechtold and Knight 1982). Man's intense and widespread modification of the mature pine-oak communities of the sandhills has fragmented fox squirrel populations and reduced the quality of most remaining habitat. As habitat quality decreases, the area required to support viable squirrel populations increases. To accommodate the squirrel's large home range and varied food resources, suitable habitat must be fairly extensive and, in addition to the drier hilltops (upland), it should include the more productive lower slopes of the sandhills, where the predominant longleaf pines and turkey oaks are interspersed with sand post oak (*Quercus stellata* var. *margaretta*), live oak (*Q. virginiana*), laurel oak (*Q. hemisphaerica*), and bluejack oak (*Q. incana*). Kantola and Humphrey (1990) suggested that the highest-quality habitat might be along the edge of longleaf pine savanna and live oak forest, because live oak acorns appear to be a major food source when turkey oak acorn crops fail.

Moore (1957) considered longleaf pine seeds and turkey oak acorns to be the primary foods of Sherman's fox squirrel. Kantola and Humphrey (1990), however, found production of these two seed crops to be extremely patchy, varying considerably from year to year and site to site. During mast failures of turkey oak, squirrels moved downslope to include live oak forest in their home ranges during the mast season. Consequently acorns of live oak appear to be a major component of the diet. Pine cones are cut from the trees from late May through October, beginning while they are still green. Acorns are harvested from both the trees and ground beginning in late September, and some are cached for later use. Other acorns and nuts, fungi, bulbs, vegetative buds, insects, and staminate pine cones also are eaten. Squirrel reproduction, and thus densities, may be expected to vary with resource abundance (Weigl et al. 1989).

Tree cavities occasionally are used for nesting, but they apparently are not as important for Sherman's fox squirrel as they are for more northern squirrels. Instead, leaf nests are used extensively. These usually are located in large oaks and often contain Spanish moss (*Tillandsia usneoides*), which provides insulation. Turkey oaks in the low slopes of the sandhills typically are larger, have more Spanish moss, and produce more acorns than those in the upland, making the low slopes a more important component of fox squirrel habitat.

VULNERABILITY OF SPECIES AND HABITAT: Populations of Sherman's fox squirrel are highly vulnerable to habitat loss. The remaining habitat continues to be threatened by development. Under some conditions, hunting also may be detrimental to local populations (Harper 1927; Moore 1953, 1954; J. Reinman, personal communication).

CAUSES OF THREAT: The already extensive loss of fox squirrel habitat is continuing, further reducing and fragmenting fox squirrel populations. Hunting pressure may threaten isolated populations that have low potential for recolonization. Many hunters who pursue this large, elusive, and variably-colored squirrel do so in search of challenge and trophy rather than solely for food.

RESPONSES TO HABITAT MODIFICATION: The pine monoculture to which much of the fox squirrel's original sandhill habitat has been converted does not supply the varied food resources necessary to support the squirrel. Agricultural, commercial, and residential developments have destroyed and fragmented the large pine-oak areas needed by fox squirrels, fragmenting their populations. Forests protected from fire undergo succession to closed-canopy forests that do not support fox squirrels. Such forests are characterized by dense undergrowth and many small turkey oaks that compete with longleaf pines by preventing their germination and impeding growth. (Acorn production also drops in such situations.) Fox squirrels benefit from prescribed burning of the sandhills every two to three years to control the understory growth and prepare the seedbed for longleaf pines.

DEMOGRAPHIC CHARACTERISTICS: Typically, two breeding seasons occur each year (Moore 1957). Kantola (1986) found that fox squirrels on the Ordway Preserve apparently did not breed in the winter season following a failure of the turkey oak acorn crop. The fairly well-defined winter breeding season usually begins with conception near the end of November; young are born in mid-January. The summer breeding season appears to be spread over a longer period, beginning with conception in May or June. Mean litter size is only 2.3 young (Moore 1957) compared with 2.5–3.2 for the midwestern fox squirrel (*S. n. rufiventer*). The young squirrels' eyes open in about 40 days. They stay in the brood nest roughly 75 days, and are weaned at about 90 days. Sexual maturity is reached at about 9 months; consequently, squirrels conceived in either the winter or summer breeding season cannot reproduce until they are 1 year old. Longevity is unknown, but captives have lived 10 years and longer (Moore 1957). Principal predators may include the bald eagle, red-tailed hawk, and man. Predation is probably not a limiting factor, however, because Sherman's fox squirrels are fairly adept at avoiding capture.

KEY BEHAVIORS: Sherman's fox squirrels are diurnal, solitary animals. Adults generally are seen together only during the breeding season. At this time they exhibit mating chases, with several males pursuing a female, competing for position to mate with her. Individuals cover large areas to exploit patchy food resources. This squirrel is not strictly territorial but has a large home range. Home range size averages 30 ha (40 ha for males, 20 ha for females; Kantola and Humphrey 1990). In contrast, the home range of midwestern fox squirrels averages only 0.8–7.0 ha (Baumgartner 1943; Bernard 1972; Adams 1976; Havera and Nixon 1978). Although home ranges of Sherman's fox squirrels overlap somewhat, core use areas of home range generally are separate. Size, position, and use of the home range may vary with seasonal food abundance, reproductive activity, and weather.

Sherman's fox squirrels use mostly leaf nests rather than cavities, and may use up to 30 nests per year. More nests occur in low slopes of the sandhills than in upland, indicating higher squirrel densities in the low slopes.

CONSERVATION MEASURES TAKEN: This subspecies is currently under review by the U.S. Fish and Wildlife Service for listing as threatened or endangered under the Endangered Species Act. It has been placed in Category 2—species known to be subject to some threat of extinction but for which more data are needed before a status designation can be made. Based on the trend of habitat loss, the state of Florida has listed Sherman's fox squirrel as a species of special concern. Hunting is still permitted in recognition that the practice is traditional and not the primary cause of threat.

CONSERVATION MEASURES PROPOSED: Habitat preservation and reclamation are the most needed conservation actions for Sherman's fox squirrel. Large areas (at least 25 km^2) of heterogeneous, natural sandhill habitat are recommended to preserve viable populations (Kantola 1986).

Such refuges should include vital lower slopes as well as upland. Management of fox squirrel habitat also must include prescribed summer burning every two to three years to regulate turkey oaks and maintain longleaf pines.

Status surveys (counting nests or squirrels) are needed to determine population levels throughout the squirrel's reported range, particularly where habitat loss has been great or hunting is common. The effects of hunting, particularly on small or closed populations, should be evaluated, and heavily hunted populations should be monitored closely.

ACKNOWLEDGMENTS: The work of Joseph C. Moore has contributed much to our knowledge of Sherman's fox squirrel. Additional data were gathered in recent research supported by the Florida Museum of Natural History and the U.S. Fish and Wildlife Service. Reviewers S. R. Humphrey, L. W. Lefebvre, and T. W. Turnipseed provided valuable comments on the manuscript. The author of the species account in the 1978 edition was Llewellyn M. Ehrhart.

Literature Cited

Adams, C. E. 1976. Measurement and characteristics of fox squirrel, *Sciurus niger rufiventer*, home ranges. Amer. Midland Nat. 95:211–215.

Allen, J. A. 1871. On the mammals and winter birds of east Florida with an examination of certain assumed specific characters in birds and a sketch of the bird fauna of eastern North America. Bull. Mus. Comp. Zool. 2:161–540.

Bangs, O. 1896. A review of the squirrels of eastern North America. Proc. Biol. Soc. Wash. 10:145–167.

Baumgartner, L. L. 1943. Fox squirrels in Ohio. J. Wildl. Manage. 7:193–202.

Bechtold, W. A., and H. A. Knight. 1982. Florida's forests. U.S. Dep. Agric., For. Serv. Resour. Bull. SE-62:1–84.

Bernard, R. J. 1972. Social organization of the western fox squirrel. Unpubl. M.S. thesis, Michigan State University, East Lansing. 41 pp.

Brady, J. R. 1977. Status of the fox squirrel in Florida. Pittman-Robertson Federal Aid Proj. W-41-24, Job XV-A-1. Project report in Florida Game and Fresh Water Fish Commission files, Gainesville.

Harper, F. 1927. The mammals of the Okefenokee Swamp Region of Georgia. Proc. Boston Soc. Nat. Hist. 38:191–396.

Havera, S. P., and C. M. Nixon. 1978. Interaction among adult female fox squirrels during their winter breeding season. Trans. Illinois State Acad. Sci. 71:24–38.

Humphrey, S. R., J. F. Eisenberg, and R. Franz. 1985. Possibilities for restoring wildlife of a longleaf pine savanna in an abandoned citrus grove. Wildl. Soc. Bull. 13:487–496.

Kantola, A. T. 1986. Fox squirrel home range and mast crops in Florida. Unpubl. M.S. thesis, University of Florida, Gainesville. 68 pp.

Kantola, A. T., and S. R. Humphrey. 1990. Habitat use by Sherman's fox squirrel (*Sciurus niger shermani*) in Florida. J. Mammal. 71:411–419.

Maynard, C. J. 1872. Catalog of the mammals of Florida with notes on their habits, distribution, etc. Bull. Essex Inst. 4:135–150.

Moore, J. C. 1953. The fox squirrel in Florida: Variation and natural history. Unpubl. Ph.D. diss., University of Florida, Gainesville. 203 pp.

Moore, J. C. 1954. Fox squirrel receptionists. Everglades Nat. Hist. 2:152–160.

Moore, J. C. 1956. Variation in the fox squirrel in Florida. Amer. Midland Nat. 55:41–65.

Moore, J. C. 1957. The natural history of the fox squirrel *Sciurus niger shermani*. Bull. Amer. Mus. Nat. Hist. 113:1–71.

Nowak, P. M., and J. L. Paradiso. 1983. Walker's mammals of the world. 4th ed. Johns Hopkins University Press, Baltimore.

Weigl, P. D., M. A. Steele, L. J. Sherman, J. C. Ha, and T. S. Sharpe. 1989. The ecology of the fox squirrel (*Sciurus niger*) in North Carolina: Implications for survival in the Southeast. Tall Timbers Bull. Res. Sta. 24:1–93.

Prepared by: Angela T. Kantola, Fish and Wildlife Enhancement, U.S. Fish and Wildlife Service, P.O. Box 25486, Denver Federal Center, Lakewood, CO 80225.

Threatened

Southeastern Beach Mouse
Peromyscus polionotus niveiventris
FAMILY CRICETIDAE
Order Rodentia

TAXONOMY: The taxon was first described by Chapman (1889) as *Hesperomys niveiventris*. Allen (1898) referred to the same form as the white-bellied Florida deermouse (*Peromyscus niveiventris*). Osgood (1909) assigned it to the genus *Peromyscus* and named it *P. polionotus niveiventris*, the beach mouse. Sherman (1937) used *P. p. niveiventris* and the common name Micco beach mouse. Hall (1981) recognized 16 subspecies of *P. polionotus*, including *niveiventris*.

DESCRIPTION: The overall appearance of *P. p. niveiventris* is that of a light, buffy *Peromyscus* with a strikingly white venter. The white hair of the venter extends well up the flanks, high on the jaw, and within 2–3 mm of the eye. Adults have buffy hair on the nasal, periorbital, and temporal regions; these hairs are buffy to the base. The dorsum is buffy from the back of the head to the tail. Hairs on the dorsum are tipped with buff and have dark gray bases. The tail is bicolored, white below and buffy above. Average measurements of 12 females and 19 males are, respectively, total length 138.6 and 134.0 mm, tail length 55.5 and 52.7 mm, length of hind foot 18.5 and 18.4 mm, length from tip of ear to notch 13.9 and 14.4 mm, and body mass 14.5 and 14.4 g.

POPULATION SIZE AND TREND: Grid-based study of *P. p. niveiventris* has been reported from the Canaveral Peninsula of Merritt Island, Brevard County (Stout 1979). Minimum numbers known to be alive on a 1.44-ha live-trapping grid (included both east and west aspects of the primary dune) in July over 5 years were as follows: 8 (1975), 16 (1976), 15 (1977), 27 (1978), and 29 (1979). The population peaked in March and April 1979, when the density reached 64 per ha (Extine and Stout 1987). A companion study of another grid located about 1 km inland revealed another population that was less numerous and more variable in abundance from month to month than the coastal one (Keim 1979; Extine 1980; Stout 1979). Eighty-one years earlier, Bangs (1898) had reported *P. p. niveiventris* to be extremely abundant from Palm Beach to Ponce (Mosquito) Inlet. Humphrey et al. (1987) found it in modest numbers from the Canaveral Peninsula north to coastal dunes opposite Mosquito Lagoon. These populations appear to be relatively stable. Additional trapping at Sebastian Inlet State Recreation Area (southern portion), Turtle Trail Public Beach Access, and Pepper Park revealed apparently small populations at Sebastian Inlet and Turtle Trail Public Beach Access. A more recent survey in 1988–89 by M. S. Robson and J. B. Miller (personal communication) corroborated the presence of *P. p. niveiventris* at Sebastian Inlet State Recreation Area and revealed its presence (1 capture in 123 trap-nights) at Ft. Pierce Inlet State Recreation Area in St. Lucie County. The local populations of Canaveral National Seashore and Cape Canaveral appear to be secure. In contrast, populations from Cape Canaveral to Ft. Pierce (St. Lucie County) appear to be highly fragmented and at low levels of abundance. Southeastern beach mice may be extirpated from their former range south of Ft. Pierce.

DISTRIBUTION AND HISTORY OF DISTRIBUTION: The type specimens were collected on the east peninsula, opposite Micco in Brevard County (Chapman 1889). Sherman (1952) reported the subspecies from Canaveral, north of Micco, and from Hillsboro Inlet, Palm Beach County. Layne (1974) reviewed the distribution in southern Florida; it has been recorded from Jupiter Inlet, Palm Beach, Lake Worth, Hillsboro Inlet, and Hollywood Beach, and may once have reached Miami Beach. The northern limit of records of *P. p. niveiventris* is from New Smyrna Beach at Ponce (Mosquito) Inlet, Volusia County. Hall (1981:669) incorrectly excluded the Volusia County portion of the range. Local distribution was confined to the sea oats (*Uniola paniculata*) zone and associated dune system. An exception was the unusually broad expanse of coastal strand and coastal scrub on the Canaveral Peninsula where *P. p. niveiventris*, at least in recent times, has occurred from the sea oats zone on the Atlantic side to the limit of xeric vegetation on the inland side adjacent to the Banana River (Stout 1979). No historic data indicate that the Atlantic coastal ridge was ever occupied. Extensive live trapping of intact vegetation on the Atlantic coastal ridge in Brevard, Martin, and Palm Beach counties has offered no evidence of *P. p. niveiventris* (I. J. Stout, personal observation). Two recent surveys have delimited the occurrence of *P. p. niveiventris*. Humphrey et al. (1987) found extant populations in Canaveral National Seashore and Cape Canaveral Air Force Station (Brevard County), the southern half of Sebastian Inlet State Recreation Area (Indian River County), Turtle Trail Public Beach Access, and Pepper Park (St. Lucie County). Local populations were not discovered on the East Peninsula of Cape Canaveral or in the northern half of Sebastian Inlet State Recreation Area (Brevard County). M. S. Robson and J. B. Miller

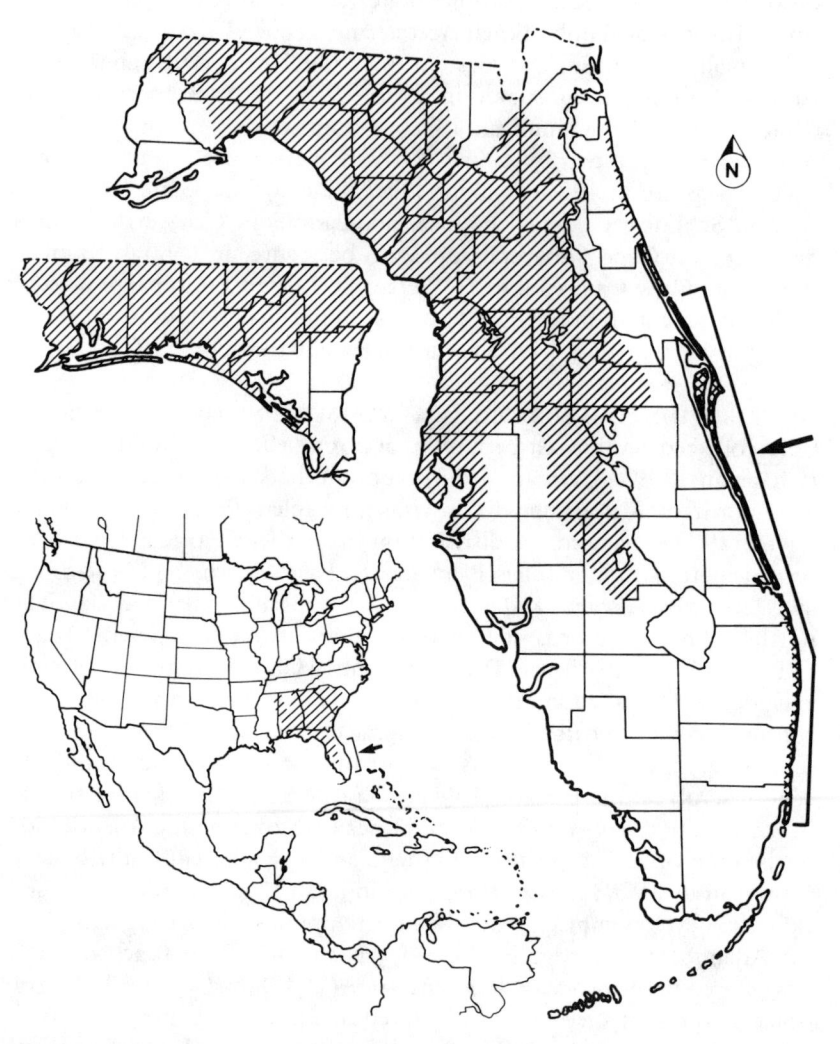

Distribution map of the southeastern beach mouse, *Peromyscus polionotus niveiventris*. Hatching represents the species; crosshatching and bracket represent the subspecies.

(personal communication) surveyed the northern and southern portions of Sebastian Inlet State Recreation Area (1988), Ft. Pierce Inlet State Recreation Area (1988), St. Lucie Inlet State Park (1988), MacArthur Beach State Park (1988), and Manalapan Beach, Palm Beach County (1989). The subspecies was found on the southern portion of Sebastian Inlet State Recreation Area and at Ft. Pierce Inlet State Recreation Area. These data suggest that about 225 km of coastline south of Ft. Pierce no longer have local populations.

GEOGRAPHICAL STATUS: This subspecies was endemic to coastal dunes along the ocean from south of Ponce Inlet in Volusia County to Hollywood Beach in Broward County. The apparent extinction of *P. p. decoloratus* (Humphrey and Barbour 1981), which occurred north of Ponce Inlet, leaves *P. p. niveiventris* isolated from other subspecies of *P. polionotus*.

HABITAT REQUIREMENTS AND HABITAT TREND: The sea oats zone of primary coastal dunes represents the principal habitat of *P. p. niveiventris*. Grassland and open sandy areas with scattered shrubs behind the foredune may be occupied in some locations (e.g., Canaveral National Seashore; I. J. Stout, personal observation). Inland from the coastline on Cape Canaveral, Brevard County, the southeastern beach mouse occurs in stands of woody shrubs dominated by oaks (*Quercus* spp.), rosemary (*Ceratiola ericoides*), and saw palmetto (*Serenoa repens*; Keim 1979; Stout 1979). South of Cape Canaveral, *P. p. niveiventris* was seldom live trapped outside the sea oats zone (Humphrey et al. 1987; M. S. Robson and J. B. Miller, personal communication). Good to excellent habitat for the subspecies exists along the Atlantic coast from the northern limit of Canaveral National Seashore, 8 km south of New Smyrna Beach, to the limit of federal land on Cape Canaveral. Some suitable habitat still exists on the East Peninsula, north and south of Sebastian Inlet, and north of Ft. Pierce Inlet. Habitat south of Jupiter may be too fragmented and isolated to support *P. p. niveiventris*. The Manalapan Beach between Lake Worth and Boynton Beach appears to be sufficiently large and undeveloped to support the subspecies, but a recent survey failed to capture a single animal (M. S. Robson and J. B. Miller, personal communication). The predicted future is for development of all private ocean-side lands within the range of the subspecies.

VULNERABILITY OF SPECIES AND HABITAT: It is very unlikely that *P. p. niveiventris* will persist anywhere in its range where development on private land occurs seaward of current setback regulations. Populations on county, state, and federal lands are less vulnerable because of existing regulations (such as the Federal Endangered Species Act of 1973). Loss of coastal dunes because of beach erosion continues to be a problem on private and public lands. Loss of the subspecies on private land will occur in spite of the existence of intact dune systems. Development brings domestic commensals (*Mus musculus, Felis catus, Canis familiaris*) increasingly in contact with *P. p. niveiventris* with inimical effects (see accounts of other subspecies of beach mice). Survival of the subspecies on state and federally owned land may be anticipated because habitat protection and management will occur as a matter of policy.

CAUSES OF THREAT: Loss of suitable habitat is the cause of local extirpation. Cumulative effects of real estate development, house cats, and house mice appear to render otherwise suitable habitat uninhabitable by *P. p. niveiventris* (see other accounts of *P. polionotus* subspecies for details). Many stretches of beach have lost the sea oats zone because of beach erosion. Sea-level rises due to global climatic changes may eliminate the remaining habitat within a century, in spite of the apparent short-term security of *P. p. niveiventris*. Alternatively, these mobile animals might be able to follow with the shifting sand as the coastline moves upslope.

RESPONSES TO HABITAT MODIFICATION: Oldfield mice occupy abandoned farmland and may achieve high densities on occasion (Smith 1968). The southeastern beach mouse on the Cape Canaveral portion of Merritt Island National Wildlife Refuge reoccupied coastal dunes where beach houses were common prior to government purchase for development of the John F. Kennedy Space Center. Foundations, remnant access roads, and partially recovered vegetation are evident where apparently normal populations of *P. p. niveiventris* exist. Powerline rights-of-way on Cape Canaveral are maintained by brush hogging weedy plants, and *P. p. niveiventris* flourish in spite of the disturbance (I. J. Stout, personal observation). At least two formerly cleared sites on Cape Canaveral are 1 km from the sea oats zone, yet both support *P. p. niveiventris* (I. J. Stout, personal observation). Elsewhere on the primary dunes, storms damage the vegetation and initiate grass and herb succession; these local conditions favor the subspecies. Conversely, *P. p. niveiventris* is rare or absent in places where woody vegetation >2 m in height is dominant.

DEMOGRAPHIC CHARACTERISTICS: Populations exhibited an annual cycle of abundance on Merritt Island during 37 months of live trapping on two permanent grids (Stout 1979). Major population increases were correlated with fall and winter breeding (Extine and Stout 1987). Interior populations were less numerous than the ones on the coastal dunes (Keim 1979; Extine 1980). Of 1061 individuals that were live trapped, 536 were males and 525 were females. Judged on external features (descended testes) males were most frequently reproductively active in August (76% descended); over 50 percent were descended in July, August, and September. Lactating females were rarely seen between February and July, whereas the proportion lactating peaked in August (61%) and again in January (64%), with a major decline (25%) in September (Stout 1979). (See other accounts of *P. polionotus* for additional details.)

KEY BEHAVIORS: No data have been reported specifically for *P. p. niveiventris*; see the species account on the Choctawhatchee beach mouse for reports on other subspecies.

CONSERVATION MEASURES TAKEN: This animal is listed as threatened by the U.S. Fish and Wildlife Service (1989). It is also listed as threatened by the Florida Game and Fresh Water Fish Commission (1990). In the future, private development of coastal dune and strand habitat will restrict *P. p. niveiventris* to public lands from Volusia to St. Lucie counties, Florida. Therefore protection by federal agencies on Canaveral National Seashore, Merritt Island National Wildlife Refuge, and the Eastern Test Range will be critical in saving the subspecies from extinction.

CONSERVATION MEASURES PROPOSED: Populations on public lands need to be protected absolutely from further loss of habitat by development. Loss of habitat to development on private lands should be avoided; however, when authorized, the loss should be mitigated by in-kind-and-amount purchase of land to augment existing public lands where the species occurs. Land management programs should strive to prevent house cats, house mice, and feral dogs from cooccurring with *P. p. niveiventris*. Suitable but currently unoccupied habitat on public lands should be evaluated with the objective of reintroducing the form where it is now extirpated. For example, Humphrey et al. (1987) suggested that animals from the southern portion of Sebastian Inlet State Recreation Area might be reintroduced on the northern portion of the tract, which appears not to have a local population.

ACKNOWLEDGMENTS: Most of my field work on the southeastern beach mouse was supported by contract No. NAS 10-8986 from NASA. I appreciate the sharing of information on the subspecies by S. R. Humphrey, M. S. Robson, and M. E. Ludlow.

Literature Cited

Allen, J. A. 1898. The mammals of Florida. Amer. Nat. 32:433–436.

Bangs, O. 1898. The land mammals of peninsular Florida and the coastal region of Georgia. Boston Soc. Nat. Hist. Proc. 28:157–235.

Chapman, F. M. 1889. Description of two apparently new species of the genus *Hesperomys* from Florida. Amer. Mus. Nat. Hist. Bull. 2:117.

Extine, D. D. 1980. Population ecology of the beach mouse, *Peromyscus polionotus niveiventris*. Unpubl. M.S. thesis, University of Central Florida, Orlando. 82 pp.

Extine, D. D., and I. J. Stout. 1987. Dispersion and habitat occupancy of the beach mouse, *Peromyscus polionotus niveiventris*. J. Mammal. 68:297–304.

Florida Game and Fresh Water Fish Commission. 1990. Official lists of endangered fauna and flora in Florida. Tallahassee, Florida. 23 pp.

Hall, E. R. 1981. The mammals of North America. 2d ed. John Wiley and Sons, New York. 1181 + 90 pp.

Humphrey, S. R., and D. B. Barbour. 1981. Status and habitat of three subspecies of *Peromyscus polionotus* in Florida. J. Mammal. 60:840–844.

Humphrey, S. R., W. H. Kern, Jr., and M. E. Ludlow. 1987. Status survey of seven Florida mammals. Unpubl. final report to U.S. Fish and Wildlife Serv., 310 University Boulevard, South Suite 120, Jacksonville. 39 pp.

Keim, M. H. 1979. Small mammal population dynamics and community structure in three east central Florida communities. Unpubl. M.S. thesis, University of Central Florida, Orlando. 144 pp.

Layne, J. N. 1974. The land mammals of south Florida. Pp. 386–413 *in* P. J. Gleason, ed., Environments of south Florida: Present and past. Miami Geol. Soc., Memoir 2:1–452.

Osgood, W. H. 1909. Revision of the mice of the American genus *Peromyscus*. North Amer. Fauna 28:1–285.

Sherman, H. B. 1937. List of the recent wild land mammals of Florida. Proc. Florida Acad. Sci. 1:102–128.

Sherman, H. B. 1952. A list and bibliography of the mammals of Florida, living and extinct. Quart. J. Florida. Acad. Sci. 15:86–126.

Smith, M. H. 1968. A comparison of different methods of capturing and estimating numbers of mice. J. Mammal. 49:455–462.

Stout, I. J. 1979. Terrestrial community analysis. Unpubl. final report to NASA/KSC, a continuation of base-line studies for environmentally monitoring space transportation systems (STS) at John F. Kennedy Space Center. Contract No. NAS 10-8986. 628 pp.

U.S. Fish and Wildlife Service. 1989. Endangered and threatened wildlife and plants; endangered status for the Anastasia Island beach mouse and threatened status for the southeastern beach mouse. Federal Register 54:20598–20602.

Prepared by: I. Jack Stout, Department of Biological Sciences, University of Central Florida, Orlando, FL 32816.

Threatened

Florida Mouse
Podomys floridanus
FAMILY CRICETIDAE
Order Rodentia

TAXONOMY: The Florida mouse was described by Chapman (1889) as *Hesperomys floridanus* from a single specimen collected at Gainesville, Alachua County. *Hesperomys macropus*, described by Merriam (1890) from specimens from Canaveral and Lake Worth, is a synonym. Chapman (1894) used the generic name *Sitomys* for this species. Use of the name *Peromyscus floridanus*, by which the species was long known, dates from Bangs (1898). Osgood (1909) placed *Peromyscus floridanus* in the monotypic subgenus *Podomys*, which Carleton (1980) elevated to full generic rank. *Podomys* appears to be most closely related to the genera *Habromys* and *Neotomodon*, which occur in southern Mexico and Guatemala (Linzey and Layne 1969; Carleton 1980).

No subspecies have been described, although differences among some populations in size, pelage coloration, and behavior are as great as or greater than those distinguishing many named mammalian subspecies (e.g., Layne 1970, 1971a, 1990; Wolfe and Layne 1968). There appears to be a general trend toward larger body size in populations along the Atlantic and Gulf coasts and at the southern end of the Lake Wales Ridge compared with that in the area of the type locality in north-central Florida.

Other common names used for the species include big-eared mouse, Florida deer mouse, and gopher mouse, the latter referring to the close association of the mouse with the burrows of the gopher tortoise (*Gopherus polyphemus*).

DESCRIPTION: In general body form and coloration, the Florida mouse resembles typical members of the genus *Peromyscus*. It has comparatively large eyes, ears, and hind feet and a tail that is about 80 percent of the head and body length. The pelage is relatively long and soft. Adults are

Florida mouse, *Podomys floridanus*. (Photo by James N. Layne)

brownish or brownish-gray on the back and upper sides and bright orange-buff on the shoulders and along the lower sides. The venter is white, frequently with a tawny patch of varying size in the middle of the chest. A buffy wash is sometimes present on the abdominal region. The tail is indistinctly bicolored, dusky above and lighter below. Juveniles are gray above and whitish on the underparts. Approximate ranges of measurements and mass of adults are: total length 179–197 mm, tail 79–90 mm, hind foot 23–27 mm, and mass 25–49 g. As an example of the magnitude of size differences that may occur between populations, mean measurements and mass of adults from Alachua County in the vicinity of the type locality (first value) and the Cedar Key area in Levy County (second value) are: total length 179.4 and 197.0 mm, tail 79.3 and 89.7 mm, hind foot 23.5 and 25.3 mm, ear from notch 19.4 and 20.8 mm, mass 25.8 and 35.5 g.

The most distinctive external diagnostic feature of the Florida mouse is the presence of only five well-developed plantar tubercles on the soles of the hind feet rather than the six typical of members of the genus *Peromyscus*. A rudimentary sixth tubercle may occur in *Podomys* (Layne 1970; Stout and Keim 1981), and some individuals of *Peromyscus polionotus* may lack the sixth tubercle (Smith 1967). Other distinctive features of the species include fragile skin on the distal portion of the tail (Layne 1972), the structure of the phallus and male accessory sex glands (Hooper 1958; Linzey and Layne 1969), and a skunk-like odor (Layne 1990).

Of the other small rodents within its range, the Florida mouse is most likely to be confused with the cotton mouse, *Peromyscus gossypinus*. The two species can easily be distinguished at any age by the difference in plantar tubercles and the fragile tail sheath of *Podomys*. Adult cotton mice do not have the bright color on the shoulders and lower sides, and their eyes and ears are noticeably smaller. Juveniles of the cotton mouse are a darker shade of gray than those of the Florida mouse.

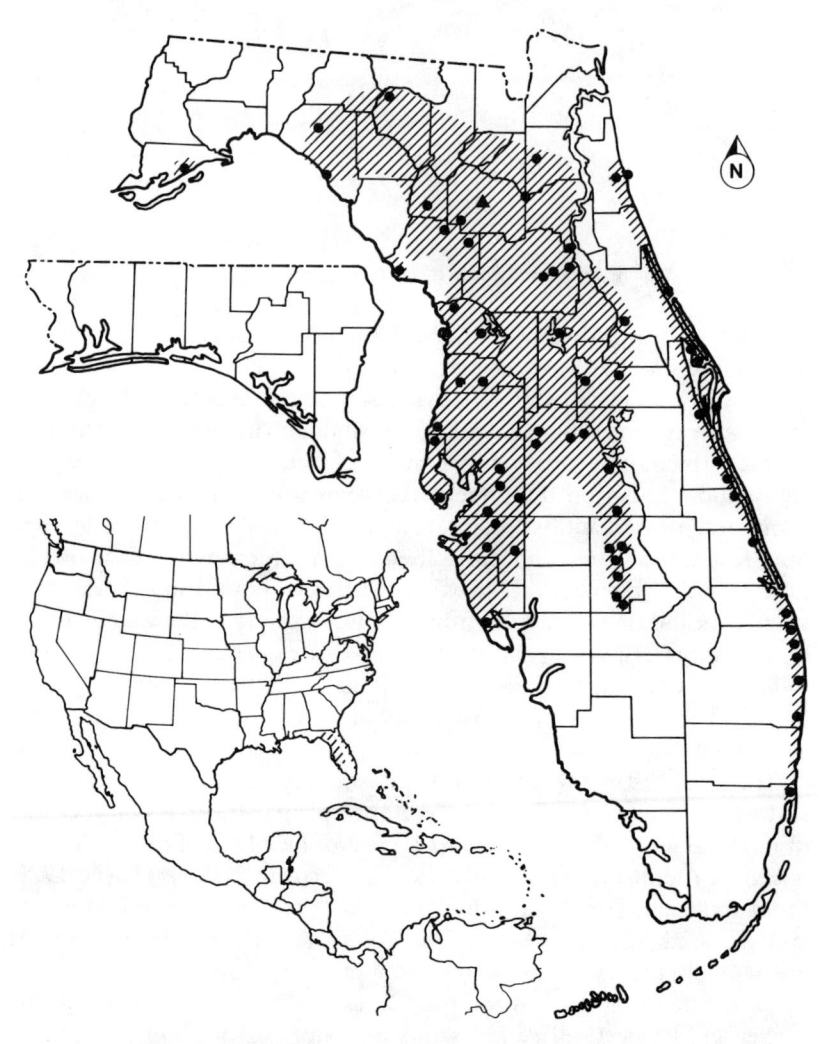

Distribution map of the Florida mouse, *Podomys floridanus*. Hatching represents the species. Dots show specific locality records, the triangle shows the type locality, and the X shows a known introduction.

POPULATION SIZE AND TREND: No estimate of the total population of the Florida mouse is available and, given the considerable variability in density of local populations within and among habitats, any such estimate would be highly speculative. Based on the long-term trend of habitat loss, however, the current total population undoubtedly is well below the presettlement level and continuing to decline.

DISTRIBUTION AND HISTORY OF DISTRIBUTION: The known range includes the northern two-thirds of the Florida peninsula and an apparently isolated area in the vicinity of Carabelle, Franklin County, in the eastern panhandle (Layne 1978). Peripheral peninsular records include Taylor, Suwannee, Alachua, Clay, and St. Johns counties on the north, Dade County (North Miami) on the east coast, Sarasota County (Englewood) on the west coast, and Highlands County (southern end of the Lake Wales Ridge) in the south-central region of the state. Within these general limits, the actual distribution is patchy, reflecting that of xeric upland vegetation, particularly scrub and sandhill associations. As the result of habitat loss, the distribution is becoming increasingly fragmented. On the Atlantic coast, the species apparently no longer reaches Dade County, the southernmost recent records being from the vicinity of Boca Raton, Palm Beach County. A population a short distance north near Boynton Beach that I surveyed in 1960 was still in existence in 1984 (R. Roberts, personal communication). The species is still relatively common in the area of the historic range limit (Highlands County) in the central part of peninsular Florida, but it has not been documented at the southwestern range limit in Sarasota County in recent years. Although the species probably no longer occurs on the Pinellas peninsula, a single specimen was recorded near Clearwater in 1984 (S. Godley, personal communication). The current status of the species at the northern range limit and in the vicinity of Carabelle is unknown.

All fossil (Pleistocene) records of the species are from north-central Florida localities within the area of the Recent range (Webb 1974).

GEOGRAPHICAL STATUS: The Florida mouse has one of the smallest geographic ranges of any species of North American mammal and is the only genus of mammal endemic to Florida, although *Neofiber* is a near-endemic. Unlike *Neofiber*, which is known from fossils in Pennsylvania, Texas, and Kansas (Birkenholz 1972), *Podomys* is not known from the fossil record outside of Florida. The Florida mouse is one of a number of Florida vertebrates, including the indigo snake (*Drymarchon corais*), burrowing owl (*Athene cunicularia*), crested caracara (*Polyborus plancus*), gopher tortoise (*Gopherus polyphemus*), and pocket gopher (*Geomys pinetis*), that have an affinity with the southwestern United States, Mexico, or Middle America. In addition to its close relation to rodent genera occurring in southern Mexico and Guatemala, *Podomys* has a narrowly host-restricted flea (*Polygenis floridanus*) that most closely resembles a species occurring in northern South America (Johnson and Layne 1961).

HABITAT REQUIREMENTS AND HABITAT TREND: The Florida mouse is narrowly restricted to fire-maintained, xeric, upland vegetation occurring on deep, well-drained sandy soils. Specific plant communities in which the species has been recorded include sand pine scrub, coastal scrub, scrubby flatwoods, longleaf pine-turkey oak (sandhill), south Florida slash pine-turkey oak (southern ridge sandhill), upland hammock, live oak (xeric) hammock, and drier pine flatwoods (Bangs 1898; Layne 1963, 1978; Stout 1979). Occasional occurrences in such atypical habitats as mesic hammock, seasonal pond margin, freshwater marsh, and oldfields probably represent transients (Layne 1990). The two major habitats of the Florida mouse are scrub, including sand pine scrub and scrubby flatwoods, and sandhill. Scrub is the primary habitat and probably more closely similar to the ancestral habitat, while sandhill vegetation is a secondary habitat that may not have been generally occupied until historic times when the original state of the habitat—a pine savanna—was converted as a result of human disturbance (logging of pines and fire suppression) to a drier, more open condition more suitable for Florida mice (Layne 1990). A major difference in the two vegetation types is in the number of species and density of oaks. Scrub has a well-developed shrub layer usually dominated by three or four species of oaks, whereas sandhill characteristically has one major species, turkey oak, usually occurring in a relatively open stand. In addition to their greater abundance, scrub oak species also tend to have higher and more consistent acorn production than turkey oak (Layne 1990). The difference in acorn production between scrub and sandhill association is reflected in numerous demographic and behavioral differences between Florida mouse populations in these habitats.

The combination of more widespread xeric vegetation and greatly expanded peninsular Florida land-mass during the Late Wisconsinan glacial stage (Watts 1980) suggests that *Podomys* may have had a larger and more continuous distribution during that period than by the end of the Holocene, when the sea had risen to its present level, accompanied by large-scale replacement of xeric habitats by more mesic or hydric associations. During historic times, there has been a continuing loss of xeric upland habitats to real estate development and agricultural use. Much of the sand pine scrub association along the Atlantic coast has been destroyed, with resultant loss of *Podomys* populations. The same is true of the more disjunct scrubs along the Gulf coast of the peninsula. Vast areas of the original sandhill vegetation along the Lake Wales Ridge have been converted to citrus groves, and only a few small remnants continue to support *Podomys* populations. An example of the magnitude of Florida mouse habitat loss in some areas is provided by data compiled for Highlands County by Peroni (1983). During the period from 1940–44 to 1981, approximately 64 percent of the xeric upland habitat suitable for *Podomys* was destroyed and an additional 10 percent was disturbed. Since 1981, the rate of clearing of the remaining scrub and sandhill habitats for development and citrus has escalated sharply. In the northern portion of the range, many former sandhill and scrub sites have been converted to pine plantations. In addition, suppression of fire and the resultant successional changes have resulted in further reduction or elimination of *Podomys* populations in many remaining sandhill and scrub habitats.

VULNERABILITY OF SPECIES AND HABITAT: As a result of its narrow habitat specificity, the Florida mouse is extremely vulnerable to habitat loss. In turn, the well-drained soils of the xeric upland habitats used by Florida mice make them especially favorable for residential and commercial development and for citrus culture.

CAUSES OF THREAT: The major threat to the Florida mouse is habitat loss. Much of the xeric upland vegetation in which the species is found has already been lost to development and agriculture, and that which remains is under increasing pressure for development. Unlike the case of wetlands, there is currently little effective government control on use of uplands. The Florida mouse is but one of a large number of plants and animals endemic to Florida's xeric upland habitats, particularly scrub. In addition to their unique biota, xeric uplands have a high recharge rate and thus are important to water resources.

RESPONSES TO HABITAT MODIFICATION: *Podomys* populations are highest in early successional stages of scrub and sandhill vegetation following fire, and decline as the habitat becomes denser and more shady and the microclimate more mesic in the absence of fire. The rate of decrease in abundance and the length of time a population can persist in an unburned site depend upon the rate of vegetation change, which in turn is related to such factors as local edaphic conditions, the time and intensity of the last fire, postfire weather conditions, and the availability of seed sources of such mesic species as red bay (*Persea borbonia*), laurel oak (*Quercus hemisphaerica*), American holly (*Ilex opaca*), and southern magnolia (*Magnolia grandiflora*), which readily invade scrub and sandhills in the absence of fire. In one scrub site in Levy County, Florida, mice have persisted for 44 years following a catastrophic fire, but absolute density and relative abundance in comparison to other small rodents (cotton mice, *Peromyscus gossypinus*, and golden mice, *Ochrotomys nuttalli*) underwent a sharp decline about 10 years after the fire. At the Archbold Biological Station in Highlands County, *Podomys* is still present in very low numbers in sandhill and scrub sites that were last burned in 1927, whereas populations in similar nearby habitats that have burned periodically are higher and more stable (Layne 1990). Jones (1989) also found that populations were higher in burned than unburned sandhill habitats on the Ordway-Swisher Preserve in Putnam County.

In scrub, optimal conditions for both the Florida mouse and scrub jay (*Aphelocoma c. coerulescens*) are similar and thus the presence of the jay in scrubs large enough to support a population is a good indicator of high-quality *Podomys* habitat. The Florida mouse is able to persist longer in unburned scrub than the scrub jay. For example, at the Archbold Biological Station the mice were still present in one scrubby flatwoods area 62 years after the last burn, whereas scrub jays had disappeared from the same site 10 years earlier (Woolfenden and Fitzpatrick 1984). Reduction of acorn yield is probably one of the causes of reduced suitability of long-unburned, overgrown scrub habitat for both Florida mice and scrub jays. In the absence of the normal fire frequency, mechanical disturbance of the vegetation (in the form of logging, partial clearing, disking, chopping, mowing, or vehicular traffic) may favor Florida mice by creating more open cover conditions. Although extensive clearing of scrub or sandhill habitat is detrimental to Florida mice, the species may persist in areas with remnants of scrub or sandhill vegetation (Humphrey et al. 1985).

DEMOGRAPHIC CHARACTERISTICS: Means and ranges of relative abundance (mice/100 trap-nights) of Florida mice in five habitats (Layne and Griffo 1961) were: sand pine scrub, 13.2 (1.2–36.1); slash pine-turkey oak (southern ridge sandhill), 18.1 (2.6–21.0); longleaf pine-turkey oak (sandhill), 3.5 (0.2–10.0); longleaf pine flatwoods, 6.4 (5.2–6.9); ecotonal (transition zone between sandhill and flatwoods and mesic, upland, and live oak hammocks), 7.4 (1.3–25.0). Arata (1959) recorded relative abundance of 0.3/100 trap-nights in a sandhill plot after a midwinter burn. Humphrey et al. (1985) took 0.26 mice/100 trap-nights on trap lines and grids and 0.45/100 in traps at entrances to gopher tortoise burrows in a natural pine savanna (sandhill) habitat compared with 0.52/100 and 7.1/100, respectively, in a remnant sandhill site (oldfield with turkey oaks). Minimum known numbers of Florida mice on a 1.44-ha grid in pine flatwoods on the Kennedy Space Center, Merritt Island, ranged from 0 to 6 (Stout 1979). Estimates of actual population densities include 1.6 to 14.1/ha in mixed scrub on Merritt Island (Ehrhart 1976), a mean of 5/ha and maximum of 28/ha in sandhill habitat in Alachua County (Layne 1990), and a mean of 9.4/ha and range of 3.9 to 16.5/ha in a sand pine scrub plot in Levy County (J. N. Layne, unpubl. data). Scrub habitat appears to generally support higher populations than sandhill. Based on data from removal-trapping transects at over 50 localities in 8 counties, sandhills had an overall mean trap success of 2.5 mice/100 trap-nights compared with 9.9 and 12.7, respectively, for sand pine scrub and scrubby flatwoods. In both habitats, numbers were highest in the winter-to-early-spring period and lowest during the summer-to-early-winter interval (Layne 1990). Scrub populations tend to have a more pronounced annual cycle and less erratic multiannual fluctuations than those in sandhills. The generally higher densities of Florida mice in scrub than sandhills suggest that the size of a preserve needed to maintain a population with a given probability of long-term survival would be smaller in scrub than in sandhill habitat. Stout et al. (1987) suggested that the minimum viable population size for the species was 100 breeding individuals.

Vulvas of females of the Alachua County population became perforate at an average of about 5 weeks of age, and spermatozoa were present in the cauda epididymis of a male 11 weeks old. Breeding is usually in fall and early winter (Layne 1978). Reported mean litter sizes in laboratory and field studies vary from 1.7 to 3.4, with a range of from 1 to 5 (Dice 1954; Layne 1966; Rood 1966; Drickhamer and Vestal 1973; Glazier 1985). Females probably seldom have more than 2 litters per breeding season. Sex ratios in 3 scrub habitats in Levy County ranged from 45 to 51 percent male, and the overall sex ratio of 592 specimens from various habitats and localities was 48 percent male (Layne 1967, 1968). On Merritt Island, males comprised 49 percent of mice trapped in mixed scrub (Ehrhart 1976) and in pine flatwoods (Stout 1979). Eisenberg (1984) stated that in sandhill habitat on the Ordway-Swisher Preserve sex ratios were strongly female-biased during the breeding season "but tend to approach equality before the animals cease reproduction and pass through the winter."

Mean survival time (known residence on live-trapping grids) of different age classes in an Alachua County sandhill population was 4.2 months for adults, 3.6 months for subadults, and 2.4 months for juveniles, compared with 2.0, 1.8, and 1.7 months, respectively, on a scrub site in Levy County (Layne 1990). About 70 percent of the individuals in a pine flatwoods grid on Merritt Island survived over a 4-week interval (Stout 1979). Eisenberg and Jones (1985) reported that 7 (5%) of 129 marked individuals in sandhill habitat survived more than a year. A male approximately 3 weeks of age at capture lived 7 years and 4 months in captivity (Keim and Stout 1987). Five individuals wild-caught as adults or subadults survived from 50 to 79 months (mean = 65) in captivity, and 4 known-age captive-born individuals lived from 48 to 64 months (mean = 59; J. N. Layne, unpubl. data).

Mean distances between successive captures of Florida mice in a scrub site were 16 m for adults, 0 for subadults (small sample size), and 5 m for juveniles, whereas comparable measures on a sandhill study area were adults 25 m, subadults 19 m, and juveniles 16 m (Layne 1990). These data suggest larger home-range size in sandhill than scrub habitat, in keeping with the apparently lower food base of the former. Mean distances between successive captures of males and females in pine flatwoods were 49.8 and 39.1 m, respectively (Stout 1979).

Predation is probably the major mortality factor in *Podomys* populations, although Goin (1944) encountered a mouse impaled on a cactus spine. Known or presumed predators include various snakes, owls, foxes, bobcats, raccoons, and house cats (Layne 1978; Maehr and Brady 1986; Wassmer et al. 1988). *Podomys* is a known host to 32 species of parasites, including 13 endoparasites and 19 ectoparasites (Layne 1963, 1967, 1971*b*; Lichtenfels 1970; Ehrhart 1976). Scrub populations tend to have a high incidence of infections of the liver-inhabiting nematode *Capillaria hepatica*, but this does not appear to be a direct cause of mortality (Layne 1968).

KEY BEHAVIORS: The Florida mouse appears to be an exclusively burrow-dwelling species, frequently using the burrows of the gopher tortoise, *Gopherus polyphemus*, constructing its own burrows and nest chambers off the main tortoise burrow (Eisenberg and Jones 1985; Layne 1990). The species may be more closely tied to gopher tortoise burrows in sandhill than in scrub habitats. The use of burrows is correlated with poorly developed nest-building behavior, which may indicate a long history of burrow dwelling (Layne 1969). One of the adaptive values of the use of burrows in relatively xeric habitats is reduction of water loss, as *Podomys* has no specialized physiological water-conservation mechanisms such as found in some desert rodents (Fertig and Layne 1963, Glenn 1970, Layne and Dolan 1975). The Florida mouse is similar to the oldfield mouse (*Peromyscus polionotus*), another species of relatively dry, open habitats, in being less arboreal than such species as the cotton mouse (*P. gossypinus*) that are characteristic of more forested habitats (Layne 1970). The fragile distal tail skin of *Podomys* also appears to be a defensive adaptation suited to a relatively terrestrial species living in open habitats. In addition to the anatomical modifications of the skin, the mouse exhibits behavior apparently functioning to attract an attacking predator's attention to the tail and to facilitate loss of the skin if the tail is seized (Layne 1972). Under experimental conditions, *Podomys* exhibited stricter nocturnality than *P. gossypinus*, which may reflect an adaption for its more open, xeric habitats (Layne 1971a). Another behavior apparently correlated with the Florida mouse's narrow habitat-specificity is a lower dispersal tendency compared with that of the more habitat-tolerant and widespread cotton mouse (J. N. Layne, unpubl. data).

CONSERVATION MEASURES TAKEN: Basic research on many aspects of the ecology, life history, behavior, physiology, and genetics of *Podomys floridanus* has been important in guiding conservation efforts for this species. It is listed as a species of special concern by the Florida Game and Fresh Water Fish Commission (1990) and is under review for listing by the U.S. Fish and Wildlife Service (1989). Protection of Florida mouse populations on sites proposed for development is one of the factors considered in Development of Regional Impact reviews by Regional Planning Councils and the Florida Game and Fresh Water Fish Commission. Scrub or sandhill habitat suitable for the species is preserved on a number of public and private lands. Such sites that are known to have populations of Florida mice include San Felasco State Preserve, Wekiwa River State Park, Cedar Key Scrub State Preserve, Merritt Island National Wildlife Refuge, Ocala National Forest, Highlands Hammock State Park, Lake Arbuckle Wildlife Management Area, Jonathan Dickinson State Park, South Fork State Park, Wingate Creek State Park, Lake Manatee State Recreation Area, Archbold Biological Station, Saddleblanket Lake Scrub Preserve of the Nature Conservancy, Tiger Creek Preserve of the Nature Conservancy, and the University of Florida's Ordway-Swisher and Welaka preserves.

Other conservation measures directly or indirectly benefiting the Florida mouse have been the development of design standards for a sand pine scrub ecosystem preserve in an urban context (Stout et al. 1987), recommendations for restoration of a former sandhill wildlife community in a decadent citrus grove (Humphrey et al. 1985), and efforts to reclaim phosphate-mined lands to xeric-type habitats suitable for Florida mice, gopher tortoises, and other xeric upland species (Godley 1989). Adams (1978) documented the successful introduction of Florida mice from Highlands County to a site in Hillsborough County.

CONSERVATION MEASURES PROPOSED: Although a substantial amount is known about the basic biology of the Florida mouse, more detailed data on various aspects of the species' ecology, life history, and population dynamics are needed to more effectively preserve and enhance remaining populations. High priority should be given to a comprehensive survey of its current distribution and population status. More intensive research also is needed on the relationships of population density and persistence to various habitat factors such as vegetation structure and successional stage, size of habitat, and dispersion of habitats. Further development and refinement of habitat management techniques are necessary. As protected scrubs and other xeric upland habitats are increasingly encroached upon by highways and development, use of fire to maintain the desired successional stages will become more difficult and thus burning techniques will need to be more precise and reliable. On the likely assumption that burning as a means of habitat management will become impossible in some areas, more attention should be directed to mechanical and chemical means of vegetation control as an analogue of fire. The genetic implications of increasingly more isolated and reduced Florida mouse populations and of the probable future need to relocate or reintroduce the species to different areas also need to be addressed.

Habitat protection by means of effective land-use regulations (including regulation of agriculture) and actual acquisition for preserves is the single most important conservation measure that should be pursued. Because it is doubtful that land-use regulations could fully guarantee the necessary degree of habitat protection in the long term, acquisition should receive highest priority. Federal, state, and local government agencies and private organizations involved in land acquisition for environmental purposes should focus more of their efforts on xeric upland habitats, particularly sand pine scrub, which is under the most immediate threat. In view of the high recharge capacity of xeric uplands, they should receive greater consideration for purchase under the Save-Our-Rivers Program of the Water Management Districts. Government agencies and private conservation organizations involved in preservation of scrub and other xeric habitats should closely coordinate their efforts to ensure establishment of a system of preserves that encompasses the full range of local and regional variants of these communities, with emphasis on such centers of endemism as the southern Lake Wales Ridge.

Literature Cited

Adams, S. R. 1978. Population studies and ecology of a native population of *Peromyscus gossypinus* and an introduced population of *Peromyscus floridanus* on Buck Island, Florida. Unpubl. M.A. thesis, University of South Florida, Tampa. 65 pp.
Arata, A. A. 1959. Effects of burning on vegetation and rodent populations in a long-leaf pine turkey oak association in north central Florida. Quart. J. Florida Acad. Sci. 22:94–104.
Bangs, O. 1898. The land mammals of peninsular Florida and the coast region of Georgia. Proc. Boston Soc. Nat. Hist. 28:157–235.
Birkenholz, D. E. 1972. *Neofiber alleni*. Mammalian Species 15:1–4.
Carleton, M. D. 1980. Phylogenetic relationships in neotomine-peromyscine rodents (Muroidea) and a reappraisal of the dichotomy within New World Cricetinae. Misc. Publ. Mus. Zool., Univ. Mich. 157:1–146.
Chapman, F. M. 1889. Preliminary descriptions of two apparently new species of the genus *Hesperomys* from Florida. Bull. Amer. Mus. Nat. Hist. 2:117.
Chapman, F. M. 1894. Remarks on certain land mammals from Florida, with a list of the species known to occur in the state. Bull. Amer. Mus. Nat. Hist. 6:333–346.
Dice, L. R. 1954. Breeding of *Peromyscus floridanus* in captivity. J. Mammal. 35:260.
Drickhamer, L. C., and B. M. Vestal. 1973. Patterns of reproduction in a laboratory colony of *Peromyscus*. J. Mammal. 54:523–528.
Ehrhart, L. M. 1976. Mammal studies: Final report to the National Aeronautics and Space Administration Kennedy Space Center (Grant No. NGR 10-019-004). Office of Graduate Studies and Research, Florida Technological University, Orlando. 182 pp.
Eisenberg, J. F. 1984. Ecology of the Florida mouse, *Podomys* (=*Peromyscus*) *floridanus*. Abstracts Amer. Soc. Mamm. 64th Ann. Meeting, Humboldt State University, Arcata, California.
Eisenberg, J. F., and C. A. Jones. 1985. Long-term residence patterns shown by *Peromyscus floridanus*. Abstracts Amer. Soc. Mamm. 65th Ann. Meeting, University of Maine at Orono.
Fertig, D. S., and J. N. Layne. 1963. Water relationships in the Florida mouse. J. Mammal. 44:322–334.
Florida Game and Fresh Water Fish Commission. 1990. Official lists of endangered and potentially endangered fauna and flora in Florida. Tallahassee. 23 pp.

Glazier, D. S. 1985. Relationship between metabolic rate and energy expenditure for lactation in *Peromyscus*. J. Comp. Biochem. Physiol. 80A:587–590.

Glenn, M. E. 1970. Water relations in three species of deer mice (*Peromyscus*). Comp. Biochem. Physiol. 33:231–248.

Godley, J. S. 1989. Comparison of gopher tortoise populations relocated onto reclaimed phosphate-mined sites in Florida. Pp. 43–58 *in* J. E. Diemer, D. R. Jackson, J. L. Landers, J. N. Layne, and D. A. Wood, eds., Gopher tortoise relocation symposium proceedings. Florida Game and Fresh Water Fish Commission, Nongame Wildlife Program Tech. Rept. 5. 109 pp.

Goin, O. B. 1944. *Peromyscus* impaled on *Opuntia*. J. Mammal. 25:200.

Hooper, E. T. 1958. The male phallus in mice of the genus *Peromyscus*. Misc. Publ. Mus. Zool., Univ. Mich. 105:1–24.

Humphrey, S. R., J. F. Eisenberg, and R. Franz. 1985. Possibilities for restoring wildlife of a longleaf pine savanna in an abandoned citrus grove. Wildl. Soc. Bull. 13:487–496.

Johnson, P. T., and J. N. Layne. 1961. A new species of *Polygenis* Jordan from Florida, with remarks on its host relationships and zoogeographic significance. Proc. Entomol. Soc. Wash. 63:115–123.

Jones, C. A. 1989. Fire and the Florida mouse (*Podomys floridanus*). Florida Sci. 52(Suppl. 1):19.

Keim, M. H., and I. J. Stout. 1987. Longevity record for the Florida mouse, *Peromyscus floridanus*. Florida Sci. 59:41.

Layne, J. N. 1963. A study of the parasites of the Florida mouse, *Peromyscus floridanus*, in relation to host and environmental factors. Tulane Stud. Zool. 11:1–2.

Layne, J. N. 1966. Postnatal development and growth of *Peromyscus floridanus*. Growth 30:23–45.

Layne, J. N. 1967. Incidence of *Porocephalus crotali* (Pentastomida) in Florida mammals. Bull. Wildl. Dis. Assoc. 3:105–109.

Layne, J. N. 1968. Host and ecological relationships of the parasitic helminth *Capillaria hepatica* in Florida mammals. Zoologica 53:107–122.

Layne, J. N. 1969. Nest-building behavior in three species of deer mice, *Peromyscus*. Behaviour 35:288–303.

Layne, J. N. 1970. Climbing behavior of *Peromyscus floridanus* and *Peromyscus gossypinus*. J. Mammal 51:580–591.

Layne, J. N. 1971a. Activity responses of two species of *Peromyscus* (Rodentia: Muridae) to varying light cycles. Oecologia 7:223–241.

Layne, J. N. 1971b. Fleas (Siphonaptera) of Florida. Florida Entomologist 54:35–51.

Layne, J. N. 1972. Tail autotomy in the Florida mouse, *Peromyscus floridanus*. J. Mammal. 53:62–71.

Layne, J. N. 1978. Florida mouse. Pp. 21–22 *in* J. N. Layne, ed., Rare and endangered biota of Florida. Vol. 1. Mammals. University Presses of Florida, Gainesville. xx + 52 pp.

Layne, J. N. 1990. The Florida mouse. Pp. 1–21 *in* C. K. Dodd, Jr., R. E. Ashton,

Jr., R. Franz, and E. Wester, eds., Burrow associates of the gopher tortoise. Proc. 8th Ann. Meeting of the Gopher Tortoise Council, Florida Museum Nat. Hist., Gainesville. 134 pp.

Layne, J. N., and P. G. Dolan. 1975. Theromoregulation, metabolism, and water economy in the golden mouse (*Ochrotomys nuttalli*). Comp. Biochem. Physiol. 52A:153–163.

Layne, J. N., and J. V. Griffo, Jr. 1961. Incidence of *Capillaria hepatica* in populations of the Florida deer mouse, *Peromyscus floridanus*. J. Parasit. 47:31–37.

Lichtenfels, J. R. 1970. Two new species of *Pterygodermatites* (*Paucipectines*) Quentin, 1969 (Nematoda: Rictulariidae) with a key to the species from North American rodents. Proc. Helminthol. Soc. Washington 37:94–101.

Linzey, A. V., and J. N. Layne. 1969. Comparative morphology of the male reproductive tract in the rodent genus *Peromyscus* (Muridae). Amer. Mus. Novitates 355:1–47.

Maehr, D. S., and J. R. Brady. 1986. Food habits of bobcats in Florida. J. Mammal. 67:133–138.

Merriam, C. H. 1890. Description of a new species of *Hesperomys* from southern Florida. Pp. 53–54 *in* Description of twenty-six new species of North American mammals. North Amer. Fauna 4:1–60.

Osgood, W. H. 1909. Revision of the mice of the American genus *Peromyscus*. North Amer. Fauna 28:1–285.

Peroni, P. A. 1983. Vegetation history of the southern Lake Wales Ridge, Highlands County, Florida. Unpubl. M.S. thesis, Bucknell University, Lewisburg, Pennsylvania. 85 pp.

Rood, J. P. 1966. Observations on the reproduction of *Peromyscus* in captivity. Amer. Midland Nat. 76:496–503.

Smith, M. H. 1967. Variation in plantar tubercles in *Peromyscus polionotus*. Quart. J. Florida Acad. Sci. 30:108–110.

Stout, I. J. 1979. Terrestrial community analysis, final report to the National Aeronautics and Space Administration John F. Kennedy Space Center, a continuation of base-line studies for environmentally monitoring space transportation systems (STS) at John F. Kennedy Space Center. Vol. 1 of 4. Contract No. NAS10-8986. 189 pp. + appendices.

Stout, I. J., and M. H. Keim. 1981. Variation in number of plantar tubercles in *Peromyscus floridanus*. Florida Sci. 44:126–128.

Stout, I. J., D. R. Richardson, R. E. Roberts, and D. F. Austin. 1987. Design of a nature preserve in a subtropical, urbanizing landscape: Application of ecologic and genetic principles. Bull. Ecol. Soc. Amer. 68:423–424.

U.S. Fish and Wildlife Service. 1989. Endangered and threatened wildlife and plants; animal notice of review. Federal Register 54:554–579.

Wassmer, D. A., D. D. Guenther, and J. N. Layne. 1988. Ecology of the bobcat in south-central Florida. Bull. Florida State Mus., Biol. Sci. 33:159–228.

Watts, W. A. 1980. The late Quaternary vegetation history of the southeastern United States. Ann. Rev. Ecol. Syst. 11:387–409.

Webb, S. D. 1974. Chronology of Florida Pleistocene mammals. Pp. 5–31 *in* S. D. Webb, ed., Pleistocene mammals of Florida. University Presses of Florida, Gainesville. 270 pp.

Wolfe, J. L., and J. N. Layne. 1968. Variation in dental structures of the Florida mouse, *Peromyscus floridanus*. Amer. Mus. Novitates 2352:1–7.

Woolfenden, G. E., and J. W. Fitzpatrick. 1984. The Florida scrub jay: Demography of a cooperative-breeding bird. Monogr. in Population Biology 20, Princeton University Press, Princeton, New Jersey. 406 pp.

Prepared by: James N. Layne, Archbold Biological Station, P.O. Box 2057, Lake Placid, FL 33852.

Threatened

Florida Black Bear
Ursus americanus floridanus
FAMILY URSIDAE
Order Carnivora

TAXONOMY: No other names. Descriptions and supporting citations for 17 subspecies of black bears can be found in Hall (1981). The Florida subspecies also has recently been referred to as *Euarctos americanus floridanus*. The type specimen, originally referred to as *Ursus floridanus*, was collected from Key Biscayne and described by Merriam (1896).

DESCRIPTION: The black bear is the largest land mammal in Florida (since the bison was extirpated). Average weights of females and males are about 82 kg and 113 kg, respectively. A 285-kg male specimen recently was recorded for Collier County (J. Schortemeyer, personal communication). Black pelage is usually entire except for a brown muzzle and occasionally a blonde chest patch. Bears in southern Florida often lose the black guard hairs on the upper limbs and body, becoming woolly-brown in appearance. This may be a physiological response to the region's subtropical climate. Front feet average 9–14 cm by 8.5–15 cm and 7.5 cm by 1.7–11.0 cm for males and females, respectively. Rear feet measure 7.5 cm by 14–22 cm for males and 6.5–9.5 cm by 13.5–18 cm for females. No other mammal in Florida (except humans) creates footprints comparable in size and dimensions to those of the black bear.

POPULATION SIZE AND TREND: Because of its secretive habits and primary occurrence in remote, forested areas, the population size of the black bear in Florida is difficult to estimate. Nonetheless, widely varying figures have been presented by a number of authors (Brady and Maehr 1985). Most estimates have been between 500 and 1000. We suspect the actual number falls somewhere within this interval. Although localized, large populations may be stable, the statewide trend is downward, and

Florida black bear, *Ursus americanus floridanus*. (Photo by Barry W. Mansell)

extirpations of small, isolated populations will undoubtedly continue to occur.

DISTRIBUTION AND HISTORY OF DISTRIBUTION: The black bear is the most widespread of North America's three bear species (Pelton 1982:504) and, historically, probably occurred in every continental state as well as in Canada and Mexico. Intensive agriculture and human settlement have eliminated the black bear from most of the Midwest and restricted current populations to remote forested tracts (Maehr 1984*a*). Brady and Maehr (1985:5) stated that "Historically, black bears occurred throughout the Florida mainland and on some coastal islands, often associated with large forested tracts. Currently, the black bear remains widespread in Florida, but its distribution is reduced and has become fragmented. Large, undeveloped woodland tracts are still the preferred habitat. The continued fragmentation of remaining bear habitat and the local extinc-

Ursus americanus floridanus 267

tions which will likely follow are an important threat to black bear existence in Florida."

GEOGRAPHICAL STATUS: A summary of local records indicates this native bear was widespread throughout mainland Florida and on some is-

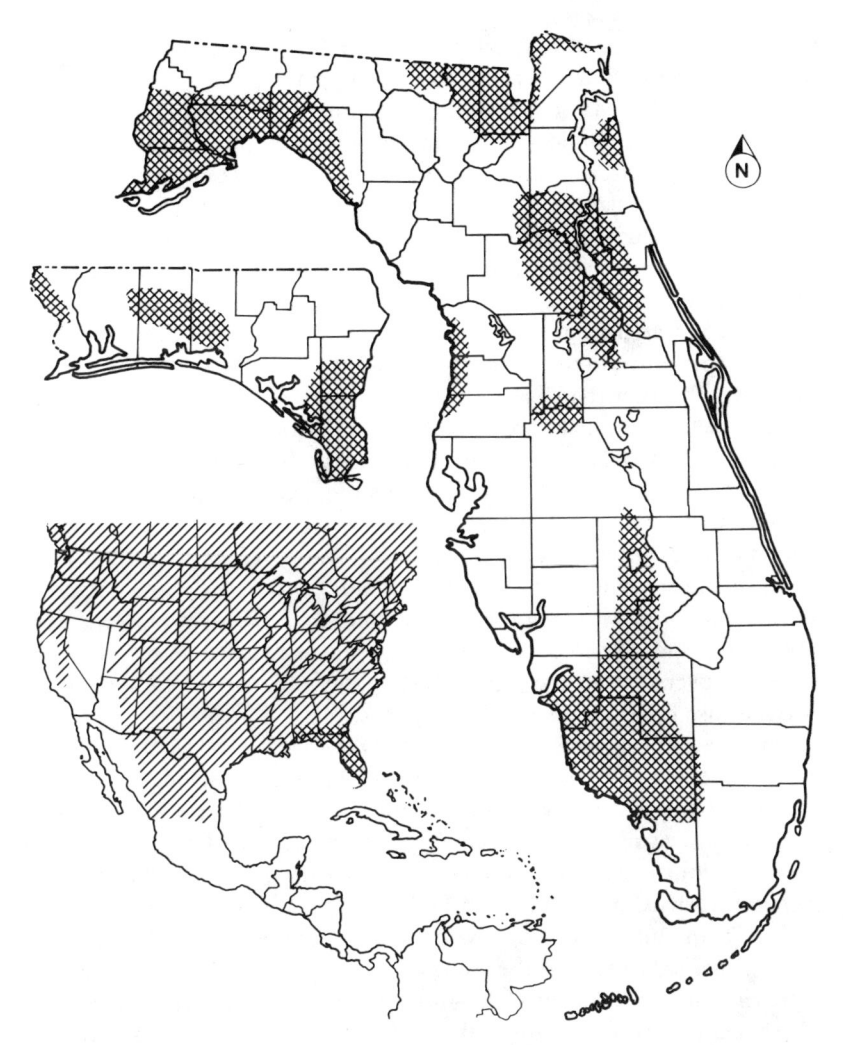

Distribution map of the Florida black bear, *Ursus americanus floridanus*. Hatching represents the species; crosshatching represents the subspecies. The main map shows current range, whereas the inset shows the original range.

lands (Brady and Maehr 1985). The patchy, fragmented distribution found today is in stark contrast to its probable range in the nearly complete cover of forest before human settlement. Black bears in northeastern Florida and extreme northwestern Florida undoubtedly cross the political borders of neighboring states (Maehr 1984a).

HABITAT REQUIREMENTS AND HABITAT TREND: Black bears use a wide variety of forested landscapes, from temperate plant communities in northwestern Florida to subtropical communities in southern Florida. Some of the more important forest types include pine flatwoods, hardwood swamp, cypress swamp, cabbage palm forest, sand pine scrub, and mixed hardwood hammock. A combination of several major forest types is typical of occupied bear range. As with black bears in other parts of their range, seasonal changes in habitat use occur in response to food availability (Pelton 1982). In the Osceola National Forest, black bears use forested wetlands in greater proportion than available, and flatwoods in lower proportion than available. Mykytka and Pelton (1989) found that swamps larger than 150 ha were important components of bear habitat in Osceola National Forest. In the Ocala National Forest, bears prefer flatwoods and avoid longleaf pine forests (Wooding and Hardisky 1988).

Cover, especially for the female's denning requirements, also is an essential habitat component. Beds usually are located in remote swamps or thickets. Nests measure 45–75 cm across and 5–17 cm deep and often occur in a nearly impenetrable tangle of vines and stems characterized by *Lyonia lucida, Lyonia ferruginea, Clethra alnifolia, Serenoa repens, Ilex glabra,* and *Smilax* spp. Shrub cover at denning sites is usually dense, with midstory and overstory sparse (J.B. Wooding, unpubl. data). Bears in Florida also may use hollow trees for denning.

The diet of black bears in Florida varies both temporally and geographically (Maehr and Brady 1982a, 1984a, 1984b) and includes a great variety of plants and animals (Maehr and DeFazio 1985). As with black bears in other parts of their range, Florida bears follow a chronology of food availability from herbaceous matter in early spring, to soft fruits in summer, to hard mast during fall (Pelton 1982:508). Major food items are the fruits and hearts of saw palmetto (*Serenoa repens*) and cabbage palm (*Sabal palmetto*) and the fruits of swamp tupelo (*Nyssa biflora*), oaks (*Quercus* spp.), blueberry (*Vaccinium* spp.), and gallberry (*Ilex glabra*). These species are found throughout the state and probably account for nearly 80 percent of the diet. Insects are the most important animal food, with the introduced honey bee (*Apis mellifera*) occurring most frequently.

Other important insects include yellow jackets (*Vespula* spp.), carpenter ants (*Campanotus abdominalis floridanus*), bessie bugs (*Odontotaenius disjunctus*), and walking sticks (*Anisomorpha buprestoides*). Vertebrates are taken infrequently and include armadillo (*Dasypus novemcinctus*), wild hog (*Sus scrofa*), and white-tailed deer (*Odocoileus virginianus*). For some plants requiring acid scarification of seeds, the black bear may act as an important agent of dispersal and germination (Maehr 1984*b*).

VULNERABILITY OF SPECIES AND HABITAT: Because of its large home range and low population density, the black bear is particularly vulnerable to habitat loss. In addition, a low reproductive rate makes it sensitive to excessive mortality. Pearson (1954) noted that overhunting probably led to the elimination of bears in Levy County. It appears that, in large enough areas, black bears can sustain regulated annual fall harvests. Individual bears are vulnerable to poaching, vehicle collisions (Wooding and Brady 1987), and killing by intolerant beekeepers (Maehr and Brady 1982*b*).

CAUSES OF THREAT: Expanding urbanization, agricultural development, and increasing use of the state's wildlands for recreation all have resulted in an accelerating rate of habitat loss. These pressures have created a graphic example of habitat fragmentation and population isolation (Brady and Maehr 1985). Illegal killing and roadkill mortality are probably a significant consideration only at the fringes of occupied range where potential colonizers are eliminated from suitable habitat and small, isolated populations. Black bears are not legally hunted in most of their Florida range; habitat loss is the major cause for concern.

RESPONSES TO HABITAT MODIFICATION: Because black bears are dependent upon extensive forested landscapes, human activities resulting in widespread conversion of forest to other land uses is a detriment. Intensive forestry practices involving even-age timber management probably reduce the ability of these lands to support bears. Where timber is managed in a mosaic of harvest chronology and stand age, habitat quality probably can be maintained (see Harris 1984). Because fall foods are extremely important in determining reproductive success in black bears (Rogers 1976), land management practices reducing mast production may decrease bear productivity. Maehr and Brady (1982*a*, 1984*b*) suggested that large-scale winter burning may reduce the diversity of foods available to bears by causing spread of saw palmetto and reduction in blueberry and runner oak (*Q. pumilla*). Summer burning may encourage the latter species and should be considered in managing areas occupied by bears.

DEMOGRAPHIC CHARACTERISTICS: According to studies in other parts of the black bear's range, females typically become sexually mature at 3–5 years of age (Pelton 1982:504). In Florida, females become sexually mature at 2 or 3 years, breed at 2.5, and produce the first litter at 3 or breed at 3.5 and produce the first litter at 4 (J. B. Wooding, unpubl. data). A female produces cubs, numbering 2–4 per litter with a 50:50 sex ratio, in alternate years during January or February. Mating appears to occur in June or July. Black bears are promiscuous, and females are induced ovulators. Implantation is delayed 5–6 months after copulation, and fetal development is completed in 6–8 weeks (Wimsatt 1963). Cubs weigh 290–454 g at birth (Alt 1987). Nutrition of the female plays a critical role in determining the age of first reproduction, litter size, and interbirth interval (Bunnell and Tait 1981).

Major mortality factors are associated with man and include legal hunting (in two areas of northern Florida), poaching, and vehicle collisions. Mortality rates are typically several times greater for males than females (Bunnell and Tait 1981). The greater vulnerability of males is attributed in part to their more extensive movements, which expose them to more danger. In Florida, there is a preponderance of males in both roadkill and hunting mortalities. Natural population regulation results from subadult dispersal and adult males killing younger bears (Bunnell and Tait 1981). In the Ocala National Forest, field sign indicated that a two-year-old, radio-collared female was killed by a large male (J. B. Wooding, unpubl. data). In Collier County, remains of a juvenile bear were found in a panther scat (Maehr et al. 1990).

Conte et al. (1983) identified 28 species of internal parasites of Florida black bears. These included 14 helminths, 12 nematodes, 1 trematode, and 1 acanthocephalan. The mean number of species per bear was 4, while the mean intensity of infection was 406 individuals per bear. The three most commonly encountered species were *Strongyloides* spp., *Macracanthorhyncus ingens*, and *Capillaria aerophila*. No serious pathogenic cases were noted.

KEY BEHAVIORS: While the black bear is probably the least likely of North American bears to engage in aggressive encounters with humans, attacks on humans occur occasionally (Whitlock 1950; Norris-Elye 1951;

Pelton et al. 1976; Dusel-Bacon 1979; Herrero 1985). Attacks by wild black bears on humans have not been reported in Florida.

Black bears are typically solitary animals. Family groups consist of a mother and cubs. The family breaks up when the cubs are 16–17 months old (Rogers 1987). Females rarely disperse from their natal home range, whereas males disperse when 2–4 years old (Rogers 1987). In 1986, a 2.5-year-old male black bear traveled 140 km in southern Florida during a possible dispersal event (Maehr et al. 1988). Such movements and successful recruitment of individual bears may be important in reducing the effects of genetic isolation on small, fragmented populations. Males are especially vulnerable to vehicle collisions during the age of dispersal. This form of mortality may become more significant as Florida's bear populations become more fragmented (Wooding and Brady 1987).

Home range and movements have been studied with the aid of radiotelemetry in the Osceola and Ocala national forests (Mykytka and Pelton 1988; Florida Game and Fresh Water Fish Commission, unpubl. data). Adult home range sizes were approximately 28 km^2 for females and 170 km^2 for males. The largest male home range was 317 km^2 in Ocala National Forest (Wooding and Hardisky 1988) and 457 km^2 in Osceola National Forest (Mykytka and Pelton 1988).

Black bears may hibernate for up to 9 months (Nelson and Beck 1984). Studies of winter movements have shown that Florida black bears bed in ground nests for periods ranging from a few days to several months. In Florida, hibernation may be restricted to females producing cubs (Wooding 1987*a*). A radio-collared female that gave birth to cubs in January 1987 remained stationary from late December 1986 to April 1987. The energy demands of pregnancy and lactation and the high level of parental care needed by neonates are accommodated by the energy-saving physiology of hibernation (Ramsey and Dunbrack 1986).

Beeyard depredation by black bears is a serious problem throughout Florida, particularly from April through June in the northern counties (Maehr and Brady 1982*b*). This seasonal peak in depredation occurs during a period of high honey production at the hives coupled with a shortage of nutritious natural foods for bears. Monetary losses resulting from a bear visit can be severe (Maehr and Brady 1982*b*). Protecting beeyards from bears is best accomplished by use of a well-maintained electric fence (Brady and Maehr 1982).

CONSERVATION MEASURES TAKEN: In the 18th century, Seminole Indians hunted black bears, using the skin, meat, and fat for various purposes (Bartram 1943). With the settlement of Florida by Europeans, bears, viewed as vermin, were hunted indiscriminately. This attitude is still held by some (Williams 1978). In the past 40 years, numerous restrictive changes in the legal status of the Florida black bear have occurred. Prior to 1950, bears were unprotected. In 1950, bears were afforded the status of a game animal with a legal hunting season running concurrently with the deer season. In 1971, the statewide hunting season was closed except in Baker and Columbia counties and on several wildlife management areas. The Florida black bear was listed as a threatened species in 1974, except in Baker and Columbia counties and Apalachicola National Forest, where they were classified as game animals. This designation remains today.

Brady et al. (1986) examined sex and age characteristics of bears legally harvested in Florida. They concluded that harvests should be reduced in Apalachicola National Forest to protect females from overharvest. In Osceola National Forest and on private land in Baker and Columbia counties, harvests did not appear to be excessive. The bear season in Apalachicola was shortened from 23 days to 15 days in 1987. Other publicly owned lands supporting black bears in Florida include Ocala National Forest, St. Marks National Wildlife Refuge, Florida Panther National Wildlife Refuge, Chassahowitzka National Wildlife Refuge, Big Cypress National Preserve, Everglades National Park, Fakahatchee Strand State Preserve, Collier-Seminole State Park, Big Cypress Seminole Indian Reservation, Eglin Air Force Base, and some smaller areas.

Measures taken to reduce conflicts between bears and beekeepers include a trap-and-release program for bears that raid apiaries protected by electric fences (Maehr 1983). Bears trapped in this program are released at the capture site because there is some evidence that this experience creates an aversion to the beeyard (Brady and Maehr 1982).

CONSERVATION MEASURES PROPOSED: Long-term conservation of the Florida black bear is dependent upon preservation of large contiguous woodlands. Large tracts in public ownership that currently support bear populations are recognized as important population centers. The continued survival of bears outside of public lands is questionable because of the speed with which Florida is being developed. A detailed plan to preserve specific areas from development and to guide public land acquisition is needed.

Bear crossing signs at identified highway crossings are being erected to reduce roadkill mortality. Other methods that should be considered in new highway construction projects through inhabited bear range include fencing and wildlife underpasses that permit safe passage. A proposed addition to the Florida Driver's Handbook discusses ways that drivers can reduce the chances of hitting an animal.

Continued efforts should be made to accurately monitor legal bear harvests. Further work is needed to develop an accurate census method that works in Florida. A track-count method using mast-survey estimates may be a promising tool for measuring population trends (Wooding 1987*b*). Finally, efforts should continue to aggressively prosecute cases involving illegal killings of bears.

ACKNOWLEDGMENTS: We thank T. Hardisky, J. Brady, V. Heller, T. Logan, and T. Hines for reviewing earlier drafts of the manuscript. We are grateful to T. Steele for her help in preparing the manuscript. Recent and ongoing black bear research efforts by the Florida Game and Fresh Water Fish Commission were made possible through initial efforts by L. Williams and continued facilitation of T. Logan. Partial funding of research activities was from Federal Aid to Wildlife Restoration W-41-R.

Literature Cited

Alt, G. L. 1987. Characteristics of bear cubs at birth. Pennsylvania Game News 58:10–13.

Bartram, J. 1943. Diary of a journey through the Carolinas, Georgia and Florida from July 1, 1765 to April 10, 1766. Annotated by F. Harper. Trans. Amer. Philosophical Soc. 33:1–120.

Brady, J. R., and D. S. Maehr. 1982. A new method of dealing with apiary-raiding black bears. Proc. Annu. Conf. Southeast. Assoc. Fish and Wildl. Agencies 36:571–577.

Brady, J. R., and D. S. Maehr. 1985. Distribution of black bears in Florida. Florida Field Nat. 13:1–7.

Brady, J. R., J. B. Wooding, and C. Moore. 1986. An analysis of harvest data from two black bear populations in north Florida. Unpubl. report, Florida Game and Fresh Water Fish Commission, Tallahassee. 36 pp.

Bunnell, F. L., and D. E. N. Tait. 1981. Population dynamics of bears—implications. Pp. 75–98 *in* C. W. Fowler and T. D. Smith, eds., Dynamics of large mammal populations. John Wiley and Sons, New York. 477 pp.

Conte, J. A., D. J. Forrester, and J. R. Brady. 1983. Parasites of Florida black bears. Proc. Helminthol. Soc. Wash. 50:252–256.

Dusel-Bacon, C. 1979. "Come quick! I'm being eaten by a bear!" Alaska 45(2): 12–13, 72–74, 76–77.

Hall, E. R. 1981. The mammals of North America. 2d ed. John Wiley and Sons, New York. 1081 + 90 pp.

Harris, L. D. 1984. The fragmented forest. University of Chicago Press, Chicago. 211 pp.
Herrero, S. 1985. Bear attacks. Lyons and Burford, New York. 287 pp.
Maehr, D. S. 1983. Black bear depredation on beeyards in Florida. Pp. 133–135 *in* D. J. Decker, ed., First Eastern. Wildl. Damage Control Conf., Ithaca, New York.
Maehr, D. S. 1984*a*. Distribution of black bears in eastern North America. Proc. Eastern. Workshop Black Bear Manage. Res. 7:74.
Maehr, D. S. 1984*b*. The black bear as a seed disperser in Florida. Florida Field Nat. 12:40–42.
Maehr, D. S., and J. R. Brady. 1982*a*. Fall food habits of black bears in Baker and Columbia counties, Florida. Proc. Annu. Conf. Southeast. Assoc. Fish and Wildl. Agencies 36:565–570.
Maehr, D. S., and J. R. Brady. 1982*b*. Florida black bear-beekeeper conflict: 1981 beekeeper survey. Amer. Bee J. 122:372–375.
Maehr, D. S., and J. R. Brady. 1984*a*. Food habits of Florida black bears. J. Wildl. Manage. 48:230–235.
Maehr, D. S., and J. R. Brady. 1984*b*. Comparison of food habits in two north Florida black bear populations. Florida Scientist 47:171–175.
Maehr, D. S., and J. T. DeFazio. 1985. Foods of black bears in Florida. Florida Field Nat. 13:8–12.
Maehr, D. S., J. N. Layne, E. D. Land, J. W. McCown, and J. C. Roof. 1988. Long distance movements of a Florida black bear. Florida Field Nat. 16:1–6.
Maehr, D. S., R. C. Belden, E. D. Land, and L. Wilkins. 1990. Food habits of panthers in southwest Florida. J. Wildl. Manage. 54:420–423.
Merriam, C. H. 1896. Preliminary synopsis of the American bears. Proc. Biol. Soc. Wash. 10:65–83.
Mykytka, J. M., and M. R. Pelton. 1988. Evaluation of four standard home range methods based on movements of Florida black bear. Biotelemetry 10:159–166.
Mykytka, J. M., and M. R. Pelton. 1989. Management strategies for Florida black bears based on home range habitat composition. Internat. Conf. on Bear Research and Management 8:161–167.
Nelson, R. A., and T. D. I. Beck. 1984. Hibernation adaptation in the black bear: implications for management. Proc. Eastern Workshop Black Bear Manage. Res. 7:48–57.
Norris-Elye, L. T. S. 1951. The black bear as a predator of man. J. Mammal. 32:222–223.
Pearson, P. G. 1954. Mammals of Gulf Hammock, Levy County, Florida. Amer. Midland Nat. 51:468–480.
Pelton, M. R. 1982. Black bear. Pp. 504–514 *in* J. A. Chapman and G. A. Feldhammer, eds., Wild mammals of North America. Johns Hopkins University Press, Baltimore, Maryland. 1147 pp.
Pelton, M. R. 1986. Habitat needs of black bears in the East. Pp. 49–53 *in* D. L. Kulhavy and R. N. Conner, eds., Wilderness and natural areas in the eastern

United States: A management challenge. Stephen F. Austin State University, Nacagodoches, Texas. 416 pp.
Pelton, M. R., C. D. Scott, and G. M. Burghardt. 1976. Attitudes and opinions of persons experiencing property damage and/or injury by black bears in the Great Smokey Mountains National Park. Bears—Their Biol. and Manage. 4:157–167.
Ramsey, M. A., and R. L. Dunbrack. 1986. Physiological constraints of life history phenomena: The example of small bear cubs at birth. Amer. Nat. 127:735–743.
Rogers, L. L. 1976. Effects of mast and berry crop failures on survival, growth, and reproductive success of black bears. N. Amer. Wildl. Conf. 41:431–438.
Rogers, L. L. 1987. Effects of food supply and kinship on social behavior, movements, and population growth of black bears in northeastern Minnesota. Wildl. Monogr. 97:1–72.
Whitlock, S. C. 1950. The black bear as a predator of man. J. Mammal. 31:135–138.
Williams, L. E., Jr. 1978. Florida black bear. Pp. 23–25 *in* J. N. Layne, ed., Rare and endangered biota of Florida. Vol. 1. Mammals. University Presses of Florida, Gainesville. xx + 52 pp.
Wimsatt, W. A. 1963. Delayed implantation in the Ursidae with particular reference to the black bear (*Ursus americanus* Pallas). Pp. 49–76 *in* A. G. Enders, ed., Delayed implantation. University of Chicago Press, Chicago. 316 pp.
Wooding, J. B. 1987*a*. Final performance rept. Black bear hibernation study W-41-35. Florida Game and Fresh Water Fish Comm., Tallahassee. 7 pp.
Wooding, J. B. 1987*b*. Final performance rept. Black bear population index and estimation development W-41-35. Florida Game and Fresh Water Fish Comm., Tallahassee. 16 pp.
Wooding, J. B., and J. R. Brady. 1987. Black bear roadkills in Florida. Proc. Annu. Conf. Southeast. Assoc. Fish and Wildl. Agencies 41:438–442.
Wooding, J. B., and T. S. Hardisky. 1988. Final performance rept. Black bear habitat study W-41-35. Florida Game and Fresh Water Fish Comm., Tallahassee. 69 pp.

Prepared by: David S. Maehr, Florida Game and Fresh Water Fish Commission, 566 Commercial Boulevard, Naples, FL 33942; John B. Wooding, Florida Game and Fresh Water Fish Commission, 4005 S. Main Street, Gainesville, FL 32601.

Species of Special Concern

Round-Tailed Muskrat
Neofiber alleni
FAMILY CRICETIDAE
Order Rodentia

TAXONOMY: The genus *Neofiber* includes one living species, *Neofiber alleni*, known as the round-tailed muskrat or Florida water rat. Five subspecies were recognized by Schwartz (1953), Birkenholz (1972), and Hall (1981), with the following type localities: *N. a. alleni* True 1884, Georgiana, Brevard County, Florida; *N. a. apalachicolae* Schwartz 1953, Apalachicola, Franklin County, Florida; *N. a. exoristus*, Schwartz 1953, 12.1 mi southeast of Waycross, Ware County, Georgia; *N. a. nigrescens* Howell 1920, Ritta, Palm Beach County, Florida; and *N. a. struix* Schwartz 1953, 21 mi west of Miami, Dade County, Florida. Subspecific differences in pelage characteristics and size are believed to be of recent origin and the result of isolation of favorable habitat within the range (Schwartz 1953). Burt (1954) questioned the utility of these slightly differentiated subspecies.

DESCRIPTION: A moderately large microtine rodent, *Neofiber* is smaller than the muskrat (*Ondatra zibethicus*) but larger than any species of *Microtus* in North America. Adult measurements are: mass 187–357 g, total length 285–381 mm, tail 99–168 mm, hind foot 40–50 mm, and ear from notch 15–22 mm (Birkenholz 1972). The underfur is dense, gray at the base to rich brown at the tip on the back and pale buff on the belly (Birkenholz 1972). The guard hairs are dark brown and glossy. Juveniles are lead gray. The skull is similar to *Ondatra* but smaller, and the molars are rootless. The fur, ears, and eyes resemble *Ondatra*; but the scaled, sparsely haired tail is cylindrical rather than laterally flattened, with a long tuft of guard hairs dorsal to the base.

POPULATION SIZE AND TREND: *Neofiber* populations in local areas fluctuate tremendously from great abundance to absence (Porter 1953). Bir-

Round-tailed muskrat, *Neofiber alleni*. (Photo by Paul W. Lefebvre, U.S. Fish and Wildlife Service)

kenholz (1963) documented an increasing population trend on Paynes Prairie, Alachua County, between 1958 and 1960. He concluded that, in 4 years, round-tailed muskrats had spread from small, circumscribed population units into all suitable habitat over approximately 52 ha. Drought in 1961 reduced the Paynes Prairie population from several thousand to a few animals within 2 months. The *Neofiber* population in the sugarcane region studied by Lefebvre (1982) appeared to have declined from former levels (Steffen 1978). Population trends appear to be related to water-level fluctuations and changes in habitat conditions. The statewide population probably has declined as a result of wetland losses.

DISTRIBUTION AND HISTORY OF DISTRIBUTION: This species is restricted to Florida and southeastern Georgia, extending at least as far west as the Choctawhatchee River in Walton County (Wassmer and Wolfe 1983). A new locality was discovered in the 4450-ha Grand Bay wetlands ecosystem, 15 km northeast of Valdosta, Georgia, in 1988 (B.J. Bergstrom, personal communication). *Neofiber* is not known from northeastern Florida, including all of Nassau and Duval counties and portions of Baker, Clay, and St. Johns counties. In the remainder of the state it occurs in patches of favorable habitat, forming localized populations. Re-

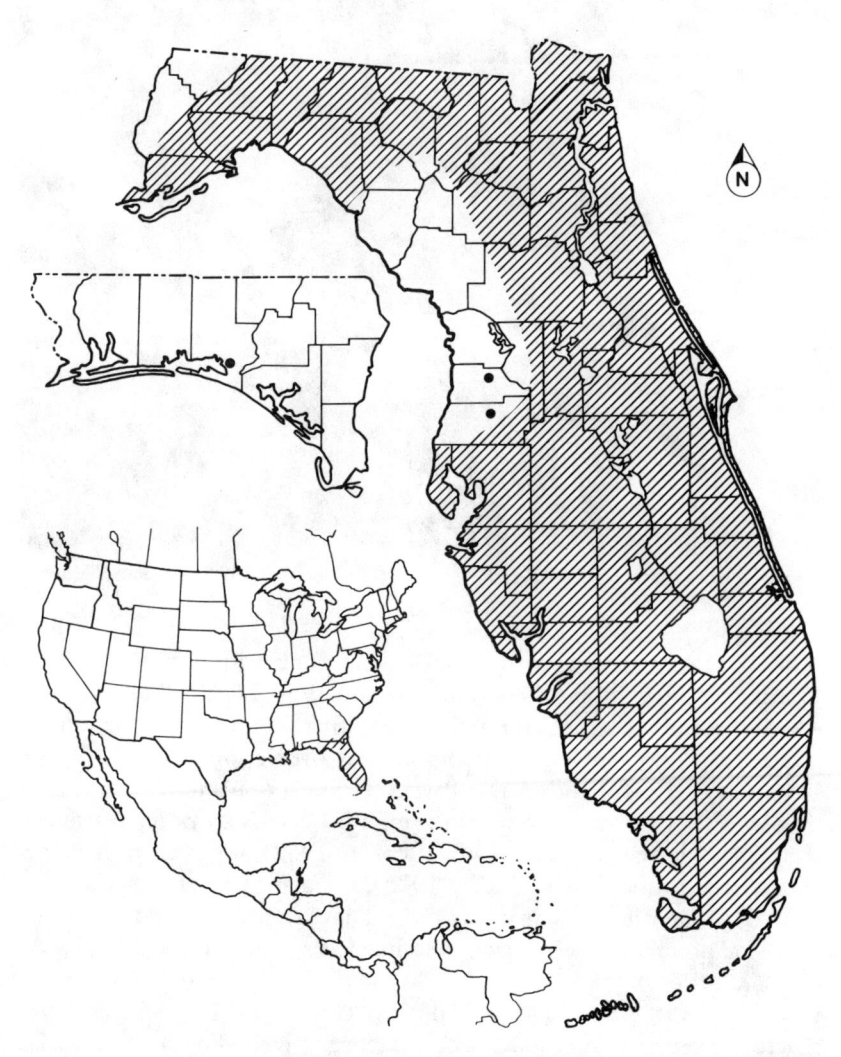

Distribution map of the round-tailed muskrat, *Neofiber alleni*. Hatching and dots represent the species; the subspecies are not shown individually.

ports from the Big Bend Gulf coast (Dixie County south to Hillsborough County) are rare, although Wassmer and Wolfe (1983) recently reported their presence in Hernando and Pasco counties. Paul (1968) provided information on a specimen from Hillsborough County and a population in Hillsborough River State Park (near the border of Hillsborough and Pasco counties). *Neofiber* is generally uncommon and spotty in occurrence in west-central Florida (Hillsborough, Polk, Manatee, Sarasota, and DeSoto counties; Layne et al. 1977). The first author found several *Neofiber* skulls in June 1987 below a barn owl (*Tyto alba*) roost on the Kissimmee River, bordering Polk and Osceola counties, 8 km north of the S-65A lock and spillway. Tilmant (1975) reported several *Neofiber* colonies in Everglades National Park, primarily in Shark River Slough, Taylor Slough, and on Cape Sable. S. R. Humphrey (personal communication) found a colony at the southern end of the Fakahatchee Strand, below U.S. 41, in a coastal brackish-water marsh (Collier County). *Neofiber* populations were documented on Merritt Island National Wildlife Refuge, Brevard County, in 1972 (Ehrhart 1984). Porter (1953) described the Loxahatchee National Wildlife Refuge in the Everglades as an excellent area for *Neofiber*. It also occurs in the Everglades Agricultural Area, where it burrows in the highly organic muck soils. In the late 1970s and early 1980s, studies conducted in the sugarcane-growing region indicated that the largest concentrations of *Neofiber* colonies were in western and northwestern Palm Beach County (Steffen et al. 1981; Lefebvre 1982). The species occurred in freshwater marshes on St. Vincent Island during the 1970s (R. Thompson, personal communication), but it was sought and not found there in 1984 (S. Christman, personal communication).

Skeletal remains of *Neofiber* have been found at 17 of 28 principal Pleistocene fossil localities in Florida (Webb 1974). Hibbard and Dalquest (1973) described an ancestral form, *Proneofiber guildayi*, from a Pleistocene deposit in Texas. A larger form (*Neofiber leonardi*) is known from the Pleistocene of Kansas and Texas (Hibbard 1943; Meade 1952). Frazier (1977) extended the range of *N. leonardi* to Florida and West Virginia, supporting Meade's (1952) speculation that the genus once ranged from Florida to the Plains region. Climatic changes probably caused a progressive range restriction through the later Pleistocene to the Recent (Birkenholz 1963; Frazier 1977). Frazier (1977) concluded that *N. leonardi* graded into *N. alleni*. *Neofiber diluvianus*, the same size as *N. alleni* but with slight differences in tooth pattern, is recorded from interglacial Pleistocene deposits of Pennsylvania (Hibbard 1955).

GEOGRAPHICAL STATUS: Because *Neofiber* is almost restricted to Florida, its survival status here essentially determines the status of this monospecific genus. Throughout most of its range, the geographical status of *Neofiber* is difficult to assess because its distribution is discontinuous and it is easily overlooked. Major swamps, ponds, cypress domes, and burrow pits in the Osceola National Forest were surveyed for round-tailed muskrats, and none were found (U.S. Fish and Wildlife Service 1978). Round-tailed muskrat houses were found in 26 percent of the 53 wetlands searched along the proposed route of the Cross Florida Barge Canal, from Yankeetown to Palatka; houses were found only in Marion and Putnam counties (Florida Game and Freshwater Fish Commission 1976). *Neofiber* burrow systems are relatively easy to locate in sugarcane fields up to 2 months after harvest (Steffen 1978; Lefebvre 1982).

HABITAT REQUIREMENTS AND HABITAT TREND: *Neofiber* typically inhabits shallow marshes with emergent vegetation. Dense stands of maidencane (*Panicum hemitomon*) and pickerelweed (*Pontederia cordata*) provide preferred habitat. *Neofiber* also inhabits salt marshes on Merritt Island National Wildlife Refuge (Ehrhart 1984) and on Cape Sable (Tilmant 1975). Marshes with water levels not exceeding 50 cm, with soft substrate deep enough to allow burrowing to water during dry periods, support the largest populations (Porter 1953; Birkenholz 1963). In sugarcane fields, where the water level is maintained at 0.60–0.75 m below the soil surface, the round-tailed muskrat builds extensive tunnel systems.

Primarily an herbivore, the round-tailed muskrat eats roots and stems of aquatic and semiaquatic vegetation. Major food plants reported for freshwater marshes are maidencane, pickerelweed, cut-grass (*Leersia*), and arrowhead (*Sagittaria*; Birkenholz 1963). In sugarcane fields, sugarcane (sprouts, stalks, and roots), sedges (*Cyperus*), and grasses (*Panicum*) appear to be the most important foods (Steffen 1978; Lefebvre 1982).

Florida has lost an enormous portion of its original wetland area (Mitsch and Gosselink 1986), which has reduced habitat for *Neofiber* and many other species. The round-tailed muskrat was present in the sugarcane growing region before the commercial cane industry began in 1929 (Howell 1920), but evidently it was not common in sugarcane fields before 1964 (Porter 1953; Steffen 1978). The sugarcane region expanded from 20,500 to 90,000 ha between 1960 and 1964 (Samol 1972) and is now approximately 162,000 ha (J. R. Orsenigo, personal communication). The relatively stable water table and deep muck soils of some sugarcane areas appear to provide a favorable habitat for *Neofiber*. The oxidation of exposed muck soils, however, limits the future availability of this habitat to *Neofiber*.

VULNERABILITY OF SPECIES AND HABITAT: Suitable *Neofiber* habitat is vulnerable to fragmentation and destruction by man and to drastic fluctuations in water levels during flood and drought periods. *Neofiber* habitat can be greatly affected by water management practices. Phenological changes in habitat, particularly reduction of emergent vegetation in the winter, influences *Neofiber* vulnerability to flooding and predation (Birkenholz 1963). The species has a relatively low reproductive potential and is vulnerable to a number of reptilian, avian, and mammalian predators. *Neofiber* is considered to be a pest in sugarcane because of the damage it does to stalks and roots (Samol 1972); however, rodent control methods are generally not specifically directed against *Neofiber*.

CAUSES OF THREAT: Draining, filling, or mining of wetlands threaten existing and potential *Neofiber* habitat. Isolation of suitable habitat by urbanization may prevent dispersal, and isolated populations would be extremely vulnerable to destruction by natural fluctuations in water levels. No records document *Neofiber* use of slime settling ponds, old flooded mine cuts, or aquatic habitats in reclaimed phosphate-mined areas (Layne et al. 1977). Prescribed winter burning of marsh on St. Vincent Island may have contributed to a *Neofiber* decline there. Some sugarcane cultivation procedures, particularly shorter (3-year) crop rotation, simultaneous replanting of large blocks of fields, and mechanical harvesting may have contributed to a decline in *Neofiber* abundance in areas of the sugarcane production region in western Palm Beach County (Lefebvre 1982).

RESPONSES TO HABITAT MODIFICATION: The round-tailed muskrat burrows extensively in drained muck soils in the Everglades Agricultural Area, indicating an ability to adapt to large-scale habitat modification. The maintenance (by canals and pumping) of a high water table in this region and the dense cover provided by maturing sugarcane are critical to this adaptation. Wetland habitat created on reclaimed phosphate-mined lands potentially might support *Neofiber* populations (Layne et al. 1977). Prescribed burns conducted when water levels were at or immediately below ground surface resulted in the least damage to *Neofiber* colonies in the Everglades, although *Neofiber* numbers declined after burning (Tilmant 1975).

DEMOGRAPHIC CHARACTERISTICS: Birkenholz (1963) concluded that *Neofiber* do not have well-marked seasonal breeding cycles under ordinary environmental conditions, and that increased reproduction was correlated with improvements in habitat conditions. *Neofiber* attains sexual maturity at approximately 90–100 days, and gestation lasts 26–29 days (Birkenholz 1963). The mean number of fetuses in 48 gravid females collected from marsh habitat (Paynes Prairie) was 2.3, and 4 or 5 litters may be produced per female per year (Birkenholz 1963). In southern Florida sugarcane fields, the mean number of fetuses in 50 females collected monthly over 2 years was 2.1 (N. Holler, unpubl. data); mean number of fetuses from 37 gravid females collected in the same habitat during two winters (December–March) was 1.8 (Lefebvre 1982). The largest numbers of juveniles and subadults appeared in May in sugarcane-field burrow systems that were live trapped monthly from May through November in 2 consecutive years (Lefebvre 1982).

In both marshes and muck fields, *Neofiber* appears to be colonial. Porter (1953) referred to the interconnecting tunnels of a burrow system as a colony. Tilmant (1975) designated a distinct cluster of houses, whether small (e.g., 5 houses) or large (>100 houses) as a colony in Everglades National Park. Birkenholz (1963) estimated that an average of 2 houses exist for every round-tailed muskrat, and that no more than 1 adult uses each house. Burrow system occupancy in sugarcane fields averaged fewer than 2 individuals at a time (Lefebvre 1982). Single adults, male/female pairs, a female with 1–2 young, or 1–2 subadults were the most common burrow-system occupants. Extended families were encountered occasionally. Adult males frequently moved among neighboring burrow systems, presumably seeking mates. Number of young per adult female tended to be lower in fields with low burrow system densities (<10/30 ha field) than in high density fields (>15 burrow systems; Lefebvre 1982).

Fourteen species of intestinal helminths have been identified in round-tailed muskrats from several sites in Florida (Forrester et al. 1987). Three primary species, a trematode (*Quinqueserialis floridensis*), cestode (*Anoplocephaloides neofibrinus*), and nematode (*Carolinensis kinsellai*), occur commonly and exclusively in *Neofiber*. The remaining less common species are primarily derived from cotton rats (*Sigmodon hispidus*). The helminths of round-tailed muskrats from sugarcane fields differed markedly from those found in muskrats from prairies and marshes, reflecting the habitat differences and perhaps the greater sharing of food and spatial resources by these two rodents in sugarcane fields (Forrester et al. 1987).

Eight species of ectoparasites were found on round-tailed muskrats collected in Alachua and Highland counties: the laelapid mites *Laelaps evansi* and *Androlaelaps farenholzi*; the chigger *Euschoengastia splendens*; the myobiid mite *Radfordia* sp.; a flea, *Polygenis gwyni*; and three new species of listrophorid mites, *Listrophorus caudatus*, *L. layni*, and *Prolistrophorus birkenholzi* (Smith et al. 1988). Two types of mites were found on *Neofiber* collected from sugarcane fields in Palm Beach County: blood sucking mites (Macronyssidae) and hair mites (Listrophoridae; Lefebvre 1982).

KEY BEHAVIORS: *Neofiber* requires access to water. Burrowing during periods of low water enables the round-tailed muskrat to occupy various habitats and to survive despite fluctuating water levels. *Neofiber* probably becomes nomadic during extreme environmental conditions such as drought or flood, although little is known about their dispersal capabilities. The greatest dispersal distances determined by recapture of a marked individual were 36 m on Paynes Prairie (Birkenholz 1963) and 320 m in sugarcane fields (Lefebvre 1982). Birkenholz (1963) found road kills near Paynes Prairie up to 800 m from the nearest known habitat. Between October 1985 and October 1989, at least 3 *Neofiber* wandered into pitfall traps around a temporary pond located in an ecotone between a sandhill and a xeric live oak community on the Ordway-Swisher Preserve (Putnam County), 400 m from the nearest permanent water body (C. K. Dodd, personal communication). The observed rate of dispersal on Paynes Prairie was no more than 270 m in a 3-month period (Birkenholz 1963). *Neofiber* probably follow waterways when moving long distances (Birkenholz 1963). The network of canals throughout the sugarcane growing region, and the ditches and subsurface drains within fields, undoubtedly aid *Neofiber* dispersal in this habitat.

Neofiber is secretive and difficult to observe. In the laboratory it displays a nocturnal and crepuscular activity pattern, with peaks occurring just before light onset and offset (Webster et al. 1980). Tightly woven spherical-to-hemispherical houses, 18–60 cm in diameter, are constructed of emergent aquatic plants. Interior chambers are about 10 cm in diameter and usually contain two plunge holes leading out beneath the surface of the water. Floating feeding stands or shelters, 10–15 cm in diameter and containing one or two plunge holes, are usually found near house sites. In sugarcane fields, sugarcane leaves are used in the construction of smaller, rudimentary houses. Houses may be an alternate strategy for wet conditions in sugarcane fields. They were not present in all burrow systems and were reinforced with fresh leaves during wet periods (Lefebvre 1982).

CONSERVATION MEASURES TAKEN: No measures specifically oriented toward *Neofiber* conservation have been taken. Preservation of large wetlands such as the Okefenokee Swamp, Paynes Prairie, and the Everglades, and many smaller ones has undoubtedly benefited this species.

CONSERVATION MEASURES PROPOSED: Continued protection of wetlands should be encouraged. The status of *Neofiber* populations in isolated wetlands regulated by the Water Management Districts should be evaluated in permit reviews. Conservation of muck soils in the Everglades Agricultural Area probably is essential to survival of the populations in this region. In protected areas such as Paynes Prairie State Preserve and Everglades National Park, impacts on *Neofiber* populations should be considered in development of fire- and water-management strategies. Although *Neofiber* is generally believed to be intolerant of salt water, it has been found in brackish water; its status in coastal marshes needs to be investigated, particularly along the northern Gulf coast. The impact of land-management practices, such as burning, phosphate mining, and crop cultivation, on *Neofiber* population dynamics also requires further study; when this species is removed from large areas, it may not readily reinvade, and long-term population declines may occur. Additional research on population ecology and experimental introduction in restored wetland habitats is needed.

ACKNOWLEDGMENTS: Helpful reviews were provided by K. J. Cook, J. N. Layne, and T. J. O'Shea. The author of the species account in the 1978 edition was James T. Tilmant.

Literature Cited

Birkenholz, D. E. 1963. A study of the life history and ecology of the round-tailed muskrat (*Neofiber alleni* True) in north-central Florida. Ecolog. Monogr. 33:255–280.
Birkenholz, D. E. 1972. *Neofiber alleni*. Mammalian Species 15:1–4.
Burt, W. H. 1954. The subspecies category in mammals. Syst. Zool. 3:99–104.
Ehrhart, L. M. 1984. Some avian predators of the round-tailed muskrat. Florida Field Nat. 12:98–99.
Florida Game and Fresh Water Fish Commission. 1976. Cross Florida barge canal restudy report wildlife study. Vol. 4. Prepared for the Department of the Army, Jacksonville District, Corps of Engineers, Jacksonville, Florida. 137 pp.
Forrester, D. J., D. B. Pence, A. O. Bush, D. M. Lee, and N. R. Holler. 1987. Ecological analysis of the helminths of round-tailed muskrats (*Neofiber alleni* True) in southern Florida. Canad. J. Zool. 65:2976–2979.
Frazier, M. K. 1977. New records of *Neofiber leonardi* (Rodentia: Cricetidae) and the paleoecology of the genus. J. Mammal. 58:368–373.

Hall, E. R. 1981. The mammals of North America. John Wiley and Sons, New York. 2 vol., xv + 1181 + 180 pp.

Hibbard, C. W. 1943. The Rezabek fauna, a new Pleistocene fauna from Lincoln County, Kansas. Univ. Kansas Sci. Bull. 29:235–247.

Hibbard, C. W. 1955. Notes on the microtine rodents from the Port Kennedy cave deposit. Proc. Acad. Nat. Sci. Philadelphia 107:87–97.

Hibbard, C. W., and W. W. Dalquest. 1973. *Proneofiber*, a new genus of vole (Cricetidae: Rodentia) from the Pleistocene Seymour formation of Texas, and its evolutionary and stratigraphic significance. Quaternary Res. 3:269–274.

Howell, A. H. 1920. Description of a new race of the Florida water-rat (*Neofiber alleni*). J. Mammal. 1:79–80.

Layne, J. N., J. A. Stallcup, G. E. Woolfenden, M. N. McCauley, and D. J. Worley. 1977. Fish and wildlife inventory of the seven-county region included in the central Florida phosphate industry areawide environmental impact study 2:664–1279. Prepared for the Fish and Wildlife Service, U.S. Department of the Interior, Washington, D.C.

Lefebvre, L. W. 1982. Population dynamics of the round-tailed muskrat (*Neofiber alleni*) in Florida sugarcane. Unpubl. Ph.D. diss., University of Florida, Gainesville. 204 pp.

Meade, G. C. 1952. The water rat in the Pleistocene of Texas. J. Mammal. 33:87–89.

Mitsch, W. J., and J. G. Gosselink. 1986. Wetlands. Van Nostrand Reinhold, New York. 539 pp.

Paul, J. R. 1968. Round-tailed muskrat in west-central Florida. Florida Acad. Sci. 30:227–229.

Porter, R. P. 1953. A contribution to the life history of the water rat, *Neofiber alleni*. Unpubl. M.S. thesis, University of Miami, Coral Gables. 84 pp.

Samol, H. H. 1972. Rat damage and control in the Florida sugarcane industry. Proc. Internat. Soc. Sugar Cane Technolog. 14:503–506.

Schwartz, A. 1953. A systematic study of the water rat (*Neofiber alleni*). Univ. of Michigan Occas. Papers Mus. Zool. 547:1–27.

Smith, M. A., J. O. Whitaker, Jr., and J. N. Layne. 1988. Ectoparasites of the round-tailed muskrat (*Neofiber alleni*) with special emphasis on mites of the family Listrophoridae. Amer. Midland Nat. 120:268–275.

Steffen, D. E. 1978. The occurrence of and damage by the Florida water rat in Florida sugarcane production areas. Unpubl. M.S. thesis, Virginia Polytechnical Institute and State University, Blacksburg. 105 pp.

Steffen, D. E., N. R. Holler, L. W. Lefebvre, and P. F. Scanlon. 1981. Factors affecting the occurrence and distribution of Florida water rats in sugarcane fields. Proc. Amer. Soc. Sugar Cane Technolog. 9:27–32.

Tilmant, J. T. 1975. Habitat utilization by round-tailed muskrats (*Neofiber alleni*) in Everglades National Park. Unpubl. M.S. thesis, Humboldt State University, Arcadia, California. 91 pp.

True, F. W. 1884. Muskrat with a round tail. Science 4:34.

U.S. Fish and Wildlife Service. 1978. Osceola National Forest phosphate extraction and processing: Impacts on federally listed threatened or endangered and other species of concern. Prepared for the Office of Biological Services, Fish and Wildlife Service, U.S. Department of the Interior, Washington, D.C.

Wassmer, D. A., and J. L. Wolfe. 1983. New Florida localities for the round-tailed muskrat. Northeast Gulf Sci. 6:197–199.

Webb, S. D., ed. 1974. Pleistocene mammals of Florida. University Presses of Florida, Gainesville. 270 pp.

Webster, D. G., R. L. Evans, and D. A. Dewsbury. 1980. Behavioral patterns of round-tailed muskrats (*Neofiber alleni*). Florida Sci. 43:1–6.

Prepared by: Lynn W. Lefebvre, National Ecology Research Center, U.S. Fish and Wildlife Service, 412 NE 16th Avenue, Gainesville, FL 32601; James T. Tilmant, South Florida Research Center, Everglades National Park, P.O. Box 279, Homestead, FL 33090.

Rare

Southeastern Big-Eared Bat
Plecotus rafinesquii macrotis
FAMILY VESPERTILIONIDAE
Order Chiroptera

TAXONOMY: Other names are Rafinesques's big-eared bat, eastern big-eared bat, eastern mule-eared bat, and eastern lump-nosed bat. Two subspecies are recognized: *P. r. macrotis* Le Conte 1831, with type locality near Riceboro, Georgia; and *P. r. rafinesquii* Lesson 1827, type locality Mount Carmel, Illinois (Handley 1959; Jones 1977).

DESCRIPTION: This is a medium-sized bat that can be recognized in Florida by very long ears (26–31 mm in five Florida Museum of Natural History [UF] specimens) and two large fleshy lumps on the snout that border the nostrils. *P. rafinesquii* can be distinguished from the closely related *P. townsendii*, which does not occur in Florida, by hairs that are bicolored with black bases. Tips of the hairs on the upper parts are gray, those of the underparts are white. The fur is long and silky, and long hairs on the feet project past the toes. Total length and forearm measurements range from 81 to 110 and 39 to 45 mm, respectively. The bats weigh 7 to 13 g; females weigh slightly more than males.

POPULATION SIZE AND TREND: Size and overall trend of the population are not known because this bat has not been studied in detail in Florida. It does not seem to be abundant anywhere in the state, is rarely seen, and is uncommon in collections. In Louisiana and Mississippi, these animals are found only in small numbers at scattered locations. Compared with other species, they are considered scarce (Jones and Suttkus 1975). The lack of breeding records for *P. rafinesquii* in Florida led Jennings (1958) to doubt that these bats have a functional population in the state.

Southeastern big-eared bat, *Plecotus rafinesquii macrotis*.
(Photo by Roger W. Barbour)

DISTRIBUTION AND HISTORY OF DISTRIBUTION: *P. r. rafinesquii* occurs in southern Indiana and Ohio, western West Virginia, Virginia, and North Carolina, the extreme northern portions of Mississippi, Alabama, and Georgia, western Arkansas and Illinois, and all of Kentucky and Tennessee. *P. r. macrotis* occurs in the southern states from Louisiana eastward across Mississippi, Alabama, Georgia, South Carolina, and the eastern half of North Carolina. In Florida, it has been collected in the panhandle and in the northern half of the peninsula (Jennings 1958; Jones 1977). G. S. Morgan (personal communication) has identified early Pleistocene *Plecotus* near Inglis in Citrus County, and Martin (1974) lists middle Pleistocene *P. rafinesquii* from near Coleman in Sumter County. *P. alleganiensis*, a form closely related to *P. rafinesquii*, is known from the Pleistocene of Maryland (Gidley and Gazin 1933).

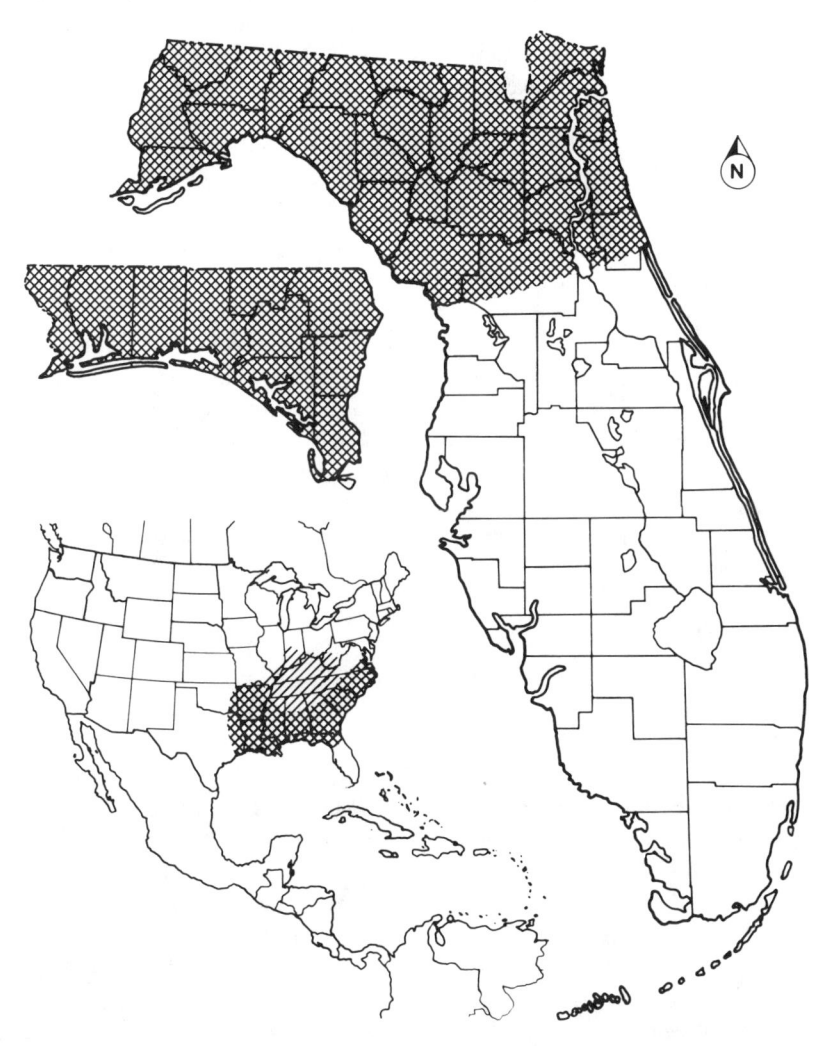

Distribution map of the southeastern big-eared bat, *Plecotus rafinesquii macrotis*. Hatching represents the species; crosshatching represents the subspecies.

GEOGRAPHICAL STATUS: Florida is the southernmost part of the range of this species. Local distribution patterns are not known.

HABITAT REQUIREMENTS AND HABITAT TREND: These animals are restricted to forested areas. In Florida they have been taken in hardwood hammocks and pine flatwoods (Moore 1949; Jennings 1958). Roosts have been found in old buildings and shacks with semilighted interiors, as well as in tree hollows and crevices. These bats enter torpor during the winter and hibernate in caves. Although they do not appear regularly in northern Florida caves (McNab 1974), one specimen (UF 13016) was taken in Florida Caverns State Park (Jackson County). Whether they migrate is not known.

VULNERABILITY OF SPECIES AND HABITAT: Because very little is known about this bat, definitive statements about its vulnerability cannot be made. Forested areas and the old buildings in them that serve as bat roosts are often destroyed as a result of human activity and changes in land use. This destruction is likely to be particularly harmful to *P. rafinesquii* because abandoned dwellings also serve as nursery sites (Jones and Suttkus 1975). Disturbance and harassment from people can also be serious threats. Insecticides and other pesticides are also likely to have a negative impact on these animals. Clearing and development of forested areas in northern Florida are increasing steadily and are likely to be deleterious to this bat.

CAUSES OF THREAT: Unknown.

RESPONSES TO HABITAT MODIFICATION: Unknown.

DEMOGRAPHIC CHARACTERISTICS: Jones (1977) reported that in northern portions of the range of *P. r. macrotis* (Louisiana and Mississippi) adults copulate in autumn and winter. Females give birth to one young in late May or early June. Lowery (1974) reported pregnant females in Louisiana in May and early June and lactating females from May to July. In Florida, male specimens had small testes in August (UF 13016) and January (UF 13298). A male collected in April (UF 13052) had large testes (4×9 mm).

Young *P. rafinesquii* begin to fly on their own at about 3 weeks of age. At 1 month they have attained adult weight. Breeding, at least in males, does not appear to take place until the second year (Jones and Suttkus

1975). These authors recaptured several marked individuals after 8 years. The longevity record is 10 years (Paradiso and Greenhall 1967).

KEY BEHAVIORS: *P. rafinesquii* has relatively large, broad wings with low wing loading (Norberg and Rayner 1987). These characteristics suggest that these bats are slow fliers with excellent maneuverability, and that they fly within, or close to, vegetation. They are also likely to hover, indicating a hovering-gleaning mode of foraging (U. M. Norberg, personal communication). The wing shape of these bats suggests that they are inefficient for fast or long-distance flight. Consequently, these bats are not believed to be able to fly continuously for long periods of time. If forced to forage in deforested areas, they are not likely to compete successfully with faster-flying bats (i.e., those with higher aspect-ratio wings) for food. Barbour and Davis (1969) describe the flight of this bat as alternating between swift and nearly hovering.

Because little published information exists on the behavior of *P. rafinesquii*, the following observations are worth reporting. In February 1988, William H. Kern received and subsequently released a healthy adult male *P. rafinesquii*. We had the opportunity to observe it fly and feed in a small room (J. J. Belwood and W. H. Kern, personal observation). In a confined area, this bat was a slow and delicate flier. Its large ears were held erect as it flew. The bat maneuvered gracefully about the room and spent most of its time flying close to the floor, furniture, and other large objects. It was extremely agile in small spaces and it occasionally hovered in place for several seconds.

This bat produced echolocation calls with frequencies that ranged from 60 to 90 kHz. *P. rafinesquii* probably uses broadband, frequency-modulated echolocation calls to navigate and prey-produced sounds to locate prey. These characteristics are tied to the habit of foraging (gleaning) in small feeding areas in cluttered environments (Simmons et al. 1979; Belwood 1988).

The captive bat occasionally landed on a wall between flying bouts. Like members of most other bat species, this *P. rafinesquii* responded visibly to sounds that were deliberately made to attract its attention. Unlike long-eared phyllostomines (J. J. Belwood, personal observation), this bat did not move its ears in the direction of sounds. Instead, it moved its whole head.

P. rafinesquii are insectivorous and emerge from roosts to feed after dark (Barbour and Davis 1969). Although diet in these animals has not been studied in detail, fecal material removed from a museum specimen (UF 13052) indicates that the bat had fed on moths (J. J. Belwood, personal observation). Brown (1978) stated that the bats can feed on insects in the air or can glean them from vegetation. The captive bat observed by Belwood and Kern was hand-fed and readily consumed 18 noctuid moths, 1 chironomid fly, and 1 green lacewing in 2 hours, but it refused a small scarab beetle (wing length 12 mm). The largest moth consumed had a wing length of 15 mm, the smallest 10 mm. The head capsules and wings of the moths and lacewing were culled before the insects were eaten. The bat weighed 6.5 g prior to being fed and 8.5 g afterwards.

In Louisiana and Mississippi, most *P. rafinesquii* roost in small to medium-sized colonies that number from 2 to 100 individuals, although some individuals roost singly (Jones and Suttkus 1975). Sexual composition in these roosts varies seasonally as a function of reproductive activities. Adults show roost-site fidelity and two or three roosting sites are often used by the same bats. *P. rafinesquii* may share roosts with members of other species (including *Pipistrellus subflavus* and *Myotis austroriparius*), and they have been captured in mist nets with other species that occur in Florida (*Lasiurus borealis, Eptesicus fuscus, Nycticeius humeralis*, and *P. subflavus*).

CONSERVATION MEASURES TAKEN: None specifically for *P. rafinesquii*. Large tracts of land in northern Florida have been publicly acquired as parks or preserves. The individual big-eared bat whose behavior is reported here was obtained at Manatee Springs State Park, Levy County. Whether other individuals of this species occur regularly on these public lands is not known.

CONSERVATION METHODS PROPOSED: Virtually nothing is known about these bats in Florida. They must be studied in detail to determine basic facts about their habitat requirements, status, and breeding potential in the state. If roosts are found in forested areas, the roosts and the areas surrounding them should be preserved. An attempt also should be made to halt the destruction or repair of older buildings and structures in such areas. An attempt should be made to educate people about the importance of these animals.

ACKNOWLEDGMENTS: This was written during a Smithsonian postdoctoral fellowship. Much of this account was taken from Jones (1977) and the original account on *P. rafinesquii* in this series (Brown 1978). Ulla Norberg examined drawings of a specimen and provided information on flight characteristics and habitat restrictions, G. S. Morgan supplied information on fossil records, and W. H. Kern allowed access to the *P.*

rafinesquii he had in captivity. Roger W. Barbour kindly supplied the photograph. Jeffery A. Gore read and commented on the manuscript.

Literature Cited

Barbour, R. W., and W. H. Davis. 1969. Bats of America. University Press of Kentucky, Lexington. 286 pp.

Belwood, J. J. 1988. Foraging behavior, prey selection, and echolocation in phyllostomine bats. Pp. 601–605 *in* P. E. Nachtigall and P. W. B. Moore, eds., Animal sonar: Processes and performance. Plenum Press, New York. 862 pp.

Brown, L. N. 1978. Southeastern big-eared bat, *Plecotus rafinesquii* Lesson. P. 35 *in* J. N. Layne, ed., Rare and endangered biota of Florida. Vol. 1. Mammals. University Presses of Florida, Gainesville. xx + 52 pp.

Gidley, J. W., and C. L. Gazin. 1933. New Mammalia in the Pleistocene fauna from Cumberland Cave. J. Mammal. 14:343–357.

Handley, C. O., Jr. 1959. A revision of American bats of the genera *Euderma* and *Plecotus*. Proc. U.S. Nat. Mus. 110:95–246.

Jennings, W. L. 1958. The ecological distribution of bats in Florida. Unpubl. Ph.D. diss., University of Florida, Gainesville. 125 pp.

Jones, C. 1977. *Plecotus rafinesquii*. Mammalian Species 69:1–4.

Jones, C., and R. D. Suttkus. 1975. Notes on the natural history of *Plecotus rafinesquii*. Occas. Papers Mus. Zool. Louisiana State Univ. 47:1–14.

Lowery, G., Jr. 1974. The mammals of Louisiana and its adjacent waters. Louisiana State University Press, Baton Rouge. 565 pp.

McNab, B. K. 1974. The behavior of temperate cave bats in a subtropical environment. Ecology 55:943–958.

Martin, R. A. 1974. Fossil mammals from the Coleman IIA fauna, Sumter County. Pp. 35–99 *in* S. D. Webb, ed., Pleistocene mammals of Florida. University Presses of Florida, Gainesville. 270 pp.

Moore, J.C. 1949. Putnam County and other Florida mammal notes. J. Mammal. 30:57–66.

Norberg, U. M., and J. M. V. Rayner. 1987. Ecological morphology and flight in bats (Mammalia: Chiroptera): Wing adaptations, flight performance, foraging strategy and echolocation. Phil. Trans. R. Soc. London B, 316:335–427.

Paradiso, J. L., and A. M. Greenhall. 1967. Longevity records for American bats. Amer. Midland Nat. 78:251–252.

Simmons, J. A., M. B. Fenton, and M. J. O'Farrell. 1979. Echolocation and pursuit of prey by bats. Science 203:16–21.

Prepared by: Jacqueline J. Belwood, Bat Conservation International, P.O. Box 162603, Austin, TX 78716.

Rare

Eastern Chipmunk
Tamias striatus striatus
FAMILY SCIURIDAE
Order Rodentia

TAXONOMY: No other names (Jones et al. 1986). The eastern chipmunks that occur in Florida are closely related to those from Alabama, Mississippi, and Louisiana, and should be recognized as *Tamias striatus striatus* (Jones and Suttkus 1979; Hall 1981).

DESCRIPTION: The eastern chipmunk is a rather typical sciurine form, with a stout body, a well-haired tail, and prominent ears with rounded tips. In adult animals, total length is 215–300 mm; length of tail is 78–113 mm. The species is distinguished by five longitudinal dark and four light stripes on the dorsum; two of the light stripes are at least twice as broad as any of the other stripes. Individuals from Florida are among the largest and most richly colored of all *T. striatus* (Jones and Suttkus 1979).

POPULATION SIZE AND TREND: Size of the population in Florida is unknown. Interviews with local residents suggest that the species has not declined recently over any large portion of its Florida range, but quantitative estimates of trends in population size are not available (Gore, in preparation).

DISTRIBUTION AND HISTORY OF DISTRIBUTION: The eastern chipmunk occurs throughout most of the eastern United States and southern Canada east of the Great Plains and from James Bay, Ontario, to Louisiana and western Florida. The species was first reported from Florida in 1962 and for many years was known only from two localities: 8 km southwest of Laurel Hill and 11.2 km southwest of Laurel Hill in Okaloosa County adjacent to the Yellow River (Stevenson 1962, 1976; Jones and Suttkus

Eastern chipmunk, *Tamias striatus striatus*. (Photo by Barry W. Mansell)

1979). A recent survey, however, found eastern chipmunks across a much larger range that includes portions of five Florida counties and watersheds of four major streams (Gore, in preparation). The new range recorded within Florida probably reflects differences in the extent of the surveys rather than a recent range expansion. Unfortunately, few data exist to determine past changes in the distribution of this species within Florida.

GEOGRAPHICAL STATUS: The current lack of specimens from southern Alabama make it impossible to determine whether the Florida population is isolated from other populations or is simply at the southern end of a narrow peninsula of distribution extending from the more northern range of the species (Jones and Suttkus 1979). Most references (e.g., Hall 1981) do not consider the Florida population to be isolated, and the wide distribution of sightings or reported sightings across western Florida and southern Alabama further suggest that eastern chipmunks are broadly distributed from central Alabama into Florida (Gore, in preparation). The southern boundary of the species' range in Florida has been delineated only crudely and factors limiting expansion to the south or east have not been identified. Neill (1957) and Wolfe et al. (1988) have discussed some of the unusual ecological conditions of the panhandle of

Florida and the patterns of distribution of numerous animals and plants that occur there.

HABITAT REQUIREMENTS AND HABITAT TREND: In Florida, eastern chipmunks are typically found in deciduous woods, but they also occur in

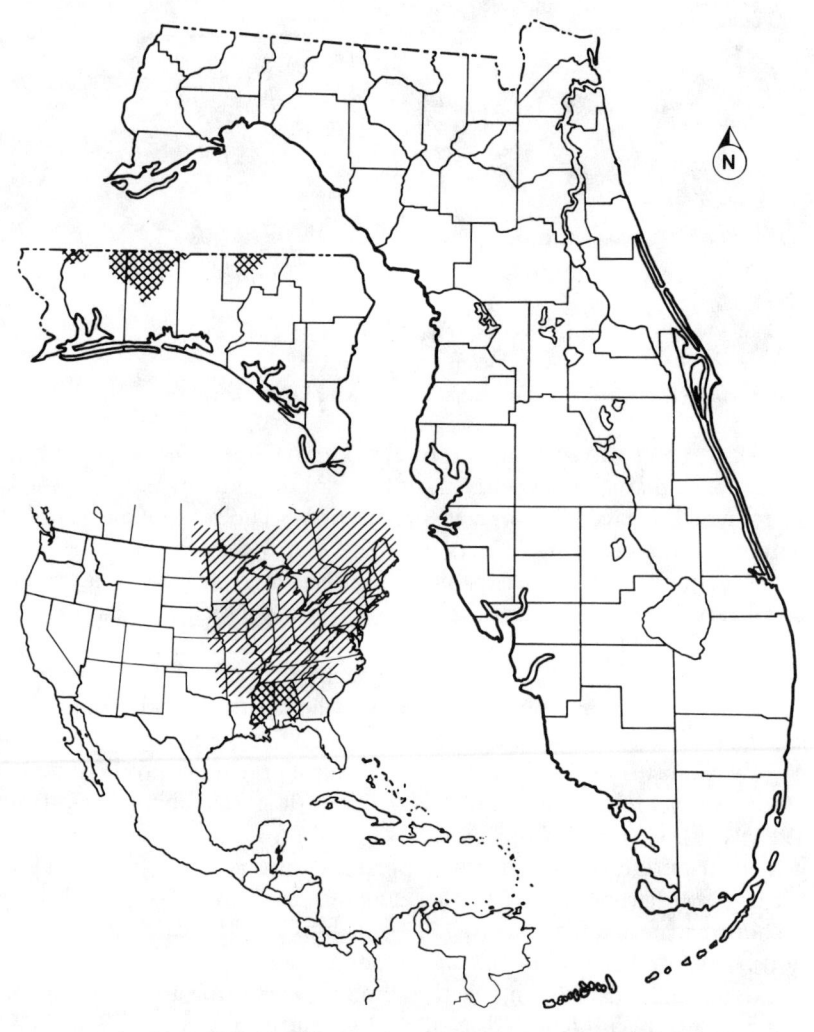

Distribution map of the eastern chipmunk, *Tamias striatus striatus*. Hatching represents the species; crosshatching represents the subspecies.

mixed deciduous-pine (*Pinus* sp.) forests and along wooded fencerows and forest edges. They may be present on residential property, if trees and shrubs are abundant. Stevenson (1962, 1976) noted that in Florida the species occurred infrequently in either pine woods or deciduous woods without an undergrowth of yaupon (*Ilex vomitoria*). Forest vegetation within the eastern chipmunk's range along the Yellow River includes *Pinus glabra, Taxodium distichum, Sabal minor, Smilax* sp., *Myrica cerifera, Carya* sp., *Carpinus caroliniana, Betula nigra, Fagus grandifolia, Quercus alba, Q. stellata, Q. virginiana, Q. nigra, Magnolia grandiflora, Illicium floridanum, Asimina parviflora, Persea borbonia, Itea virginica, Hamamelis virginiana, Liquidambar styraciflua, Crataegus marshallii, C. lacrimata, Amelanchier arborea, Prunus angustifolia, Gleditsia triacanthos, Rhus radicans, Cyrilla racemiflora, Ilex cassine, I. vomitoria, I. opaca, Euonymus americanus, Acer floridanum, A. drummondii, Aesculus pavia, Parthenocissus quinquefolia, Vitis rotundifolia, Hypericum* sp., *Cornus florida, Nyssa aquatica, Kalmia latifolia, Rhodendron canescens, R. austrinum, Vaccinium* sp., *Diaspyros virginiana, Symplocos tinctoria, Halesia deptera, Styrax grandifolia, Fraxinus* sp., *Chionanthus virginica, Gelsemium sempervirens, Callicarpa americana, Lonicera sempervirens,* and *Viburnum dentatum* (Jones and Suttkus 1979).

Eastern chipmunks were originally found in Florida only in the forested floodplain and ravines along the Yellow River (Stevenson 1962; Jones and Suttkus 1979). Gore (in preparation) reported individuals from several localities well away from large streams, although the majority of observations were within 100 m of permanent streams.

VULNERABILITY OF SPECIES AND HABITAT: The eastern chipmunk is vulnerable to habitat loss, specifically from large-scale clearing or logging of forests. In Florida, this loss of habitat may be offset somewhat by the abandonment of cutover or tilled pine forests and their succession into deciduous forests. Unfortunately, the net effect of current land use practices on eastern chipmunk populations in Florida remains unknown.

CAUSES OF THREAT: Habitat loss is assumed to be the most serious threat to the species in Florida. Although eastern chipmunks often live alongside humans, they cannot tolerate clearing of their forest habitat and its conversion into tilled land, pine plantations, or developed property.

RESPONSES TO HABITAT MODIFICATION: The eastern chipmunk responds unfavorably to clearing of vegetation and related land use changes. Individuals do not disperse widely (Snyder 1982) and therefore are unlikely to escape disturbance within their home ranges or to quickly reinhabit disturbed areas. Frequent flooding of burrows is not tolerated by these animals.

DEMOGRAPHIC CHARACTERISTICS: No demographic information is available for the animals that live in Florida. General information on the life history of the eastern chipmunk is available in several works (e.g., Hamilton and Whitaker 1979; Snyder 1981).

KEY BEHAVIORS: Eastern chipmunks seem unusually secretive and difficult to observe in Florida. Stevenson (1962:110) reported that "chipmunks were heard frequently, but seen rarely, for a period of two years before a specimen could be secured." Following the report by Stevenson (1962) of the presence of eastern chipmunks in western Florida, numerous searches were made in the area, but no animals were obtained until October 1971 (Jones and Suttkus 1979).

At the southern edge of their range, eastern chipmunks are active outside of their burrows throughout the year. Jones and Suttkus (1979) either observed or collected individuals in each month of the year; most animals were seen in October, and the fewest were observed during July. They also noted that eastern chipmunks in this region were most active in the fall, especially when cold fronts were present, and that animals seemed to range greater distances from the entrances to burrows in the fall than at other times of the year.

CONSERVATION MEASURES TAKEN: The eastern chipmunk has been recognized as a rare species by the Florida Committee on Rare and Endangered Plants and Animals (Jones 1978). It is also protected as a species of special concern by the Florida Game and Fresh Water Fish Commission (1989).

CONSERVATION MEASURES PROPOSED: Conservation of deciduous forest habitats throughout the species' range in Florida should be encouraged. In particular, future management plans for the Blackwater State Forest should address conservation of habitat specifically for the eastern chipmunk. Education should be promoted to enhance the public's awareness of the presence and importance of eastern chipmunks in western Florida.

ACKNOWLEDGMENTS: J. N. Layne, J. L. Wolfe, and S. R. Humphrey have provided information and participated in discussions about eastern chipmunks in Florida.

Literature Cited

Florida Game and Fresh Water Fish Commission. 1989. Official lists of endangered and potentially endangered fauna and flora in Florida. Tallahassee. 19 pp.
Gore, J. A. In press. Distribution of the eastern chipmunk (*Tamias striatus*) in Florida. Florida Sci. 53.
Hall, E. R. 1981. The mammals of North America. John Wiley and Sons, New York. 2 vols., xv + 1181 + 180 pp.
Hamilton, W. J., Jr., and J. O. Whitaker, Jr. 1979. Mammals of the eastern United States. 2d ed. Comstock Publishing Associates, Ithaca, New York. 346 pp.
Jones, C. 1978. Eastern chipmunk. Pp. 35–36 *in* J. N. Layne, ed., Rare and endangered biota of Florida. Vol. 1. Mammals. University Presses of Florida, Gainesville. xx + 52 pp.
Jones, C., and R. R. Suttkus. 1979. The distribution and taxonomy of *Tamias striatus* at the southern limits of its geographic range. Proc. Biol. Soc. Washington 91:828–839.
Jones, J. K., Jr., D. C. Carter, H. H. Genoways, R. S. Hoffman, D. W. Rice, and C. Jones. 1986. Revised checklist of North American mammals north of Mexico, 1986. Occas. Pap. Mus. Texas Tech Univ. 107:1–22.
Neill, W. T. 1957. Historical biogeography of present-day Florida. Bull. Florida State Mus., Biol. Sci. 2:175–220.
Snyder, D. P. 1982. *Tamias striatus*. Mammalian Species 168:1–8.
Stevenson, H. M. 1962. Occurrence and habits of the eastern chipmunk in Florida. J. Mammal. 43:11–111.
Stevenson, H. M. 1976. Vertebrates of Florida. University Presses of Florida, Gainesville. xix + 607 pp.
Wolfe, S. H., J. A. Reidenauer, and D. B. Means. 1988. An ecological characterization of the Florida panhandle. U.S. Fish and Wildlife Service, Biol. Rep. 88(12):1–277.

Prepared by: Clyde Jones, Department of Biological Sciences and the Museum, Texas Tech University, Lubbock, TX 79409; Cheri A. Jones, Mississippi Museum of Natural History, 111 N. Jefferson Street, Jackson, MS 39202; Jeffery A. Gore, Florida Game and Fresh Water Fish Commission, 6938 Highway 2321, Panama City, 3 FL 32409.

Rare

Lower Keys Population of Rice Rat
Oryzomys palustris natator (in part)
FAMILY CRICETIDAE
Order Rodentia

TAXONOMY: The population of rice rats in the Lower Keys was described by Spitzer and Lazell (1978) as a species, the silver rice rat (*Oryzomys argentatus*), based on two specimens from Cudjoe Key, Monroe County. Goodyear and Lazell (1986) evaluated additional measurements from a sample of nine specimens. Humphrey and Setzer (1989), reviewing geographic variation in rice rats of the United States, were unable to distinguish the Lower Keys sample at either the species or subspecies level. They considered the available sample from the Lower Keys to be inadequate because it was too small and included both sexes and various ages, a crushed and unmeasurable skull, and individuals reared in captivity and fed a laboratory diet that may have affected growth of the skulls. Concluding that the genus had been overly split on the basis of random variation in small samples, Humphrey and Setzer considered rice rats from the Lower Keys to be a disjunct population of a subspecies, *O. p. natator*, ranging throughout peninsular Florida. They suggested that reliable data on this population should be obtained from a larger sample of adult males, from colorimetry of the pelage, or from karyotypic or electrophoretic studies. They also suggested that evidence might eventually be brought forward to justify recognition as a subspecies on the basis of color alone.

DESCRIPTION: The marsh rice rat has a sparsely haired tail that is about the same length as the head and body. In general, color of the pelage is brown, brownish-gray, or gray dorsally, buffy to pale gray on the sides, and grayish white below. The tail is bicolored, brown above and white below. The skull of old individuals is distinguished by a prominent supraorbital ridge. The palate extends posteriorly beyond the molar toothrow.

The marsh rice rat is extraordinarily variable in pelage color and size of the body and skull (Merriam 1901; Goldman 1918; Paradiso 1960; Wolfe 1985; Humphrey and Setzer 1989). Merriam (1901:276) cautioned that for taxonomic work "it is very necessary to use skulls of fully adult and even old individuals, as in many forms the specific characters do not develop until rather late in life. The males are larger than the females and usually have the characters more accentuated." Skull measurements vary according to age and sex, which have progressively greater influences on more southerly populations (Humphrey and Setzer 1989). *O. p. natator* is significantly larger than the continental subspecies of marsh rice rat, *O. p. palustris*, in all 12 skull characters measured by Humphrey and Setzer (1989). The most important variables for differentiating the two subspecies were alveolar length of the upper molars (averaging 4.9 mm in *natator*), alveolar length of the lower molars (4.9 mm), breadth across the condyles (7.3 mm), and breadth of the braincase (13.0 mm). External measurements for the subspecies *natator* ($n = 179$) average 140 mm for length of the head and body and 34 mm for length of the hind foot.

The small sample of *natator* from the Lower Keys had less variation in skull measurements than most other samples, possibly resulting from sampling error or low genetic variation (Humphrey and Setzer 1989). Pelage color of the Lower Keys population was described as being brownish-gray dorsally and distinctively silver-gray on the sides (Spitzer and Lazell 1978). Although most specimens of other populations are darker, the palest specimen in a sample from Sanibel Island, Lee County was approximately as pale as a Lower Keys specimen (Humphrey and Setzer 1989).

POPULATION SIZE AND TREND: Few data exist on population numbers, and there are none on trend. Based on the numbers of catches per trap-night, Goodyear (1987) considered densities to be low at eight of the nine Lower Keys sites where rice rats have been found—capture rates there ranged from 0.07 to 1.8 percent. She had a much higher trapping success of 8.6 percent on Johnston Key. Goodyear concluded that rice rats were not abundant in the Lower Keys.

DISTRIBUTION AND HISTORY OF DISTRIBUTION: Marsh rice rats are widespread in the Lower Keys. They have been found on Cudjoe Key, Big Torch Key, Middle Torch Key, Johnston Key, Little Pine Key, Raccoon Key, Saddlebunch Key, Summerland Key, and the Water Keys. Trapping at Annette Key, Big Pine Key, Boca Chica Key, Crab Key, Little Torch Key, and Porpoise Key have not yielded captures (Goodyear 1987).

Goodyear (1983, 1984) considered rice rats likely to be present on numerous keys where they have not been recorded, based on presence of suitable areas of appropriate habitat. Rice rats are absent from the Upper Keys (Schwartz 1952; Goodyear 1987) and have not been taken from the

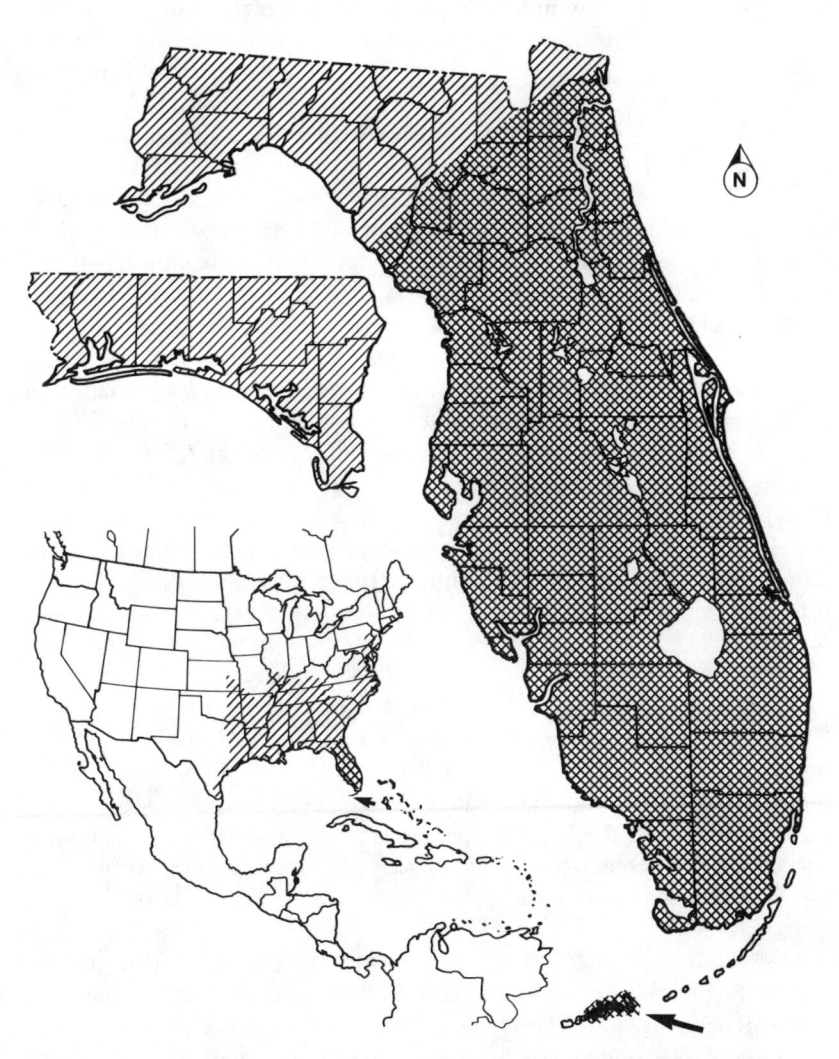

Distribution map of the Lower Keys population of rice rat, *Oryzomys palustris natator* (in part). Hatching represents the species; crosshatching represents the subspecies; and the arrow indicates the population.

Middle Keys sampled to date (Goodyear 1987). After the original captures at the type locality on Cudjoe Key were reported, Barbour and Humphrey (1982) were unable to capture rice rats in freshwater habitats there or on nearby keys, and they mistakenly concluded that the population might be extirpated. Goodyear (1987) also could no longer find animals on Cudjoe Key.

Rice rats have been present in Florida at least since early in the Rancholabrean land-mammal age (Webb and Wilkins 1984; the Sangamonian interglacial of the late Pleistocene, about 122,000 yBP). At that time, sea level was about 8 m higher than at present, and all of the Keys were submerged. Subsequently, sea level dropped to 100–130 m lower than at present during the Wisconsinan glaciation, reaching its most extreme only 18,000 yBP (Kennett 1982:268–273). At that time, all of the present-day Keys were connected directly with the mainland. In Recent times, minor oscillations of sea level have occurred ranging from +4 m to −2 m between 6000 and 1600 yBP (Fairbridge 1974; Tanner et al. 1989). All this activity would have had obvious effects on the availability of habitat to rice rats in the Lower Keys. The most recent events would have completely inundated the Lower Keys and then extended Everglades habitat over most of Florida Bay, favoring colonization from the mainland <2000 yBP (Humphrey and Setzer 1989).

GEOGRAPHICAL STATUS: The Lower Keys population of the marsh rice rat is a peripheral, disjunct population of an abundant, widespread subspecies. The subspecies occupies salt and freshwater marshes all over peninsular Florida, and the species is widely distributed from southern New Jersey to eastern Kansas to southern Texas. Many species of vertebrates share this pattern, being distributed more widely in peninsular Florida or the eastern United States with a disjunct but undifferentiated population in the Lower Keys (including the American alligator, *Alligator mississippiensis*; striped mud turtle, *Kinosternon bauri*; Big Pine Key ringneck snake, *Diadophis punctatus acricus*; red rat snake, *Elaphe guttata guttata*; Florida brown snake, *Storeria dekayi victa*; and Florida ribbon snake, *Thamnophis sauritus sackeni*). Others have the same distributional pattern but with the keys populations considered as distinct taxa (including the Lower Keys marsh rabbit, *Sylvilagus palustris hefneri*; Key deer, *Odocoileus virginianus clavium*; Florida Keys mole skink, *Eumeces egregius egregius*; and Key silverside, *Menidia conchorum*).

HABITAT REQUIREMENTS AND HABITAT TREND: Although the type locality of the Lower Keys population of the marsh rice rat was a freshwater marsh, subsequent trapping (Barbour and Humphrey 1982; Goodyear 1987) showed that this is not its primary habitat. Instead, the primary habitat is thoroughly described by Goodyear (1987) as the upland-to-marine interface, including what is locally known as the buttonwood (*Conocarpus erectus*) transition, through salt flats and coastal strand, to the upslope face of black mangrove (*Avicennia germinans*) forest. She termed this gradient "salt marsh," which it resembles in having a moist and periodically flooded surface of open mud and scattered grasses but from which it differs in being partly wooded. *O. palustris* is more abundant in tidal marsh than in other habitats (Wolfe 1985, 1990).

Details of habitat use are based on intensive radiotracking of a single male rice rat, with sparse but consistent data on several others (Goodyear 1983). The home range (minimum convex polygon) of this individual over a 2-month period was 22.8 ha. Strikingly smaller home ranges have been reported in live-trapping studies of *O. palustris* (Negus et al. 1961; Birkenholz 1963), but these results are not comparable. Use of the zones within the habitat gradient depends on the tide level. The buttonwood transition is used most extensively when the mangrove edge, coastal strand, and salt flats are flooded. Saltwort (*Batis maritima*) is a major part of the diet and hence an indicator of good habitat. Sea ox-eye daisy (*Borrichia fruitescens*) is another indicator of suitable habitat. The animals usually nest in grass tussocks of *Sporobolus* or *Distichlis* in the salt flats, where they are found in the daytime. The nests are spheres of grass and a few leaves or stems of other plants, about 15 cm in diameter. Nests are usually near ground level, slightly elevated by the grass tussocks in which they are located. The animals also nest in disturbed areas or on spoil piles where slash or trash is piled, and sometimes nests are made of trash. At night the animals forage in various portions of the gradient. The edge of the mangrove forest is used for foraging only. During high tides, the animals stay in place in the taller vegetation or on beach ridges or spoil piles. The diet includes seeds of saltwort, seeds of coconut palm (*Cocos nucifera*), and isopods. Crabs and seeds of buttonwood also may be eaten. Frequent observations of rice rats on spoil piles were associated with nesting or with presence of dried, split coconut seeds, which were seen being carried or eaten on three occasions. The extensive literature on diet of *O. palustris* has been reviewed by Wolfe (1982a).

VULNERABILITY OF SPECIES AND HABITAT: Goodyear (1984) considered the Lower Keys population of the marsh rice rat to be vulnerable to competition by black rats (*Rattus rattus*) and predation by raccoons (*Procyon lotor*), both of which have high densities in coastal areas and in close association with human habitation. The habitat of the population is vulnerable to development for human use.

CAUSES OF THREAT: The primary, immediate threat is filling of the buttonwood transition and salt flats with marl for construction of residential and commercial buildings. This causes both a loss of the absolute amount of habitat and fragmentation of the remaining habitat into smaller, disconnected parcels. A greater but longer-term threat is the sea-level rise predicted to result from global warming. Most of the Lower Keys will go under water, with sea level rising 1–2 m by the year 2100 under the most likely scenario (Hoffman et al. 1983; Hoffman 1984).

RESPONSES TO HABITAT MODIFICATION: Largely unknown. The male studied carefully by Goodyear (1983) used spoil piles and canal banks extensively, but these were small incursions into a home range otherwise comprising native habitats. The spoil piles were attractive because they provided dry nesting sites during high tides, often contained piles of trimmings from coconut palms that were a good matrix for nest building, and often included coconut seeds that provided food. It is unlikely that a more complete conversion of the upland-to-marine interface, as usually accompanies real estate development, would allow rice rats to persist.

DEMOGRAPHIC CHARACTERISTICS: Among free-ranging animals, Goodyear (1983) recorded a pregnant female on 6 December and a birth of at least four young on 17 November. For a captive female, she recorded successive litters of three, five, and four young.

O. palustris has a gestation period of 21–28 days (Svihla 1931; Park and Nowosielski-Slepowron 1972). Recent reports of average litter size are 4–6 (Conaway 1954), 4.8 (Negus et al. 1961), and 3.6 (Park and Nowosielski-Slepowron 1972). Breeding may be year-round but is influenced greatly by population levels and the environment (Goldman 1918; Worth 1950; Negus et al. 1962; Park and Nowosielski-Slepowron 1972). Both sexes reach reproductive maturity at 50–60 days; the estrous cycle is 6–9 days, and there is a postpartum estrus (Svihla 1931; Conaway 1954). Densities have been reported as 0.1–3.3/ha, 2–25/ha, and >50/ha (Negus et al. 1961; Wolfe 1985; Smith and Brieze 1979). The mortality rate is about 80 percent in a 3-month period, and maximum longevity in the field is 24 months (Wolfe 1985).

Goodyear (1983) recorded a few cases of nest disturbance and predation on Lower Keys rice rats, apparently by raccoons. Snakes also are potential predators, but owls are the best-documented predators of the species (Wolfe 1982a). Competition from black rats also may be an important pressure on the rice rats (Goodyear 1984); one of two sites inhabited by rice rats alone was the population with highest density, and most sites with rice rats absent or in very low numbers had black rats at their highest densities. Joule and Jameson (1972) conducted experiments indicating that the presence of rice rats adversely affected populations of cotton rats (*Sigmodon hispidus*) but that cotton rats did not affect rice rats. Predators, competitors, and a long list of parasites of *O. palustris* are reviewed by Wolfe (1982a).

KEY BEHAVIORS: One radiotracked animal used at least 16 different nesting sites in a 2-month period (Goodyear 1983). Nests were not simply used and abandoned. Often the animal alternated use of nests. In an extreme example, it moved back and forth twice between two nests located 1 km apart over the course of 4 days. The animal also returned to reuse nest sites after long periods of disuse. It often moved long distances in the predawn hours of the morning, radically shifting location before nesting. This animal usually traveled by walking or running, but it did not hesitate to swim, it climbed at least 4 m high in vegetation, and occasionally it hopped bipedally when carrying food or attempting to pass through dense vegetation (Goodyear 1983). The excellent swimming ability of *O. palustris* is well documented (Svihla 1931; Hamilton 1946; Esher et al. 1978). Rice rats have survived storm tides of 3 m in salt marshes of the northern Gulf of Mexico (Wolfe 1982b).

CONSERVATION MEASURES TAKEN: This animal is listed as endangered by the Florida Game and Fresh Water Fish Commission (1990). The U.S. Fish and Wildlife Service (1990) has proposed to list this population as endangered. About half of the documented distribution of the population is in the National Key Deer Wildlife Refuge or Great White Heron National Wildlife Refuge (Goodyear 1984).

CONSERVATION MEASURES PROPOSED: The most important conservation action is fee simple acquisition of the known capture sites identified by Goodyear (1984). Areas protected to benefit marsh rice rats in the Lower Keys should greatly exceed 20 ha, the minimum size of one home range (Goodyear 1984). Additional surveys should be conducted in areas identified as potential range by Goodyear (1984). Considering the threat of sea-level rise, acquisitions also should be focused on the highest ground

Oryzomys palustris natator (in part)

in the Lower Keys or even in the Middle Keys, in anticipation of its colonization by rice rats when the land-sea interface moves upslope.

ACKNOWLEDGMENTS: The current state of knowledge of this population is attributable mainly to difficult and intensive field studies by Numi C. Goodyear (the author of the species account in the 1978 edition). I thank James N. Layne and Charles A. Woods for helpful criticisms of the manuscript.

Literature Cited

Barbour, D. B., and S. R. Humphrey. 1982. Status of the silver rice rat (*Oryzomys argentatus*). Florida Sci. 45:106–122.
Birkenholz, D. E. 1963. Movement and displacement in the rice rat. Quart. J. Florida Acad. Sci. 26:269–274.
Conaway, C. H. 1954. The reproductive cycle of rice rats (*Oryzomys palustris palustris*) in captivity. J. Mammal. 35:263–266.
Esher, R. J., J. L. Wolfe, and J. N. Layne. 1978. Swimming behavior of rice rats and cotton rats. J. Mammal. 59:551–558.
Fairbridge, R. W. 1974. The Holocene sea-level record in south Florida. Pp. 223–232 *in* P. J. Gleason, ed., Environments of south Florida: Past and present. Miami Geol. Soc. Memoir 2:1–452.
Florida Game and Fresh Water Fish Commission. 1990. Official lists of endangered and potentially endangered fauna and flora in Florida. Tallahassee. 23 pp.
Goldman, E. A. 1918. The rice rats of North America. N. Amer. Fauna 43:1–100.
Goodyear, N. C. 1983. Aspects of the biology of the silver rice rat *Oryzomys argentatus*. Unpubl. M.S. thesis, University of Rhode Island, Kingston. 101 pp.
Goodyear, N. C. 1984. Final report on the distribution, habitat, and status of the silver rice rat *Oryzomys argentatus*. Unpubl. report to U.S. Fish and Wildlife Service, 3100 University Boulevard, South Suite 120, Jacksonville, Florida. 49 pp.
Goodyear, N. C. 1987. Distribution and habitat of the silver rice rat, *Oryzomys argentatus*. J. Mammal. 68:692–695.
Goodyear, N. C., and J. D. Lazell, Jr. 1986. Relationships of the silver rice rat *Oryzomys argentatus* (Rodentia: Muridae). Postilla 198:1–7.
Hamilton, W. J., Jr. 1946. Habits of the swamp rice rat, *Oryzomys palustris palustris* (Harlan). Amer. Midland Nat. 36:730–736.
Harlan, R. 1837. Description of a new species of quadruped, of the order Rodentia, inhabiting the United States. Amer. J. Sci. and Arts 31:385–386.
Hoffman, J. S. 1984. Estimates of future sea level rise. Pp. 79–103 *in* M. C. Barth and J. G. Titus, eds., Greenhouse effect and sea level rise. Van Nostrand Reinhold Co., New York. 325 pp.

Hoffman, J. S., D. Keyes, and J. G. Titus. 1983. Projecting future sea level rise. U.S. GPO No. 055-000-0236-3. U.S. Government Printing Office, Washington, D.C.

Humphrey, S. R., and H. W. Setzer. 1989. Geographic variation and taxonomic revision of rice rats (*Oryzomys palustris* and *O. argentatus*) of the United States. J. Mammal. 70:557–570.

Joule, J., and D. L. Jameson. 1972. Experimental manipulation of population density in three sympatric rodents. Ecology 53:653–660.

Kennett, J. P. 1982. Marine geology. Prentice-Hall, Inc., Englewood Cliffs, New Jersey. 813 pp.

Merriam, C. H. 1901. Synopsis of the rice rats (genus *Oryzomys*) of the United States and Mexico. Proc. Washington Acad. Sci. 3:273–295.

Negus, N. C., E. Gould, and R. K. Chipman. 1961. Ecology of the rice rat, *Oryzomys palustris* (Harlan) on Breton Island, Gulf of Mexico, with a critique of social stress theory. Tulane Studies Zool. 8:93–123.

Paradiso, J. L. 1960. Size variation in the rice rat. J. Mammal. 41:516–517.

Park, A. W., and B. J. A. Nowosielski-Slepowron. 1972. Biology of the rice rat (*Oryzomys palustris natator*) in a laboratory environment. Z. Saugetierk. 37:42–51.

Schwartz, A. 1952. The land mammals of southern Florida and the Upper Florida Keys. Unpubl. Ph.D. diss., University of Michigan, Ann Arbor. 180 pp.

Smith, A. T., and M. J. Brieze. 1979. Population structure of Everglades rodents: Responses to a patchy environment. J. Mammal. 60:778–794.

Spitzer, N. C., and J. D. Lazell, Jr. 1978. A new rice rat (genus *Oryzomys*) from Florida's Lower Keys. J. Mammal. 59:787–792.

Svihla, A. 1931. Life history of the Texas rice rat (*Oryzomys palustris texensis*). J. Mammal. 12:238–242.

Tanner, W. F., S. Demirpolat, F. W. Stapor, and L. Alvarez. 1989. The "Gulf of Mexico" late Holocene sea level curve. Trans. Gulf Coast Assoc. Geol. Soc. 39:553–562.

U.S. Fish and Wildlife Service. 1990. Endangered and threatened wildlife and plants; proposed endangered status for the Lower Keys population of the rice rat (silver rice rat). Federal Register 55:43002–43008.

Webb, S. D., and K. T. Wilkins. 1984. Historical biogeography of Florida Pleistocene mammals. Pp. 370–383 *in* H. H. Genoways and M. R. Dawson, eds., Contributions in Quaternary vertebrate paleontology: A volume in memorial to John E. Guilday. Spec. Publ. Carnegie Mus. 8:1–538.

Wolfe, J. L. 1982*a*. *Oryzomys palustris*. Mammalian Species 176:1–5.

Wolfe, J. L. 1982*b*. Storm effects on rice rats inhabiting coastal marshes. Gulf Research Report 7:169–170.

Wolfe, J. L. 1985. Population ecology of the rice rat (*Oryzomys palustris*) in a coastal marsh. J. Zool. 205A:235–244.

Wolfe, J. L. 1990. Environmental influences on the distribution of rice rats (*Oryzomys palustris*) in coastal marshes. Florida Sci. 6 53:81–84.

Worth, C. G. 1950. Observations on the behavior and breeding of captive rice rats and woodrats. J. Mammal. 31:421–426.

Prepared by: Stephen R. Humphrey, Florida Museum of Natural History, University of Florida, Gainesville, FL 32611.

Rare

Southeastern Weasel
Mustela frenata olivacea
FAMILY MUSTELIDAE
Order Carnivora

TAXONOMY: The southeastern weasel was originally described by Howell (1913) as *Mustela peninsulae olivacea* and subsequently changed to *M. frenata olivacea* (Hall 1951). Hall noted that the distinction between *M. f. olivacea* and the Florida weasel (*M. f. peninsula*) is inadequate because so few specimens were available from the southern half of Florida.

DESCRIPTION: The southeastern weasel is very similar in appearance to the Florida weasel, and differs by having finer, softer pelage, smaller skull measurements, and slightly smaller overall size (Hall 1951). The pelage is chestnut brown dorsally and yellowish white below, except on the chin and upper lips, which are white. The terminal one-third of the tail is black and a dark spot may be present at the angle of the mouth. Weasels in Florida do not molt into a white winter plumage. This species is sexually dimorphic; males are larger than females.

POPULATION SIZE AND TREND: Almost nothing is known about weasel populations in Florida, and weasels appear to be extremely rare based on the scarcity of weasel captures or sightings in the state. In contrast, records of southeastern weasels are numerous from Alabama and Georgia (Hall 1951).

DISTRIBUTION AND HISTORY OF DISTRIBUTION: The southeastern weasel ranges from southern North Carolina west to eastern Mississippi and south to the northern third of peninsular Florida.

GEOGRAPHICAL STATUS: This weasel is apparently widespread throughout the southeastern United States but uncommon in Florida.

Mustela frenata olivacea 311

HABITAT REQUIREMENTS AND HABITAT TREND: Based on the limited information available, weasels in Florida do not prefer any habitat type and probably can be expected to occur wherever food and shelter are adequate. Weasels in Florida have been observed in a variety of habitats including sand pine scrub, sandhill, pine flatwoods, cypress swamps, and tropical hammock (Brown 1972).

VULNERABILITY OF SPECIES AND HABITAT: Because weasels in Florida may occur at low densities, viable populations may require large areas of suitable habitat. While habitat requirements of weasels in Florida are poorly understood, weasels appear to use a wide variety of habitats. There is little doubt that the extensive loss of habitats of many types occurring throughout Florida negatively affects weasel populations to some degree, but specific effects are undocumented.

CAUSES OF THREAT: Because weasels are seldom encountered by humans and are legally protected in Florida, humans probably cause little direct mortality. Indirect effects associated with habitat loss and modification by humans are likely to affect weasels, but so little is known about weasels in Florida that the extent of this threat is difficult to assess.

RESPONSES TO HABITAT MODIFICATION: Unknown.

DEMOGRAPHIC CHARACTERISTICS: Data specific to weasels in Florida are lacking, but studies on *M. frenata* from other areas of the United States suggest that weasel numbers may vary with season, habitat, and prey availability (Polderboer et al. 1941; Fitzgerald 1977). Sex ratios reported for *M. frenata* are heavily biased toward males (Hall 1951). This bias probably results from differential selective factors between the sexes (e.g., body size, movements) causing males to be captured more frequently than females; sex ratios actually are probably nearly equal (Hall 1951).

The breeding habits of weasels in Florida are poorly understood. It appears that litters are born in late fall to early winter following an unusually long gestation period associated with delayed embryonic implantation of perhaps nine months (Moore 1944; Brown 1973). Females can produce one litter annually, and a litter size of three has been reported for *M. f. olivacea* (Moore 1944).

KEY BEHAVIORS: Much has been written on the weasel's ferocious disposition, ability to kill animals larger than itself, and apparent lust for killing. These behaviors are predatory adaptations allowing weasels to subdue a wide variety of prey and efficiently gather food that may be stored for later use (Hall 1951). The bulk of the weasel's diet consists of small mammals, mainly rodents, but also includes a great variety of other animals (Hall 1951).

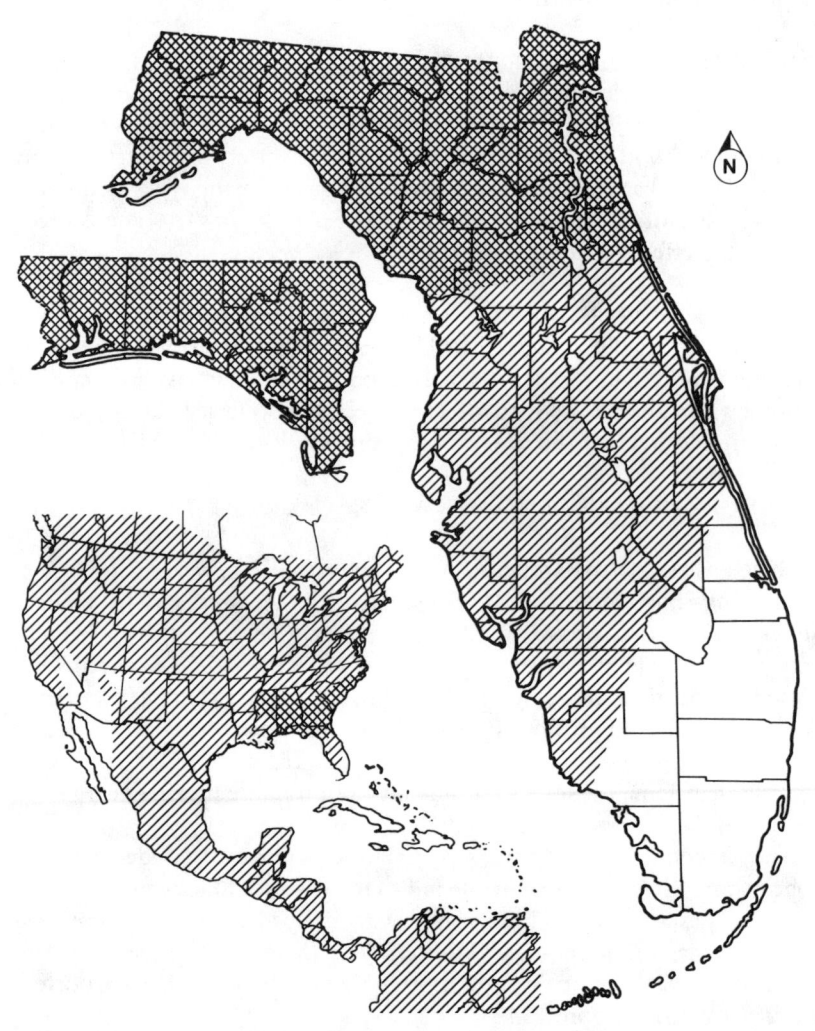

Distribution map of the southeastern weasel, *Mustela frenata olivacea*. Hatching represents the species; crosshatching represents the subspecies.

Mustela frenata olivacea

Home ranges of weasels in Florida are not known but probably vary with habitat, season, food availability, sex, and population levels. Weasels from northern latitudes have been reported to travel long distances regularly (Quick 1944), and weasels in Florida also probably use large areas to search for food and mates. Homesites of southern weasels include hollow trees, burrows, and the burrows of other animals, including pocket gophers (*Geomys pinetis*) and gopher tortoises (*Gopherus polyphemus*; Moore 1944; Frank and Lips 1989).

CONSERVATION MEASURES TAKEN: The southeastern weasel is listed as a fully protected furbearer by the Florida Game and Freshwater Fish Commission and is protected throughout the state.

CONSERVATION MEASURES PROPOSED: Data on weasels in Florida are anecdotal and sparse. An up-to-date study on the ecology of weasels in Florida would provide information required to assess the status and guide management decisions for this little-known mammal. State game officers, park rangers, and managers of public lands should be made aware of the rarity of weasel records in Florida, and any observations of weasels should be reported to knowledgeable authorities. A taxonomic study examining the differences between this weasel and the Florida weasel would be useful to determine the genetic uniqueness of these subspecies.

ACKNOWLEDGMENTS: The author of the species account in the 1978 edition was Larry N. Brown.

Literature Cited

Brown, L. N. 1972. Florida's rarest carnivore. Florida Wildlife 25(11):4–6.

Fitzgerald, B. M. 1977. Weasel predation on a cyclic population of the montane vole (*Microtus montanus*) in California. J. Anim. Ecol. 46:367–397.

Frank, P. A., and K. R. Lips. 1989. Gopher tortoise burrow use by long-tailed weasels and spotted skunks. Florida Field Nat. 17(1):20–22.

Hall, E. R. 1951. American weasels. Univ. Kansas Publ. Mus. Nat. Hist. 4:1–46.

Howell, A. H. 1913. Description of a new weasel from Alabama. Proc. Biol. Soc. Washington 26:139–140.

Moore, J. C. 1944. Life history notes on the Florida weasel. Proc. Florida Acad. Sci. 7:247–263.

Polderboer, E. B., L. W. Kuhn, and G. O. Hendrickson. 1941. Winter and spring habits of weasels in central Iowa. J. Wildl. Manage. 5:115–263.

Quick, H. F. 1944. Habits and economics of New York weasels in Michigan. J. Wildl. Manage. 8:71–78.

Prepared by: Philip A. Frank, Department of Wildlife and Range Sciences, University of Florida, Gainesville, FL 32611.

Rare

Florida Weasel
Mustela frenata peninsulae
FAMILY MUSTELIDAE
Order Carnivora

TAXONOMY: Originally described as *Putorius peninsulae* by Rhoads (1894), this taxon was revised two times before the current name *M. f. peninsulae* was accepted (Hall 1951). In his consideration of American weasels, Hall was able to examine only 10 specimens of *M. f. peninsulae* and considered the taxonomic distinction between this weasel and the southeastern weasel (*M. f. olivacea*) unsatisfactory. Perhaps enough material is now available from southern Florida for a more detailed evaluation of these two subspecies.

DESCRIPTION: The Florida weasel is very similar in appearance to the southeastern weasel but differs by having coarser pelage, larger cranial measurements, and slightly larger overall size (Hall 1951). See the species account for the southeastern weasel (*M. f. olivacea*) for a description of this weasel.

POPULATION SIZE AND TREND: See the species account on the southeastern weasel, *M. f. olivacea*.

DISTRIBUTION AND HISTORY OF DISTRIBUTION: Hall (1951) credits the range of this subspecies as extending south through the entire peninsula, but this weasel has been recorded only once south of Lake Okeechobee (Brown 1972). While the scarcity and secretive habits of this weasel undoubtedly have prevented detection of this animal in areas where it actually occurs, the lack of records from southeastern Florida and the Everglades indicates its absence there despite the availability of presumably acceptable habitat.

315

Florida weasel, *Mustela frenata peninsulae*. (Photo by Stephen R. Humphrey)

GEOGRAPHICAL STATUS: The Florida weasel is endemic to peninsular Florida and is restricted to the central third of the peninsula.

HABITAT REQUIREMENTS AND HABITAT TREND: See the species account for the southeastern weasel, *M. f. olivacea*.

VULNERABILITY OF SPECIES AND HABITAT: See the species account for the southeastern weasel, *M. f. olivacea*.

CAUSES OF THREAT: See the species account for the southeastern weasel, *M. f. olivacea*.

RESPONSES TO HABITAT MODIFICATION: Unknown.

Mustela frenata peninsulae

DEMOGRAPHIC CHARACTERISTICS: See the species account for the southeastern weasel, *M. f. olivacea*.

KEY BEHAVIORS: See the species account for the southeastern weasel, *M. f. olivacea*.

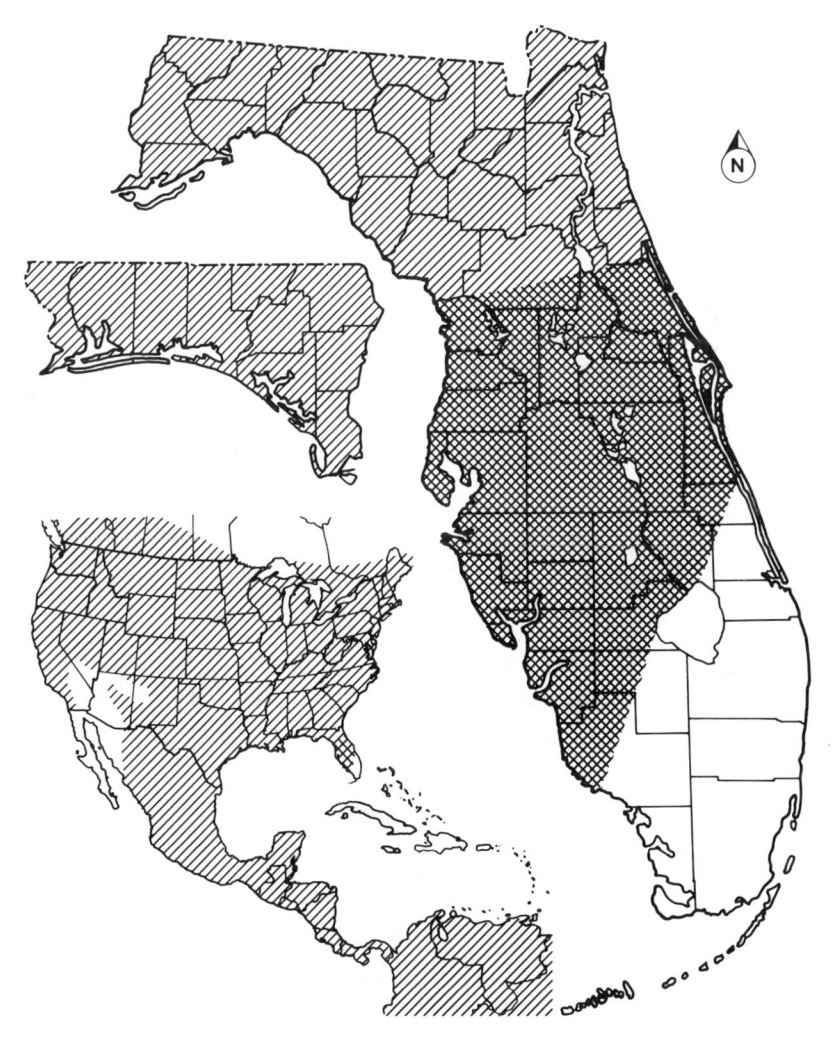

Distribution map of the Florida weasel, *Mustela frenata peninsulae*. Hatching represents the species; crosshatching represents the subspecies.

CONSERVATION MEASURES TAKEN: The Florida weasel is listed as a fully protected furbearer by the Florida Game and Fresh Water Fish Commission and is protected throughout the state. As of this writing the Florida Game and Fresh Water Fish Commission is conducting a status survey of the Florida weasel.

CONSERVATION MEASURES PROPOSED: Based on the scarcity of records, the Florida weasel may be the state's rarest carnivore (Brown 1972). A taxonomic study examining the distinctiveness between the Florida weasel and the southeastern weasel is needed to determine the degree of difference of the endemic Florida weasel. Also, see the species account for the southeastern weasel, *M. f. olivacea*.

ACKNOWLEDGMENTS: The author of the species account in the 1978 edition was Larry N. Brown.

Literature Cited

Brown, L. N. 1972. Long-tailed weasel (*Mustela frenata*) in subtropical Florida. J. Mammal. 53:407.

Hall, E. R. 1951. American weasels. Univ. Kansas Publ. Mus. Nat. Hist. 4:1–466.

Rhoads, S. N. 1894. Contributions to the mammalogy of Florida. Proc. Acad. Nat. Sci. Philadelphia, pp. 152–161.

Prepared by: Philip A. Frank, Department of Wildlife and Range Sciences, University of Florida, Gainesville, FL 32611.

Rare

Southern Florida Population of Mink
Mustela vison mink (in part)
FAMILY MUSTELIDAE
Order Carnivora

TAXONOMY: The population of mink in southernmost Florida was described as a subspecies by Hamilton (1948) on the basis of a single road-killed specimen obtained in the Big Cypress Swamp. However, because it closely resembles the population from the piedmont of the Carolinas and northern Georgia, it was considered a disjunct population of the subspecies *M. v. mink* (Humphrey and Setzer 1989). This conclusion was based mainly on study of the carnassial pair of teeth, because most specimens of this population are obtained as roadkills with crushed skulls that cannot be measured for most other characters useful in taxonomic work. Obtaining a better sample is nearly impossible because the best method for capturing these secretive animals (use of leg-hold traps) has been illegal in Florida since 1973.

DESCRIPTION: Like other members of the weasel family, mink have long, slender bodies. The head is flattened and not much larger in diameter than the body, and the ears are small and rounded. The animals from southern Florida are small with dark brown, silky fur; fur of mink in northern Florida is much paler. The distal half of the tail is blackish brown. Some specimens have a white chin spot, and a few have a white chest patch. Mink are sexually dimorphic, with males larger than females (Birney and Fleharty 1966; McNab 1971). Length of the fourth upper premolar of *M. v. mink* averages 7.55 mm measured along the outer cusp and 8.20 mm measured along the inner cusp in males, and 6.85 and 7.45 mm in females (Humphrey and Setzer 1989).

POPULATION SIZE AND TREND: The collector of the first specimen of this population asserted that mink were rare in this region (Hamilton

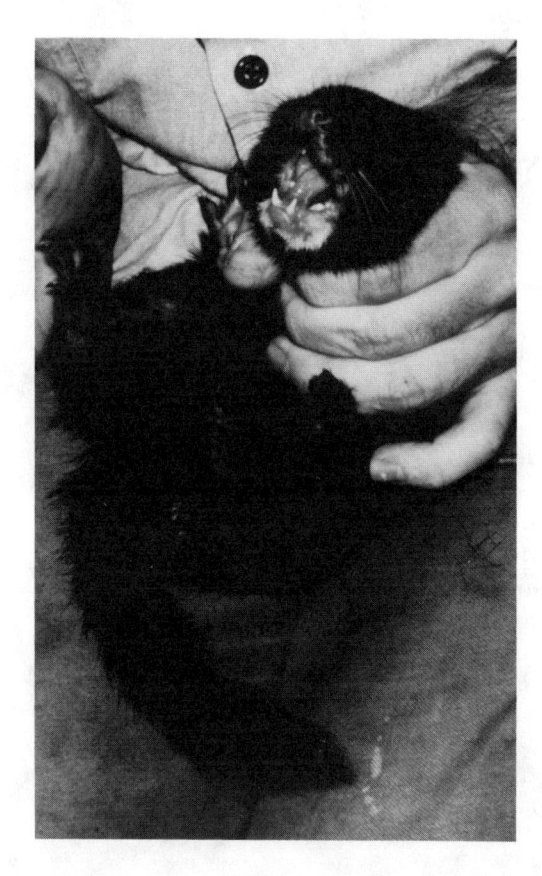

Southern Florida population of mink, *Mustela vison mink* (in part). (Photo by Terry L. Zinn)

1948), an opinion repeated by subsequent authors (Layne 1974; Brown 1978). However, mink are relatively common in portions of the Everglades and Big Cypress Swamp (Allen and Neill 1952; Smith 1980; Humphrey and Zinn 1982). There are no estimates of size or density of the population, and there are no counts indicating a decline in population numbers.

DISTRIBUTION AND HISTORY OF DISTRIBUTION: This population occurs in the shallow freshwater marshes of the Everglades and Big Cypress Swamp. Current distribution is limited to southern portions of Collier and Dade counties, and presumably northern and eastern Monroe County. Allen and Neill (1952) reported buying more than 100 mink pelts from residents of the vicinity of Okeechobee, Okeechobee County, in 1935–36. They stated that the animals had been trapped in and about Lake Okee-

Mustela vison mink (in part) 321

chobee. No subsequent information exists about occurrence of mink at Lake Okeechobee. Furthermore, no information at all exists about mink in the northern Everglades. Possibly the distribution of mink once included these areas but has been reduced. The range documented with

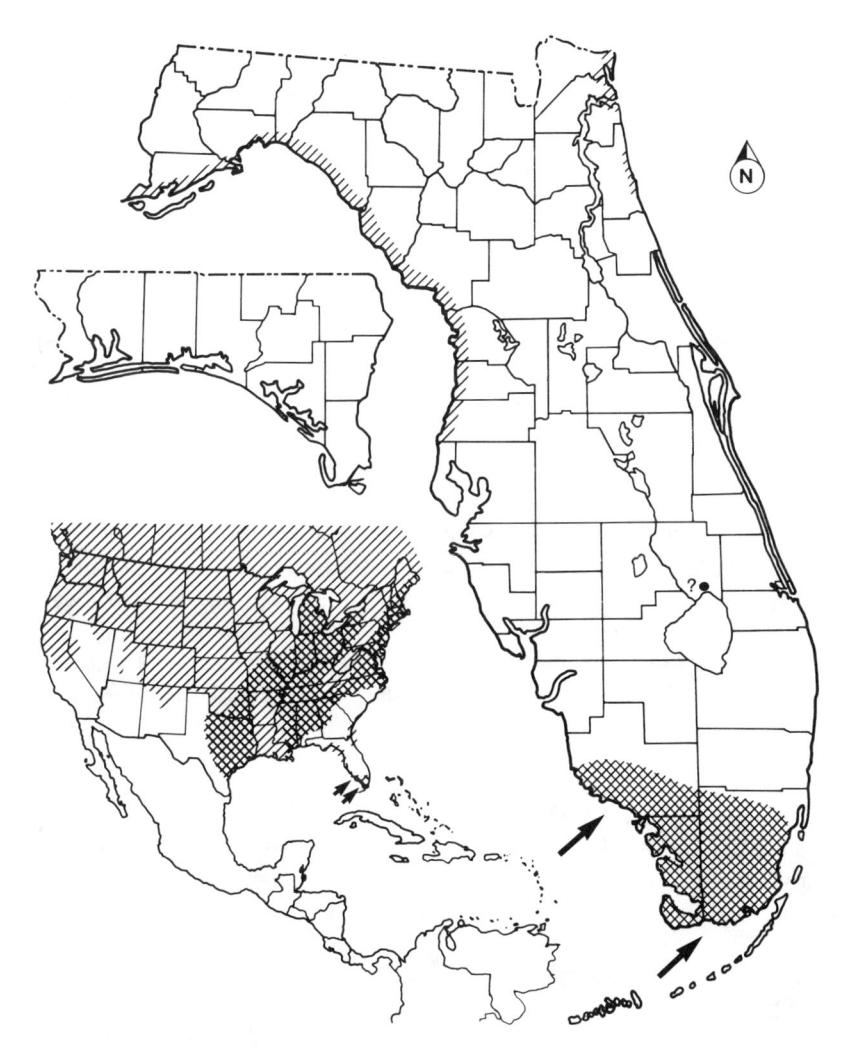

Distribution map of the southern Florida population of mink, *Mustela vison mink* (in part). Hatching represents the species; crosshatching and the dot represent the population.

museum specimens extends only as far north as Alligator Alley (U.S. Highway 84 or Interstate Highway 75) and as far east as the Florida Turnpike.

Two other subspecies of mink occur in northern Florida, in the salt marshes along the Gulf and Atlantic coasts. The absence of mink from the widespread freshwater wetlands elsewhere in Florida and southern Georgia was reported by Harper (1927), Hamilton (1948), and Layne (1974) and has been confirmed in a century of sampling by fur buyers, and by the lack of roadkills on highways across marshlands. Humphrey and Setzer (1989) showed this hiatus in mink distribution to include most of the coastal plain of Georgia and the Carolinas.

The modern mink species is known from fossil records beginning in the early Pleistocene (Kurten and Anderson 1980). Three fossil mink mandibles exist from the Rancholabrean land-mammal age (late Pleistocene) of northern Florida.

To hypothesize a past connection of the southernmost and Gulf coastal mink populations of Florida, Layne (1974) speculated that at a time of lower sea level or cooler climate both mink and salt marshes were distributed continuously along the west coast of Florida. The lack of differentiation of the southern Florida mink from mink of the piedmont, however, led Humphrey and Setzer (1989) to pose an alternative scenario. In this view, the southern Florida population was repeatedly extirpated as peninsular Florida repeatedly shortened to a headland during Pleistocene interglacial periods. The southern population may have been reestablished by colonization from the continental population during glacial times, as the peninsula again extended to present-day length or beyond. In this way the southern Florida population of the subspecies *mink* would have retained its similarity to the continental population.

GEOGRAPHICAL STATUS: The southern Florida population is an undifferentiated, disjunct population of the southeastern (continental) mink. Possibly the distribution has contracted since the 1930s by loss from the northern part of the Everglades, but a field survey is needed to ascertain whether mink are absent from that area.

HABITAT REQUIREMENTS AND HABITAT TREND: In the Big Cypress Swamp, southern Florida mink are widely dispersed in all types of shallow wetlands. Even short-hydroperiod spikerush (*Eleocharis*) marshes are occupied during the late wet season, when water levels are high. Mink retreat from drying marshlands to long-hydroperiod swamp forests as the dry season progresses. Salt marshes between the mangroves and freshwater habitats also are used during the wet season, when flooding is not confined to tidal channels (Humphrey and Zinn 1982). Some authors (Layne 1974; Smith 1980) judge that southern Florida mink are more numerous in the Big Cypress Swamp than in the Everglades. Smith (1980) reported a concentration of tracks, scats, and well-defined trails on levees and next to canals in the Everglades. Dens of this population have been seen in a hollow cypress stump and under an abandoned car.

Contents of stomachs, intestines, and scats of roadkilled southern Florida mink indicate a diet of mostly crayfish, small mammals, and fish, plus some birds, snakes, and insects (Smith 1980).

The peculiar distribution of mink in Florida results partly from discontinuous coastal habitats. Some other mechanism, however, must prevent mink from occupying vast areas of seemingly suitable freshwater marshes in peninsular Florida. Perhaps in relatively deep waters they incur heavy predation by alligators (*Alligator mississippiensis*) and snapping turtles (*Chelydra serpentina*), which usually are excluded from saltmarshes by hypersaline water and from shallow marshes by seasonal drying. Another possibility might be some infectious parasite or disease present in most freshwater environments of Florida.

VULNERABILITY OF SPECIES AND HABITAT: This amphibious animal is sensitive to hydrologic manipulation of the wetlands on which it depends. Statewide, its habitat is so large and flat that it is difficult for man to modify, yet substantial modification has occurred and more is likely.

CAUSES OF THREAT: The virtual replumbing of the Big Cypress Swamp and Everglades to drain wetlands, control hydroperiod, and apportion water among competing interests surely has affected the southern Florida mink population, but no consequences are documented clearly. Alteration of Lake Okeechobee and drainage of its associated watersheds was begun by industrialist Hamilton Disston in 1882 and continued in 1906 by the Everglades Drainage District, totaling 708 km of canals and levees by 1929 (Carter 1974; Gleason 1984). Historically, when the water level in Lake Okeechobee rose over 4.5 m above mean sea level, a sheet of water up to 51 km wide overflowed the southern rim of the lake into the northern Everglades (Parker 1974). But when floodwaters from a 1928 hurricane killed 1836 people south of the lake, the U.S. Army Corps of Engineers built Hoover Dike, preventing recurrences (Gentry 1974). Ultimately these engineering works allowed about 800,000 ac of wetlands south of Lake Okeechobee to be developed for intensive agriculture, for which the water table is kept just below the land surface. Unanticipated consequences of the drainage system include saltwater intrusion of the aquifer, protracted muck fires during the droughts of 1931–32, 1943–45, and 1970–71, loss of fish and wildlife habitat, rapid oxidation of peat soils brought into contact with air for agriculture, and overdrainage during droughts and underdrainage during floods (Carter 1974; DeGrove 1984). A series of drainage projects in the Big Cypress Swamp, beginning in 1915 and including the Tamiami, Barron River, Turner River, Golden Gate, and Faka-Union canals, has in places reduced the hydroperiod from 5–7 months to 2–2.5 months and lowered the water level 0.6 m (Carter 1974; Gleason 1984).

Currently the greatest cause of threat is massive conversion of privately owned portions of the Big Cypress Swamp to citrus culture. Concerns include loss of habitat, biological concentration of pesticides and heavy metals, and pollution of surface waters with excess nitrates from fertilizer. But probably the greatest problem is increased demand on finite surface-water resources, to support both citrus production and the burgeoning human population, driven partly by the citrus economy.

RESPONSES TO HABITAT MODIFICATION: No data exist on effects of these losses and modifications of habitat on mink numbers or distribution. Probably at least the most extreme changes have been deleterious to mink. Smith (1980) speculated, however, that the network of roads and levees in the Everglades has increased the carrying capacity of that region for mink. Smith (1980) reported that mink do not avoid sites of human activity that contain suitable habitat. A habitat suitability model has been developed for mink (Allen 1986), by which responses of southern Florida mink to development could be predicted.

DEMOGRAPHIC CHARACTERISTICS: Roadkills are frequent and may be a significant agent of mortality where roads cross wetlands. Humphrey and Zinn (1982) suggested that mating occurs from September to November, when water is widely dispersed. Southern Florida mink are suckling young in March and April (Hamilton 1948; Smith 1980), when water is most concentrated. This timing indicates mating earlier than for northern populations, which copulate during winter, have delayed implantation of the embryo, and give birth in May. Northern populations have one litter of young annually, usually numbering 3 to 4, and a nursing period of about five weeks. Thereafter the young accompany the mother on foraging trips.

KEY BEHAVIORS: Although mink generally are considered to be noctur-

Mustela vison mink (in part)

nal and crepuscular, southern Florida mink occasionally are seen in the daytime (Smith 1980). The animals are secretive and seldom seen even where common. Their presence can be determined quickly by use of chalk-dusted trackboards and anal scent attractant, especially during September–November (Humphrey and Zinn 1982).

CONSERVATION MEASURES TAKEN: This animal is listed (as *M. v. evergladensis*) as endangered by the Florida Game and Fresh Water Fish Commission (1990), and it is under review for possible listing by the U.S. Fish and Wildlife Service (1989). Major tracts of mink habitat are under public ownership, including Everglades National Park, Big Cypress National Preserve, Fakahatchee Strand State Preserve, Florida Panther National Wildlife Refuge, and Collier-Seminole State Park. Although it is not yet known whether mink are present there, the U.S. Army Corps of Engineers conserved 900,000 ac of Everglades marsh by constructing levees around Water Conservation Areas 1–3 from 1950 to 1962; these were areas judged unfit for agriculture (Soil Conservation Service 1948). This action also retained a portion of the historical watershed for Everglades National Park, a source of water guaranteed in 1970 when the U.S. Congress required release of 315,000 ac/ft per year into the park. In 1989 Congress added 107,600 more ac to Everglades National Park and ordered the U.S. Army Corps of Engineers to restore the natural flow of water to the park.

CONSERVATION MEASURES PROPOSED: A survey of mink distribution should be conducted in northern portions of the Big Cypress Swamp and Everglades, including the marshes of Lake Okeechobee. Plans to enlarge and consolidate existing public ownerships by fee title acquisition should be implemented. These areas include the Big Cypress National Preserve additions, unpurchased portions of Fakahatchee Strand State Preserve, and Bird Rookery Swamp. Removal of surface water by citrus, agricultural, and urban users from the supply needed to maintain habitat for mink and other wetland wildlife should become the basis for a human carrying capacity for the region. Stringent water-quality standards should be established and enforced for citrus and agriculture to prevent pollution of the surface waters of the Big Cypress Swamp and Everglades. Research is needed on the responses of mink to habitat modification.

ACKNOWLEDGMENTS: The work of Terry L. Zinn and Daniel Cary while graduate students has made a major contribution to current knowledge of this animal. Kenneth C. Alvarez and James A. Kushlan were instrumental in encouraging their work. Mark E. Ludlow and Mark S. Robson suggested important improvements to the manuscript.

Literature Cited

Allen, A. W. 1986. Habitat suitability index model: Mink. U. S. Fish Wildl. Serv., Biol. Report 82(10.127):1–23.

Allen, E. R., and W. T. Neill. 1952. Notes on the abundance of the Everglades mink. J. Mammal. 33:113–114.

Birney, E. C., and E. D. Fleharty. 1966. Age and sex comparisons of wild mink. Trans. Kansas Acad. Sci. 69:139–145.

Brown, L. N. 1978. Everglades mink. Pp. 26–27 in J. N. Layne, ed., Rare and endangered biota of Florida. Vol. 1. Mammals. University Presses of Florida, Gainesville. xx + 52 pp.

Carter, L. J. 1974. The Florida experience—land and water policy in a growth state. Johns Hopkins University Press, Baltimore. 355 pp.

DeGrove, J. M. 1984. Pp. 22–27 in P. J. Gleason, ed., Environments of south Florida present and past II. 2d ed. Miami Geol. Soc., Coral Gables. 551 pp.

Florida Game and Fresh Water Fish Commission. 1990. Official lists of endangered and potentially endangered fauna and flora in Florida. Tallahassee. 23 pp.

Gentry, G. G. 1974. Hurricanes in south Florida. Pp. 73–81 in P. J. Gleason, ed., Environments of south Florida: Present and past. Miami Geol. Soc. Memoir 2:1–452.

Gleason, P. J. 1984. Saving the wild places—a necessity for growth. Pp. viii–xxv in P. J. Gleason, ed., Environments of south Florida present and past II. 2d ed. Miami Geol. Soc., Coral Gables. 551 pp.

Hamilton, W. J., Jr. 1948. A new mink from the Florida everglades. Proc. Biol. Soc. Washington 61:139–140.

Harper, F. 1927. The mammals of the Okefinokee Swamp region of Georgia. Proc. Boston Soc. Nat. Hist. 38:191–396, pl. 4–7.

Humphrey, S. R., and H. W. Setzer. 1989. Geographic variation and taxonomic revision of mink (*Mustela vison*) in Florida. J. Mammal. 70:241–252.

Humphrey, S. R., and T. L. Zinn. 1982. Seasonal habitat use by river otters and Everglades mink in Florida. J. Wildl. Manage. 46:375–381.

Kurten, B., and E. Anderson. 1980. Pleistocene mammals of North America. Columbia University Press, New York. 442 pp.

Layne, J. N. 1974. The land mammals of south Florida. Pp. 386–413 in P. J. Gleason, ed., Environments of south Florida: Present and past. Miami Geol. Soc. Memoir 2:1–452.

McNab, B. K. 1971. On the ecological significance of Bergmann's rule. Ecology 52:845–854.

Parker, G. G. 1974. Hydrology of the pre-drainage system of the Everglades in southern Florida. Pp. 18–27 *in* P. J. Gleason, ed., Environments of south Florida: Present and past. Miami Geol. Soc. Memoir 2:1–452.

Smith, A. T. 1980. An environmental study of Everglades mink (*Mustela vison*). South Florida Research Center, Everglades National Park. Report T-555:1–17.

Soil Conservation Service. 1948. Soils, geology, and water control in the Everglades region. Agric. Experiment Station, Univ. of Florida, Bull. 442:1–168.

U.S. Fish and Wildlife Service. 1989. Endangered and threatened wildlife and plants; animal notice of review. Federal Register 54:554–579.

Prepared by: Stephen R. Humphrey, Florida Museum of Natural History, University of Florida, Gainesville, FL 32611.

Status Undetermined

Sherman's Short-Tailed Shrew
Blarina carolinensis shermani
FAMILY SORICIDAE
Order Insectivora

TAXONOMY: This subspecies was described by Hamilton (1955) as *B. brevicauda shermani* from specimens collected 2 mi north of Ft. Myers, Lee County. The specific location was 1 mi north of the Edison Bridge crossing the Caloosahatchee River and 0.25 mi east of U.S. Route 41. It is named after Harley B. Sherman, a former professor at the University of Florida who contributed much to the knowledge of Florida mammals. The species assignment of this subspecies is uncertain. At the time *shermani* was described, two other subspecies of *B. brevicauda*, *carolinensis* and *peninsulae*, were recognized in Florida. Subsequently, *B. b. carolinensis* was elevated to the species level (Genoways and Choate 1972) and *B. b. peninsulae* was treated as a subspecies of *B. carolinensis* (Hall 1981). At the present time, *B. c. carolinensis* is considered to occur in the panhandle and northern peninsular region of Florida and *B. c. peninsulae* in the southern part of the peninsula (Hall 1981; George et al. 1982). Recent authors have referred the subspecies *shermani* both to *brevicauda* (Hamilton and Whitaker 1979) and to *carolinensis* (Hall 1981; George et al. 1982). Hamilton and Whitaker (1979) tentatively assigned *shermani* to *B. brevicauda* on the basis of its relatively large size (hind foot $\geq$13 mm). Knowledge of its karyotype is needed for positive species identification, but to date suitable material for karyotypic analysis of this population has not been available (H. H. Genoways, personal communication). George et al. (1982) recognized three chromosomal groups, corresponding to *B. brevicauda*, *B. carolinensis* from northern Florida and elsewhere in the southeast, and *B. c. peninsulae*. *B. brevicauda* has a fundamental number of chromosomes (FN) of 48 and diploid (2N) numbers of 50, 49, or 48. *B. carolinensis* is characterized by FN = 45 or 44 and 2N = 46, 39, 38, or 37 and *B. c. peninsulae* by FN = 52 and 2N = 52, 51, or 50. The karyotype of *peninsulae* is distinct enough to suggest that it may be a separate species (George et al. 1982). Thus the *shermani* population may prove to be a subspecies of *brevicauda, carolinensis,* or *peninsulae,* if the specific status of the last taxon should be substantiated by further study. Therefore, the name *B. c. shermani* is used provisionally in this account.

DESCRIPTION: Sherman's short-tailed shrew is a medium-sized, dark-colored subspecies (Hamilton 1955). It is larger in body proportions and skull size and the adult winter pelage is darker and lacks the brownish tinge of other *Blarina* populations in Florida. Mean measurements and body mass of 27 specimens from the type locality are: total length 109.0 mm, tail 23.5 mm, hind foot 14.1 mm, and mass 13.8 g. For comparison, mean measurements and mass of a series of *B. c. peninsulae* from Highlands and Collier counties (first value) and of *B. c. carolinensis* from Alachua and Putnam counties (second value) are: total length 97.0 and 92.2 mm, tail 21.3 and 21.0 mm, hind foot 12.0 and 12.5 mm, and mass 9.9 and 8.0 g. The skull is robust, with a relatively broad cranium and rostrum. Mean skull measurements of 25 specimens are: condylobasal length 21.0 mm, cranial breadth 10.8 mm, interorbital breadth 5.5 mm, palatal length 9.0 mm, maxillary breadth 7.3 mm, and maxillary tooth row 7.9 mm. For comparison, average measurements of 7 skulls of *B. c. peninsulae* (first value) and 17 skulls of *B. c. carolinensis* (second value) from the localities noted above are: condylobasal length 20.0 and 19.3 mm, cranial breadth 10.5 and 10.3 mm, interorbital breadth 5.3 and 5.2 mm, palatal breadth 8.7 and 8.1 mm, maxillary breadth 6.7 and 6.6 mm, and maxillary tooth row 7.5 and 7.0 mm. All of the above measurements are from Hamilton (1955).

POPULATION SIZE AND TREND: Hamilton (1955) collected the type series of 27 specimens in 4000 trap-nights (snaptraps), suggesting that the shrew was not common. Subsequent efforts to document its occurrence and status at or in the immediate vicinity of the type locality have been unsuccessful. I attempted to collect specimens at the type locality in 1956 (25 livetrap-nights). In summer 1982 and spring 1983, Genoways (personal communication) trapped for five nights in the area with live-traps and pitfalls. In December 1984 and January 1985, Humphrey et al. (1986) sampled the area of the type locality and other localities to a distance of about 10 mi to the east and west with livetraps placed in lines or grids (adjusted trap-nights = 1298). Although short-tailed shrews are relatively difficult to capture anywhere in Florida, the lack of success in the above surveys, given the level of trapping effort expended, suggests that this subspecies is either very rare or has been extirpated in the vicinity of the type locality. The status of other *shermani* populations, if any, is unknown.

DISTRIBUTION AND HISTORY OF DISTRIBUTION: This shrew is known only from the type locality, although Hamilton (1955) suggested that it

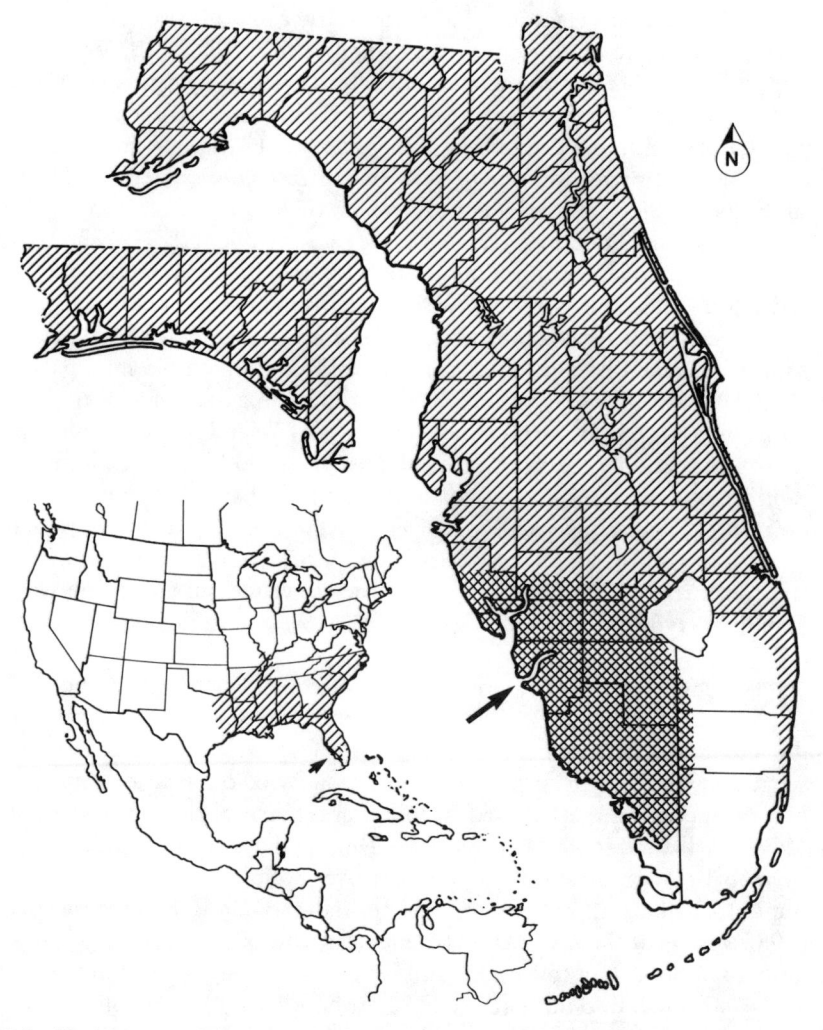

Distribution map of Sherman's short-tailed shrew, *Blarina carolinensis shermani*. Hatching represents the species; crosshatching represents the subspecies.

might have a more general distribution on the west coast of peninsular Florida. However, he failed to collect any in 200 trap-nights in Ft. Myers south of the Caloosahatchee River in 1939 and 1940 or on Pine Island, Lee County, in 1954. The localities nearest to the type locality from which *Blarina* has been collected are on the south side of the Caloosahatchee River about 9 mi E and 13 mi ENE of the type locality. Four specimens in the Archbold Biological Station collection (ABS 63-66) were collected by D. Radtke on the north side of the Orange River 0.5 mi northwest of Buckingham, Lee County, in November 1980, January and March 1981, and May 1983. Three specimens in the Florida Museum of Natural History (UF 20909-20911) were collected by Kenneth T. Wilkins and Henry W. Setzer 9 mi east of Ft. Myers off Bateman Road near Hickey Creek on 9 December 1982. Mean total length, tail, and hind foot measurements of the shrews from these two localities (Buckingham: 109.0 mm × 21.0 mm × 13.3 mm; Hickey Creek area: 98.0 mm × 22.0 mm × 13.7 mm) are distinctly larger than those given by Hamilton (1955) for *B. c. carolinensis* and intermediate between those for *B. c. peninsulae* and *shermani*. These limited data suggest that the specimens from south of the Caloosahatchee River are intergrades between *shermani* and *peninsulae*, although more detailed analysis may show they are referable to *shermani*.

GEOGRAPHICAL STATUS: On the basis of present limited knowledge, Sherman's short-tailed shrew has one of the most restricted ranges of any Florida mammal. If *shermani* should prove to be a subspecies of *B. brevicauda*, it would represent another case, similar to that of *Microtus pennsylvanicus dukecampbelli*, of a widely disjunct, relict Florida population of a more northern species.

HABITAT REQUIREMENTS AND HABITAT TREND: The sole information on the habitat of this subspecies is contained in the original description. Specimens were collected in drainage ditches with dense grass and in mole (*Scalopus aquaticus*) runways (Hamilton 1955). The type locality and vicinity have undergone intensive development, resulting in extensive loss of potential shrew habitats. Humphrey et al. (1986) mentioned the occurrence of remnants of undeveloped oak woodland and drainage ditches within 0.5 mi of the type locality in 1984–85. The specimens referred to above from near Buckingham were killed by house cats around buildings in mesic hammock habitat, and those from the vicinity of Hickey Creek were collected in slash pine-palmetto flatwoods with dense grass cover between the palmetto clumps (K. T. Wilkins, field notes). Elsewhere in

Florida, *Blarina* is typically found in moist forests or dense herbaceous habitats.

VULNERABILITY OF SPECIES AND HABITAT: Short-tailed shrews in Florida appear to be more habitat-restricted than in more northern parts of the range, and the types of habitats they prefer have been significantly reduced or degraded through urban or suburban development or agricultural practices.

CAUSES OF THREAT: Habitat loss or disturbance is the major threat to Sherman's short-tailed shrew. Because shrews are highly vulnerable to predation by house cats (*Felis catus*), an increase in free-ranging cats in developed areas may seriously affect shrew populations persisting in any remaining habitat fragments.

RESPONSES TO HABITAT MODIFICATION: Changes in habitats leading to reduction of cover and drying of soils would presumably be detrimental to shrew populations.

DEMOGRAPHIC CHARACTERISTICS: Except for the fragmentary data provided by Hamilton (1955), nothing is known of the life history or population dynamics of this form. His specimens were collected in February and March 1954 and were all adults, the males with enlarged testes and the females with turgid vaginas and well-developed uterine horns but no visible embryos.

KEY BEHAVIORS: The general behavior and biology of Sherman's short-tailed shrew are assumed to be generally similar to other populations of *Blarina*.

CONSERVATION MEASURES TAKEN: Other than the known efforts in 1956, 1979–80, and 1984–85 to capture specimens in the vicinity of the type locality, no specific measures have been directed toward its conservation. It was listed as "status undetermined" in the first edition of the FCREPA series (Layne 1978), and this classification still seems appropriate.

CONSERVATION MEASURES PROPOSED: Further intensive efforts should be made to determine whether this population is still extant in the vicinity of the type locality, and, if so, to obtain more detailed information on its distribution, habitat requirements, and life history and ecology. Based on the present evidence of its rarity and limited distribution combined with the continuing loss of potential habitats, any local population discovered would warrant strenuous efforts to ensure its protection. New surveys should emphasize use of pitfalls, preferably with drift fences, as these tend to be more effective than live or snap trapping in taking shrews. In addition, an effort should be made to encourage residents with house cats living in the vicinity of the type locality to save any shrews brought in by their cats. A priority in further sampling should be to obtain live specimens for karyotyping so that the specific status of this subspecies can be established. Further collecting to obtain additional specimens of *Blarina* from the area between the type locality in Lee County and the Lake Wales Ridge of the peninsula is needed to allow more accurate delineation of the range limits of *shermani*.

ACKNOWLEDGMENTS: I thank Kenneth T. Wilkins for permission to cite his field notes on file in the Florida Museum of Natural History; Laurie Wilkins for providing information on specimens in the Florida Museum of Natural History mammal collection; Hugh H. Genoways for information on his work in the Ft. Myers area; Stephen R. Humphrey for access to an unpublished report with data on a status survey of *shermani*; and Dieter Radtke for his interest and efforts in obtaining short-tailed shrew specimens from the Buckingham area.

Literature Cited

Genoways, H. H., and J. R. Choate. 1972. A multivariate analysis of systematic relationships of the short-tailed shrews (genus *Blarina*) in Nebraska Syst. Zool. 21:106–116.

George, S. B., H. H. Genoways, J. R. Choate, and R. J. Baker. 1982. Karyotypic relationships within the short-tailed shrews, genus *Blarina*. J. Mammal. 63: 639–645.

Hall, E. R. 1981. The mammals of North America. 2d ed. John Wiley and Sons, New York 1:1–600 + 90.

Hamilton, W. J., Jr. 1955. A new subspecies of *Blarina brevicauda* from Florida. Proc. Biol. Soc. Washington 68:37–40.

Hamilton, W. J., Jr., and J. O. Whitaker, Jr. 1979. Mammals of eastern United States. 2d ed. Cornell University Press, Ithaca, New York. 346 pp.

Humphrey, S. R., R. W. Repenning, and H. W. Setzer. 1986. Status survey of five Florida mammals. Unpubl. final report to U.S. Fish and Wildlife Serv., 3100 University Boulevard, South Suite 120, Jacksonville, Florida. 38 pp.

Layne, J. N. 1978. Sherman's short-tailed shrew. Pp. 42–43 *in* J. N. Layne, ed.,

Rare and endangered biota of Florida. Vol. 1. Mammals. University Presses of Florida, Gainesville. xx + 52 pp.

Prepared by: James N. Layne, Archbold Biological Station, P.O. Box 2057, Lake Placid, FL 33852.

Status Undetermined

Southeastern Brown Bat
Myotis austroriparius
FAMILY VESPERTILIONIDAE
Order Chiroptera

TAXONOMY: *M. austroriparius* was described by Miller and Allen (1928). Although several subspecies have been described, mainly on the basis of fur color, color and mensural characters of populations do not change gradually across the landscape; instead the species appears to be composed of a mosaic of demes, and subspecies cannot be discerned (LaVal 1970). Other common names include southeastern myotis, Mississippi myotis, and little brown bat; the last name is widely accepted as the common name of another species, *Myotis lucifugus*.

DESCRIPTION: The fur has a dull appearance, somewhat wooly texture, and little color contrast between the hair tips and bases. Color is quite variable, from gray to orange or russet above, and from tan to white below. The foot measures 10 mm in length and the hairs between the toes extend to or beyond the claw tips. The ear is 15 mm long, and the forearm is 36–41 mm long. The calcar is not keeled. The skull has a globose braincase and usually a sagittal crest.

POPULATION SIZE AND TREND: Unknown for Florida. The species originally was, and in some places still is, abundant. Uncertain accuracy of population estimates and ignorance of seasonal movements among caves preclude evaluation of trends from the scanty data available. The population in the lower Ohio River Valley has become rare and is near extinction (Barbour and Davis 1969).

In peninsular Florida, the following estimates have been published: 100,000 in Sweetgum Cave, Citrus County (LaVal 1970); 90,000 in Grant's Cave, Alachua County (Rice 1957); 30,000 in Hog Sink, Alachua County (Rice 1957); 2500 in Bat Cave, Alachua County (Rice

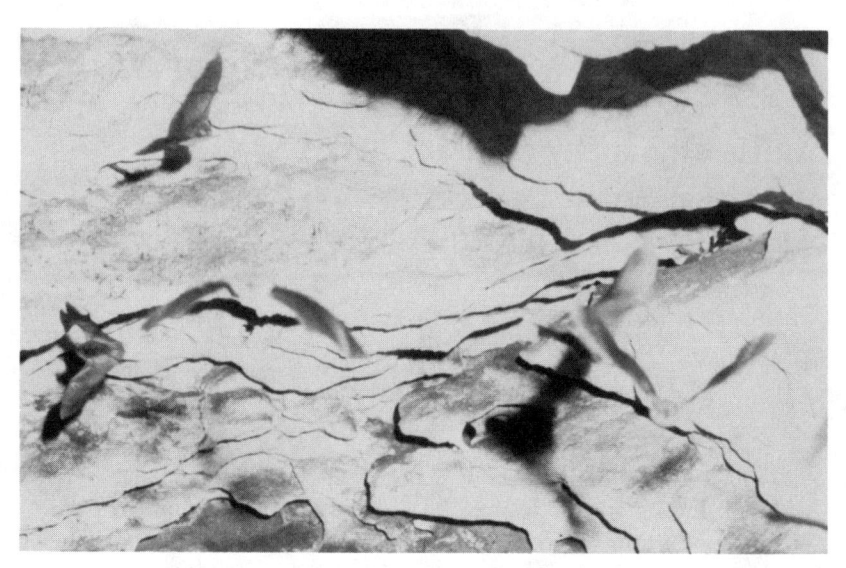

Southeastern brown bat, *Myotis austroriparius*. (Photo by Terry L. Zinn)

1957); 6000 in Robert's Cave, Gilchrist County (Rice 1957); 1000 in the Devil's Den, Levy County (Rice 1957); 2500 in the Devil's Head and Horns, Suwannee County (Rice 1957); and 15,000 adults and young in Sunday Sink, Marion County (Zinn and Humphrey 1981). All the estimates by Rice were of adults present before the birth of young in May. Currently, the population in Grant's Cave is as reported by Rice, and the population is gone from Bat Cave. The Hog Sink property is the subject of a proposed 5200-unit housing development, and most of these published population estimates are quite out of date.

In the Florida panhandle, Wenner (1984) reported the following estimates of summer populations from caves in Jackson County: 50,000–100,000 in Judge's Cave, >35,000 in Snead's Cave, 2000 in Girard's Cave, 200 in Gerome's Cave, and 2 in Old Indian Cave (150 during winter). These estimates include both *M. austroriparius* and *M. grisescens*, although in each case *M. austroriparius* was probably the most common species observed. Observations of emerging bats made in April and May, 1987 to 1989, provided estimates of <50,000 in Judge's Cave, >50,000 in Snead's Cave, 3000 in Girard's Cave, and >2000 in Old Indian Cave (J. A. Gore, personal observation). At least three caves (Fear's, Miller, and Mud) that previously had supported populations of *M. austroriparius* (Rice 1955; Wenner 1984) are now virtually devoid of bats (Wenner 1984; J. A. Gore, personal observation).

Myotis austroriparius

DISTRIBUTION AND HISTORY OF DISTRIBUTION: This species ranges from peninsular Florida (as far south as Manatee County), north to southeastern North Carolina and southern Indiana and Illinois, and west to southeastern Oklahoma and northeastern Texas. Rice (1957) considered a hiatus to occur between the Suwannee and Aucilla rivers, separating the peninsular and panhandle populations of Florida.

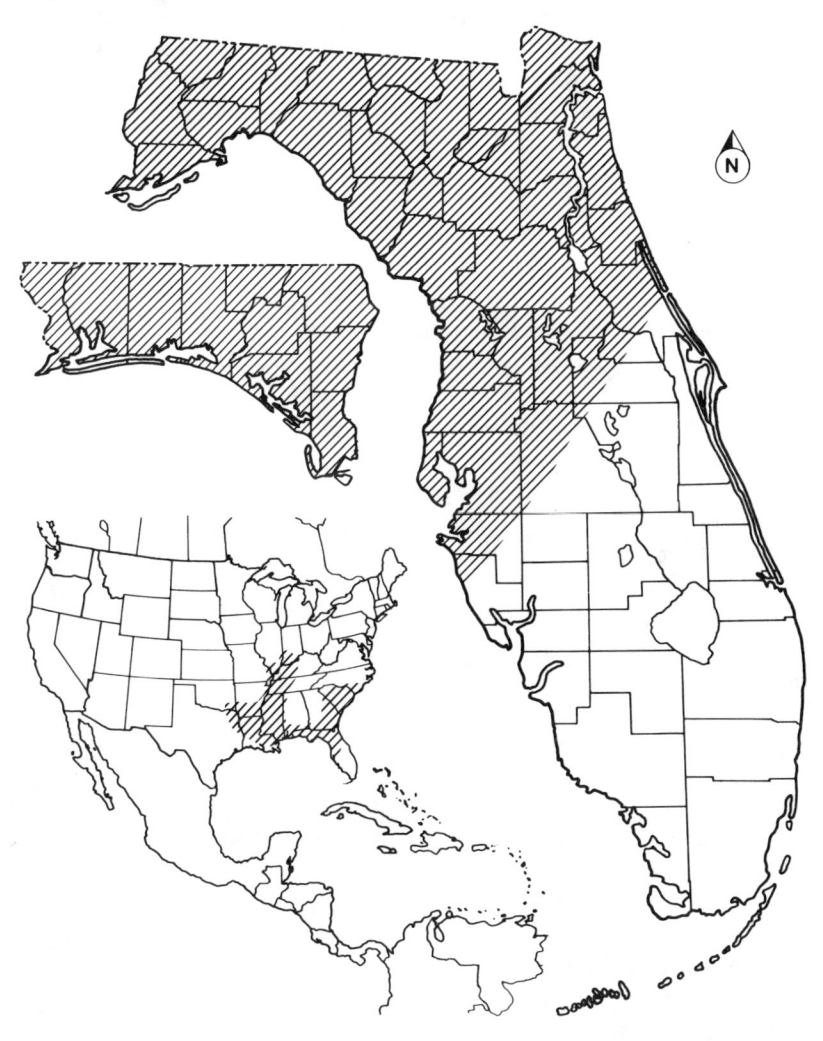

Distribution map of the southeastern brown bat, *Myotis austroriparius*. Hatching represents the species.

GEOGRAPHICAL STATUS: The Florida range is a significant part of the species' overall range. Rice (1955) reported that the species has two ecologically distinct populations in Florida, differing in their adaptation to climate. He stated that those in the panhandle mate in autumn and hibernate through the winter, whereas those in the peninsula are active during the winter, hibernating for only a few weeks if the weather becomes particularly cold, and they mate in the spring. One of us (J. A. Gore), however, has observed the evening emergence of thousands of bats from Snead's and Gerome's caves in December, January, February, and March. Probably those in the panhandle respond to temperature as do those in the peninsula.

HABITAT REQUIREMENTS AND HABITAT TREND: *M. austroriparius* lives in caves, buildings, and presumably trees, but in Florida this is primarily a cave bat. Because caves are the location of birth and rearing of most young, they are essential to the survival of the species. Most caves used as nursery sites have large, high ceilings and contain bodies of ground water. Rice (1957) considered these characteristics important as protection against climbing predators and in providing high humidity. When the water level is low, the bats may desert the cave (Rice 1957; Bain 1981). Snead's Cave is completely dry and has a low ceiling; Judge's, Gerome's, and Girard's caves contain water but most of the bats do not roost over it; and Gerome's Cave also has a low ceiling. Zinn (1977) monitored temperatures in the nursery at Sunday Sink for 24 hours during lactation. Whereas the outside air ranged from 19 to 30°C, and the cave's water was stable at 23°C and its air and rock at 22°C, the metabolic heat of the bats maintained the roost air at 28°C for most of the diel period. Roost air temperature dropped 3° while the adults were out foraging. Zinn pointed out that the domed roof of the nursery roost trapped the metabolically heated air, and that the water beneath it enabled the adults to drink without leaving the cave, which may have been especially important during lactation. Even though hibernation is abbreviated in the peninsula, it still occurs, and caves may also be essential for survival during winter.

Only a few *M. austroriparius* in Florida have been reported breeding or wintering in noncave situations. Sherman (1937) found a few females with young in a colony of free-tailed bats (*Tadarida brasiliensis*). Foster et al. (1978) studied a nursery population of about 1100 adults in the chimney of a house. Zinn (1977) reported a culvert used as a winter roost. Bain (1981) reported small numbers of *M. austroriparius* in populations of evening bats, *Nycticeius humeralis*, in Hague and Melrose, Alachua County; he also reported a culvert used as a winter roost and summer male roost in Gainesville, Alachua County. Fargo (1929) reported *M. austroriparius* wintering in mangrove trees on Indian Key, Tampa Bay. Probably the use of noncave sites is frequent but the total numbers of animals is low relative to those in caves.

M. austroriparius appears to prefer to forage over water (Rice 1957; Zinn 1977; Zinn and Humphrey 1981). In dry areas, bats having access to both longleaf pine and live oak habitats feed around the live oaks (S. R. Humphrey, personal observation) The bats feed selectively on small beetles, moths, and mosquitoes and other aquatic insects (Zinn 1977; Zinn and Humphrey 1981).

VULNERABILITY OF SPECIES AND HABITAT: Concentration of large numbers at a few sites makes the species quite vulnerable to natural disturbances such as flooding, or to changes in human land use or recreation.

CAUSES OF THREAT: The threats to cave populations are physical blocking of cave entrances, direct killing of the bats, or disturbance of the cave sufficient to make the site unusable by the bats. Rice (1955) found 11,000 *M. austroriparius* in Mud Cave, but the site later became a public dump, and the cave was filled with trash. Loss of the Bat Cave population was a direct result of regular visits to the cave by spelunkers (W. H. Oldacre, personal communication). In the late 1970s, several people contracted histoplasmosis in a Citrus County cave after throwing fresh guano containing spores of the fungus *Histoplasma capsulatum* at the bat colony overhead. On the basis that the bats were at fault, this incident resulted in demands that public health officials destroy the bat population. Most caves located in rural areas can be expected to be surrounded by increasingly dense housing as the human population of Florida continues its rapid growth, creating real or perceived conflicts between the people and the animals.

RESPONSES TO HABITAT MODIFICATION: Not known in any detail. There is evidence that disturbance in nursery caves will cause the bats to abandon the site.

DEMOGRAPHIC CHARACTERISTICS: Females first bear young when 1 year old. Virtually all females in the nursery populations are reproductive. This species is unique in its genus in having 2 young in the litter, and the young are born at an unusually early stage of development (Sherman 1930). Rice (1957) calculated that 190 young are born to 100 adult females, based on a sample of 1489 pregnant females in which 1343 had 2 fetuses and 146 had 1 and a sample of 112 females roosting with 193 newborn young. Foster et al. (1978) reported a single case of triplets. Preweaning mortality of the young is relatively high, about 12 percent, because of certain death of young falling into the water below the roost (Foster et al. 1978). Because adults are active almost all year, adult mortality also may be relatively high.

The major predators of this species are rat snakes (*Elaphe obsoleta* and *E. guttata*), opossums (*Didelphis virginiana*), owls, and cockroaches (Rice 1957; Foster et al. 1978). These bats have heavy infestations of mites (*Spinturnix* sp. and *Ichoronyssus quadridentatus*) and parasitic flies (*Basilia boardmani* and *Trichobius major*), but parasites are not considered a source of mortality (Rice 1957).

KEY BEHAVIORS: In the Florida panhandle, *M. austroriparius* intermittently hibernates in caves or flies nightly to forage, depending on cave and outside temperature (Rice 1957; J. A. Gore, personal observation); hibernation can occur as early as late October. Populations in peninsular Florida are active for most of the winter, feeding nightly. In most winters, these bats hibernate in caves for only about a month, from mid-February to mid-March (S. R. Humphrey, personal observation). It is not correct that most of these bats are absent from caves in winter, as stated by Rice (1957); instead they are even more highly concentrated than usual in a few caves. Compared with panhandle *M. austroriparius*, peninsular bats have little fat deposition in winter (McNab 1974). The mating period is estimated as late October to early December in the panhandle and mid-February to mid-April in the peninsula (Rice 1957). When peninsular populations emerge from hibernation, they engage in the swarming flight typical of mating, near cave entrances (Zinn and Humphrey 1981). Thereafter, peninsular populations remain at or disperse to nursery populations in mid-March (Rice 1957). Rice reported a few recoveries of banded bats showing movements of up to 43 km.

Nursery populations are about 80 percent females prior to birth of young in May. Most adult males roost elsewhere in separate caves or in small groups in buildings, culverts, and other sites. The young are born in the first three weeks of May, mostly in the second week, and they are able to fly five to six weeks later, beginning to leave their natal sites in July. The juveniles wander widely after they are able to fly, and probably encounter much accidental mortality.

CONSERVATION MEASURES TAKEN: Several caves in Jackson County that are used by *M. austroriparius* are protected because they are also used by the endangered *M. grisescens*. Old Indian and Judge's caves are publicly owned and their entrances have been gated or fenced. Snead's, Girard's, Gerome's, and Grant's caves are on private land but are posted against trespass; Grant's is also fenced. In 1987 the Florida Game and Fresh Water Fish Commission conducted a survey of caves in Jackson County to locate bat populations, determine their status, and assess needs and means of protecting populations from disturbance. *M. austroriparius* is under review for possible listing by the U.S. Fish and Wildlife Service (1989).

CONSERVATION MEASURES PROPOSED: A survey and mark-recapture program should be undertaken to locate cave populations, estimate their numbers, and determine the combination of sites required for the annual cycle of each major population. With this information available, it would be possible to devise a protection program that would ensure the survival of the major populations in perpetuity.

ACKNOWLEDGMENTS: Jacqueline J. Belwood and Lynn W. Lefebvre suggested important improvements to the manuscript.

Literature Cited

Bain, J. R. 1981. Roosting ecology of three Florida bats: *Nycticeius humeralis*, *Myotis austroriparius*, and *Tadarida brasiliensis*. Unpubl. M.S. thesis, University of Florida, Gainesville. 130 pp.
Barbour, R. W., and W. H. Davis. 1969. Bats of America. University Press of Kentucky, Lexington. 286 pp.
Fargo, W. G. 1929. Bats of Indian Key, Tampa Bay, Florida. J. Mammal. 10:203-205.
Foster, G., S. R. Humphrey, and P. P. Humphrey. 1978. Survival rate of young southeastern brown bats, *Myotis austroriparius*, in Florida. J. Mammal. 59:299-304.
LaVal, R. K. 1970. Infraspecific relationships of bats of the genus *Myotis austroriparius*. J. Mammal. 51:542-552.
McNab, B. K. 1974. The behavior of temperate cave bats in a subtropical environment. Ecology 55:943-958.
Miller, G. S., Jr., and G. M. Allen. 1928. The American bats of the genera *Myotis* and *Pizonyx*. Bull. U.S. National Mus. 144:1-128.
Rice, D. W. 1955. Status of *Myotis grisescens* in Florida. J. Mammal. 36:289-290.

Rice, D. W. 1957. Life history and ecology of *Myotis austroriparius* in Florida. J. Mammal. 38:15–32.

Sherman, H. B. 1930. Birth of the young of *Myotis austroriparius*. J. Mammal. 11:495–503.

Sherman, H. B. 1937. Breeding habits of the free-tailed bat. J. Mammal. 18:176–187.

U.S. Fish and Wildlife Service. 1989. Endangered and threatened wildlife and plants; animal notice of review. Federal Register 54:554–579.

Wenner, A. S. 1984. Current status and management of gray bat caves in Jackson County, Florida. Florida Field Nat. 12:1–6.

Zinn, T. L. 1977. Community ecology of Florida bats with emphasis on *Myotis austroriparius*. Unpubl. M.S. thesis, University of Florida, Gainesville. 87 pp.

Zinn, T. L., and S. R. Humphrey. 1981. Seasonal food resources and prey selection of the southeastern brown bat (*Myotis austroriparius*) in Florida. Florida Sci. 44:81–90.

Prepared by: Stephen R. Humphrey, Florida Museum of Natural History, University of Florida, Gainesville, FL 32611; Jeffery A. Gore, Florida Game and Fresh Water Fish Commission, 6938 Highway 2321, Panama City, FL 32409.

Status Undetermined

Big Brown Bat
Eptesicus fuscus fuscus
FAMILY VESPERTILIONIDAE
Order Chiroptera

TAXONOMY: Sometimes called the barn bat, house bat, or brown bat. Hall (1981) recognized 11 subspecies. Two of these, *E. f. fuscus* (Palisot de Beauvois) and *E. f. osceola* Rhoads, occur in Florida. The boundary between the two subspecies in Florida, however, is poorly defined (Hall 1981) and the status of *E. f. osceola* as a distinct subspecies is uncertain (Burnett 1983). Therefore, the present account considers all *Eptesicus* in Florida as one subspecies, *E. f. fuscus*.

DESCRIPTION: This is one of the larger bats in Florida. Wingspan ranges from about 320 to 350 mm and forearm length ranges from 42 to 51 mm. Adults normally weigh 16 to 19 g, up to 26 g if pregnant, or 30 g if fat for the winter (Scudder and Humphrey 1978). The ear is relatively short (14–15 mm) for a bat of this size, and the tragus is rounded at the tip. The calcar is keeled, and the tail extends only a few mm beyond the interfemoral membrane. The bare ears, wings, and interfemoral membrane are blackish in color. The fur is brown, but the color varies widely among individuals from reddish-brown to dark brown.

POPULATION SIZE AND TREND: This species is regarded as common or abundant over much of the United States, but it is conspicuously uncommon in the far southeastern part of the country and is apparently absent from southern Florida (Barbour and Davis 1969). Size of the population in Florida is unknown.

DISTRIBUTION AND HISTORY OF DISTRIBUTION: The big brown bat is widely distributed across most of North America. Its range extends from southeastern Alaska across southern Canada and south through Central

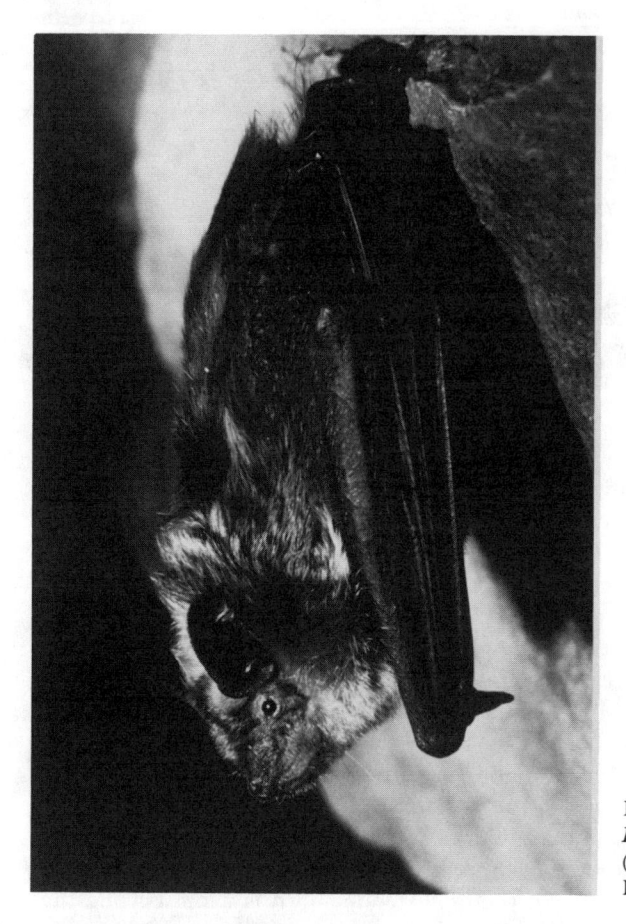

Big brown bat, *Eptesicus fuscus fuscus*. (Photo by Stephen R. Humphrey)

America and the Caribbean Islands. Its distribution within Florida is not well known, but it ranges across the northern portion of the state and perhaps as far south as Lake Okeechobee (Barbour and Davis 1969; Scudder and Humphrey 1978; Hall 1981).

GEOGRAPHICAL STATUS: Florida represents the southeastern edge of the range of *E. f. fuscus*, the widespread eastern subspecies. Although *E. f. osceola* is not considered as a separate subspecies in this account, current data suggest that the subspecies, if valid, is probably endemic to peninsular Florida (Hall 1981).

HABITAT REQUIREMENTS AND HABITAT TREND: Big brown bats eat

Eptesicus fuscus fuscus

insects (especially beetles) and forage over a variety of mostly open habitats (Barbour and Davis 1969; Black 1972). Individuals sometimes roost alone, but small colonies (usually fewer than 100 adults) are more common. This species is often found roosting in buildings, but it uses a variety of sites including trees, rock crevices, mines, caves, storm sewers,

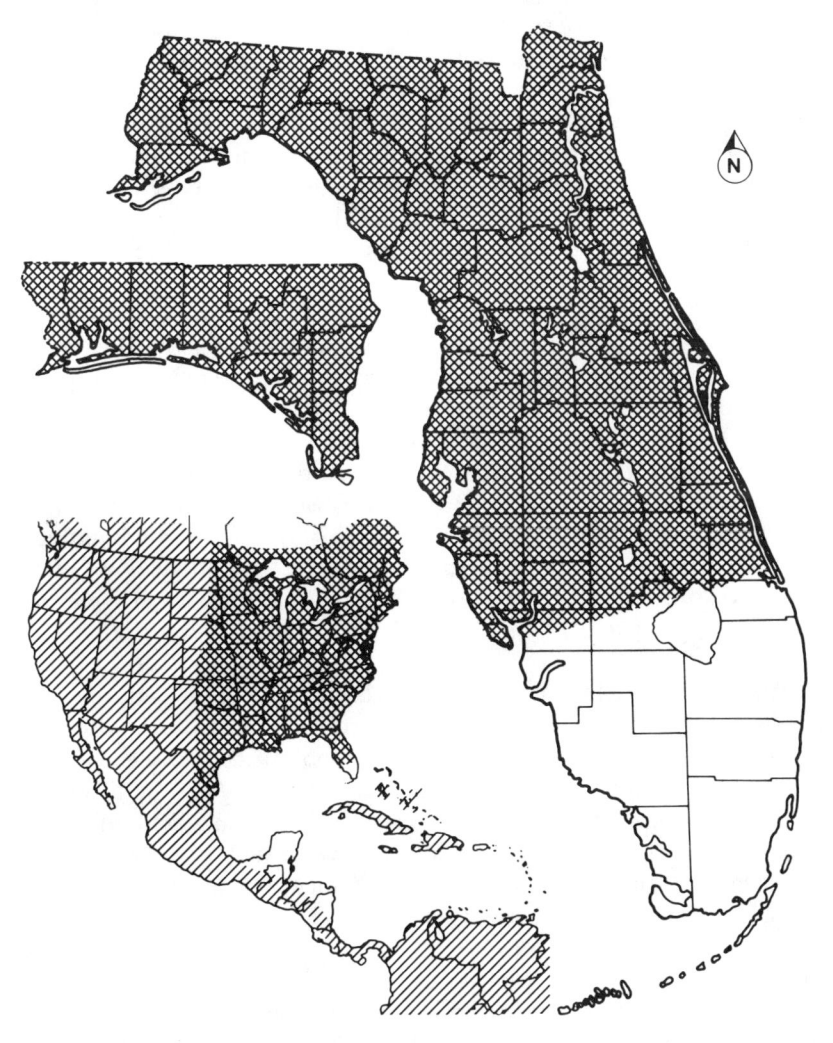

Distribution map of the big brown bat, *Eptesicus fuscus fuscus*. Hatching represents the species; crosshatching represents the subspecies.

and bridges (Barbour and Davis 1969). Buildings are frequently the sites of summer roosts, and in the mild climate of Florida buildings may also be the primary winter roosts. Mumford and Whitaker (1982) believed that this species often roosted in tree cavities in winter in Indiana; trees may also be important winter roosts in parts of Florida. Caves are not important roosting sites in Florida. Because we know so little about the habitat requirements of this species in Florida, it is not reasonable to estimate trends in habitat availability.

VULNERABILITY OF SPECIES AND HABITAT: This species may be less vulnerable than many other colonial bats because it uses a wide variety of foraging and roosting habitats and forms only small colonies. Nevertheless, the apparent small size of the population in Florida and our ignorance of the factors limiting its distribution in the state indicate at least a degree of vulnerability.

CAUSES OF THREAT: Disturbance of roosts and extermination of bats by humans are likely the greatest threats. Effects of pesticides on populations of this insect-eating bat are not known.

RESPONSES TO HABITAT MODIFICATION: Brigham and Fenton (1987) found that nursery colonies of big brown bats that were evicted from their roosts in buildings quickly found new roosts in nearby buildings. However, they also found that the evicted bats were about three times less successful in rearing young to flight age than bats that were not evicted.

DEMOGRAPHIC CHARACTERISTICS: This bat breeds in autumn, but fertilization is delayed until March or April. Each pregnant female typically produces two young in May or early June (later at more northern latitudes; Barbour and Davis 1969). Prenatal mortality may be as high as 34 percent (Kunz 1974). About 40 percent of the young die in their first year and more than half by the second year (Goehring 1972). Mortality is greatly reduced after the first few years and a few individuals live 20 years (Beer 1955; Barbour and Davis 1969; Goehring 1972).

KEY BEHAVIORS: Big brown bats are relatively tolerant of cold temperatures; they are active during rather cold weather and usually hibernate at colder sites than other bats (Barbour and Davis 1969). This species, therefore, is not expected to be torpid in Florida except in the coldest weather.

The big brown bat is rather sedentary and only short-distance shifts among roosts are common (Beer 1955; Goehring 1972; Mumford and Whitaker 1982). When disturbed, roosting individuals quickly retreat out of reach or even abandon the roost altogether. With repeated handling, this species becomes tame (Barbour and Davis 1969).

In Florida, big brown bats commonly appear with colonies of free-tailed bats, *Tadarida brasiliensis* (Scudder and Humphrey 1978). The only *Eptesicus* colonies I am aware of, one in Walton County and one in Baker County, contain larger numbers of *Tadarida*.

CONSERVATION MEASURES TAKEN: None specifically for the big brown bat. State laws and regulations prohibiting the wanton destruction of wildlife have been enforced, however, to prevent the needless destruction of bat colonies in buildings. Several government agencies and private conservation groups now provide information to the public about bats and, in particular, advise people on how to evict unwanted bats from buildings and how to construct artificial roosts or bat houses. Because big brown bats so often roost in man-made structures, these measures may benefit them more than most other species.

CONSERVATION MEASURES PROPOSED: The measures already taken should be continued and increased in scope as necessary. Before the need for species-specific conservation measures can be evaluated, more information must be collected on the distribution, abundance, and biology of this species in Florida.

Literature Cited

Barbour, R. W., and W. H. Davis. 1969. Bats of America. University Press of Kentucky, Lexington. 286 pp.
Beer, J. R. 1955. Survival and movements of banded big brown bats. J. Mammal. 36:242–248.
Black, H. L. 1972. Differential exploitation of moths by the bats *Eptesicus fuscus* and *Lasiurus cinereus*. J. Mammal. 53:598–601.
Brigham, R. M., and M. B. Fenton. 1987. The effect of roost sealing as a method to control maternity colonies of big brown bats. Canadian J. Public Health 78:47–50.
Burnett, C. D. 1983. Geographic and secondary sexual variation in the morphology of *Eptesicus fuscus*. Annals Carnegie Mus. Nat. Hist. 52:139–162.
Goehring, H. H. 1972. Twenty-year study of *Eptesicus fuscus* in Minnesota. J. Mammal. 53:201–207.

Hall, E. R. 1981. The mammals of North America. 2d ed. John Wiley and Sons, New York. 1:1–600 + 90 pp.

Kunz, T. H. 1974. Reproduction, growth, and mortality of the vespertilionid bat, *Eptesicus fuscus*, in Kansas. J. Mammal. 55:1–13.

Mumford, R. E., and J. O. Whitaker, Jr. 1982. Mammals of Indiana. Indiana University Press, Bloomington.

Scudder, S. J., and S. R. Humphrey. 1978. Big brown bat. Pp. 32–34 *in* J.N. Layne, ed., Rare and endangered biota of Florida. Vol. 1. Mammals. University Presses of Florida, Gainesville. xx + 52 pp.

Prepared by: Jeffery A. Gore, Florida Game and Fresh Water Fish Commission, 6938 Highway 2321, Panama City, FL 32409.

Status Undetermined

Northern Yellow Bat
Lasiurus intermedius floridanus
FAMILY VESPERTILIONIDAE
Order Chiroptera

TAXONOMY: *Lasiurus intermedius* was described by Allen in 1862. The type locality for the species is Matamoros, Tamaulipas, Mexico. Hall and Jones (1961) recognized three subspecies. *L. i. intermedius* occurs from southern Texas to Nicaragua; *L. i. insularis* occurs in Cuba; *L. i. floridanus* occurs from South Carolina to southern Texas and is the subspecies that occurs in Florida. This subspecies was originally described as *Dasypterus floridanus* by G. S. Miller, Jr., in 1902. The type locality for *L. i. floridanus* is Lake Kissimmee, Oceola County, Florida. Other common names used for this bat are the Florida bat, the Florida yellow bat, eastern yellow bat, greater yellow bat, and big yellow bat.

DESCRIPTION: The northern yellow bat is Florida's second largest resident bat, after the mastiff bat, *Eumops glaucinus*. The anterior half of the uropatagium (tail membrane) is thickly furred. The pelage is buffy yellow with dark tips on the hairs, giving the bat a slightly sooted appearance. Like other members of the genus *Lasiurus*, female *L. intermedius* have four mammae, unlike other vespertilionids, which have two. The additional pair of mammae accommodate the large litters (2–4 young) that occur in this genus. The baculum of the male is short and J-shaped. The penis is spiny and distally enlarged (Handley 1960). Average measurements for Florida specimens of northern yellow bats from the Florida Museum of Natural History are: for males, total length 125.2 mm ($n = 24$, range 116–131), tail length 53.7 mm ($n = 25$, range 51–60), hind foot length 9.9 mm ($n = 24$, range 8–11), forearm length 48.4 mm ($n = 18$, range 47–51), mass 17.0 g ($n = 26$, range 10–19.7); for females, total length 122 mm ($n = 3$, range 112–129), tail length 51.3 mm ($n = 3$, range 45–59), hind foot length 12 mm ($n = 2$), forearm length 50.1 mm ($n = 7$, range 49–52), mass 22.3 g ($n = 8$, range 18.2–26.1).

Northern yellow bat, *Lasiurus intermedius floridanus*. (Photo by Stephen R. Humphrey)

The rostrum is relatively long and the sagittal crest is strong compared to other *Lasiurus* species. The dental formula of the yellow bat is I 1/3, C 1/1, P 1/2, M 3/3. The upper P1 is always absent in the yellow bats, *L. intermedius* and *L. ega* (Handley 1960). The calcar is slightly keeled (Lowery 1974).

POPULATION SIZE AND TRENDS: Unknown. Little is known about the natural history of this species and nothing is known about its population

Lasiurus intermedius floridanus

size. They roost in trees and clumps of Spanish moss. This makes them difficult to census and study. The abundance of Spanish moss in Florida was greatly reduced by a fungus outbreak and has not recovered. This might have affected population levels of the yellow bat.

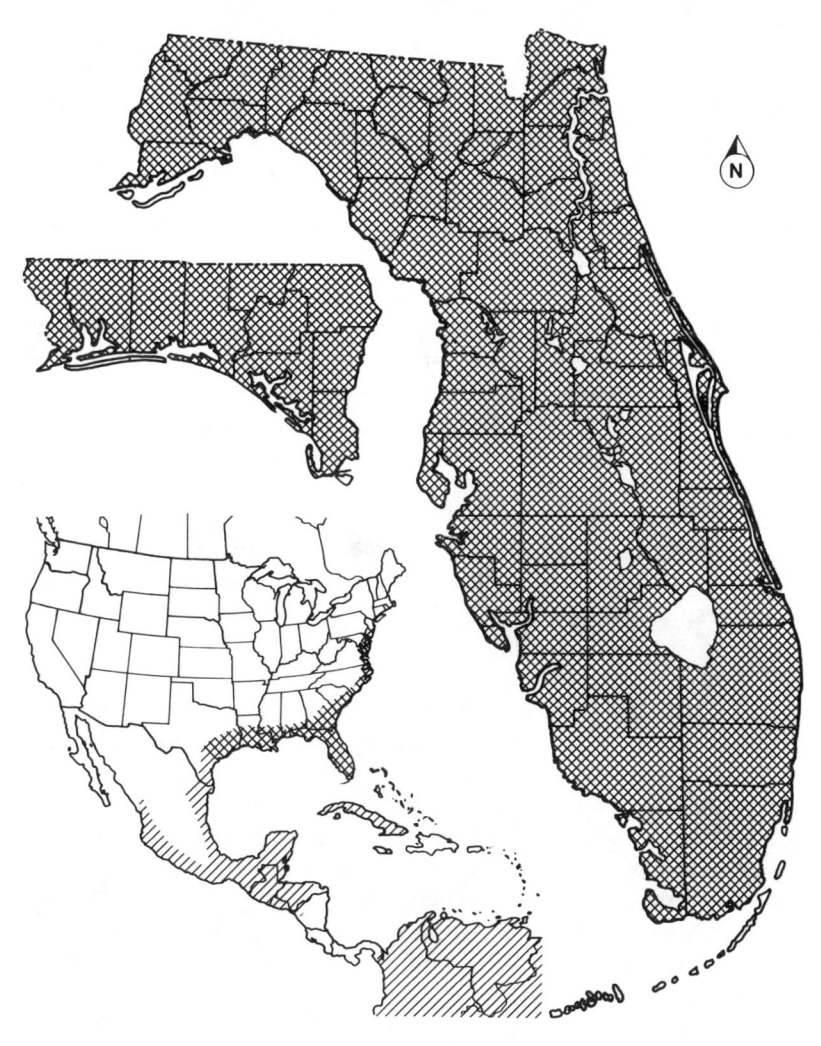

Distribution map of the northern yellow bat, *Lasiurus intermedius floridanus*. Hatching represents the species; crosshatching represents the subspecies.

DISTRIBUTION AND HISTORY OF DISTRIBUTION: The subspecies of northern yellow bat that occurs in the southeastern United States, *L. i. floridanus*, is found along the Atlantic and Gulf coastal plain from South Carolina to southeastern Texas (Webster et al. 1980). Although records of yellow bats have come from as far north as New Jersey (Koopman 1965) and Virginia (Rageot 1955), it is unlikely that this species normally occurs north of South Carolina. In Florida, the northern yellow bat may be found throughout the state but is most abundant in dry uplands. This is the dominant bat species in the central peninsula including Hillsborough, Polk, Hardee, and Highland counties (Jennings 1958). Fossil *L. intermedius* are reported from late Pleistocene sites in Florida at Devil's Den, Levy County; Reddick, Marion County; Arredondo II (Martin 1972); Haile XIB, Alachua County (Webb 1974); and Glyptodont Site, Pinnellas County (Morgan 1985).

GEOGRAPHICAL STATUS: The state of Florida is a major portion of the range of *L. i. floridanus*. This subspecies is the smallest of the three in terms of measurements and sagittal crest development. The Cuban subspecies, *L. i. insularis*, is the largest, with a well-developed sagittal crest. Hall and Jones (1961) showed that the size difference between *floridanus* and *insularis* is comparable to that between *Lasiurus borealis* and *Lasiurus cinereus*. The close proximity of Cuba to southern Florida suggests the possibility that any large yellow bats taken in southern Florida, with a forearm length approaching 60 mm, might represent a wanderer of the Cuban subspecies.

HABITAT REQUIREMENTS AND HABITAT TRENDS: The northern yellow bat in the southeastern United States is strongly associated with the epiphytic bromeliad, Spanish moss (*Tillandsia usneoides*; Barbour and Davis 1969). This bat regularly roosts and rears its young in the clumps of Spanish moss that festoon many of the trees of Florida (Jennings 1958). They have been collected from Spanish moss during all months of the year but were most abundant in June. Jennings (1958) stated that yellow bats were most often associated with longleaf pine-turkey oak sandhill habitats and live oak hammocks. They forage primarily over open areas such as fields, pastures, airports, golf courses, marshes and lake borders, and the open savanna-like habitat of sandhills (Jennings 1958). Yellow bats also roost and rear their young in the skirt of old fronds on tall palm trees. In this situation, often many females with young roost in the same tree (Davis 1974; Matt Crowder, personal communication). A structurally similar roost was reported for *L. i. intermedius* in Veracruz, Mexico. Yellow bats were found roosting among the dry corn stalks hanging on the side of a shed. Approximately 45 bats were seen and 15 were collected. The "colony" was apparently made up of females and their volant young. The observation took place on 22 July 1955 (Baker and Dickerman 1956). An assemblage of male yellow bats was reported roosting in tall palm trees in Jefferson County, Florida (Barbour and Davis 1969).

Sherman (1939) examined the stomach contents of a yellow bat taken in August near Gainesville and found the following food items: Homoptera, damselfly (Odonata, suborder Zygoptera), fly (Diptera, Anthomyiidae), bark beetle (Coleoptera, Scolytidae), predaceous diving beetle (Coleoptera, Dytiscidae), unidentified Coleoptera, and a flying ant (Hymenoptera, Formicidae, Myrmicinae).

VULNERABILITY OF SPECIES AND HABITAT: Little is known about the habits of this bat, and its vulnerability is difficult to assess. The loss of habitat, especially sandhill and oak hammock, to residential development and planting of citrus groves has probably had a serious effect on this species in Florida. It is also possible that the practice of removing old palm fronds has decreased potential roost sites for yellow bats in suburban areas.

CAUSES OF THREAT: Specific effects of habitat loss have not been measured. A former threat to yellow bats was the harvest of Spanish moss for processing into furniture stuffing (Jennings 1958; Lowery 1974). This loss of bat roosts was eliminated when development of superior synthetic materials caused the collapse of the cottage industry of moss collecting.

RESPONSES TO HABITAT MODIFICATION: Unknown.

DEMOGRAPHIC CHARACTERISTICS: The northern yellow bat mates in the fall in Florida. Jennings (1958) found that males had enlarged epididymides from July through December. Sperm were found in the uteri of two females collected in December. One of these females was a young of the year. A copulating pair of yellow bats fell to the ground and was captured on 23 November 1957 (Jennings 1958). The young are born from the last week in May to the end of June. Lactation may last well into July, while some young are foraging by the first week of July (Jennings 1958). The mean litter size based on full-term embryo counts was 3.5 ($n = 10$, range 3–4), and the mean postpartum litter size observed was 2.7

($n = 9$, range 1–4) (Jennings 1958). Young are left at the roost site while the mother is foraging, but she will carry her offspring away if the roost is disturbed.

The only recorded predator of the northern yellow bat is the barn owl, based on the presence of skulls in owl pellets (Silva-Taboada 1979). It is probable that other species of owls and hawks also prey on these bats. The following mites (Acari) have been identified from *L. intermedius: Olabidocarpus americanus* (Chirodiscidae), *Steatonyssus radovskyi* (Macronyssidae), and *Pteracarus chalinolobus* (Myobiidae; Whitaker and Wilson 1974).

KEY BEHAVIORS: Jennings (1958) stated that northern yellow bats in Florida roost singly but have a tendency to aggregate in the same tree. This phenomenon was also described for northern yellow bats in palm trees (Davis 1974).

The northern yellow bat has been taken throughout the year in southern Louisiana (Lowery 1974) and Florida (Jennings 1958) and does not appear to migrate in these states. Moore (1949) observed yellow bats foraging over a saw mill in Putnam County, Florida. They arrived later than seminole bats, 20–35 minutes after sundown. They flew at 15–20 feet above the ground, hawking insects that rose from piles of sawdust and lumber scraps.

CONSERVATION MEASURES TAKEN: None. The collapse of the Spanish moss industry in the southeastern United States has eliminated one potential threat of human disturbance. The practice of harvesting the clumps of Spanish moss from the trees not only destroyed roost sites but was the direct cause of death to numerous bats.

CONSERVATION MEASURES PROPOSED: There should be strict enforcement of Florida Administrative Code, Rule 39-4.001, the general prohibition on the taking, killing, or poisoning of wildlife except under specific permits. These permits should be strictly limited to protect bats from unnecessary persecution. Because the limited information we have on northern yellow bats in Florida seems to indicate that they prefer sandhills and xeric live oak hammocks, these dwindling habitats should be more actively protected by state purchase or owner agreements.

ACKNOWLEDGMENTS: Most of the natural history information about the *Lasiurus* bats in Florida comes from the Ph.D. dissertation of William L. Jennings. This is an excellent source of bat natural history not only for

Florida but for any species that is common in the southeastern United States. I thank Laurie Wilkins for providing records of the *Lasiurus intermedius* specimens from the Florida Museum of Natural History and Jacqueline J. Belwood for her review of the manuscript.

Literature Cited

Allen, H. 1862. Descriptions of two new species of Vespertilionidae, and some remarks on the genus *Antrozous*. Proc. Acad. Natural Sci. Philadelphia, pp. 246–248.

Baker, R. H., and R. W. Dickerman. 1956. Daytime roost of the yellow bat in Veracruz. J. Mammal. 37:443.

Barbour, R. W., and W. H. Davis. 1969. Bats of America. University Press of Kentucky, Lexington. 286 pp.

Davis, W. B. 1974. The Mammals of Texas. Texas Parks and Wildlife Dept., Bulletin 41:1–294.

Hall, E. R., and J. K. Jones, Jr. 1961. North American yellow bats, "Dasypterus," and a list of the named kinds of the genus *Lasiurus* Gray. Univ. Kansas Publ., Mus. Nat. Hist., 14:73–98.

Handley, C. O., Jr. 1960. Descriptions of new bats from Panama. Proc. U.S. National Mus. 112:459–479.

Jennings, W. L. 1958. The ecological distribution of bats in Florida. Unpubl. Ph.D. diss., University of Florida, Gainesville. 125 pp.

Koopman, K. F. 1965. A northern record of the yellow bat. J. Mammal. 46:695.

Lowery, G. H., Jr. 1974. The mammals of Louisiana and its adjacent waters. Louisiana State University Press, Baton Rouge. 565 pp.

Martin, R. A. 1972. Synopsis of Late Pliocene and Pleistocene bats of North America and the Antilles. Amer. Midland Nat. 87:326–335.

Miller, G. S., Jr. 1902. Twenty new American bats. Proc. Acad. Natural Sci. Philadelphia 54:389–412.

Moore, J. C. 1949. Putnam County and other Florida mammal notes. J. Mammal. 30:57–66.

Morgan, G. S. 1985. Fossil bats (Mammalia: Chiroptera) from the Late Pleistocene and Holocene Vero fauna, Indian River County, Florida. Brimleyana 11:97–117.

Rageot, R. H. 1955. A new northernmost record of the yellow bat, *Dasypterus floridanus*. J. Mammal. 36:456.

Sherman, H. B. 1939. Notes on the food of some Florida bats. J. Mammal. 20:103–104.

Silva-Taboada, G. 1979. Los murcielagos de Cuba. Acad. Cien. Cuba, Havana. 423 pp.

Webb, S. D. 1974. Pleistocene mammals of Florida. University Presses of Florida, Gainesville. 270 pp.

Webster, W. D., J. K. Jones, Jr., and R. J. Baker. 1980. *Lasiurus intermedius*. Mammalian Species 132:1–3.

Whitaker, J. O., Jr., and N. Wilson. 1974. Host and distribution lists of mites (Acari), parasitic and phoretic, in the hair of wild mammals of North America, north of Mexico. Amer. Midland Nat. 91:1–67.

Prepared by: William H. Kern, Jr., Department of Entomology and Nematology, University of Florida, Gainesville, FL 32611.

Status Undetermined

Brazilian Free-Tailed Bat
Tadarida brasiliensis cynocephala
FAMILY MOLOSSIDAE
Order Chiroptera

TAXONOMY: The species was described by Geoffroy Saint-Hilaire in 1824. Other names are guano bat, Mexican free-tailed bat, and free-tailed bat. Nine subspecies are recognized. See Wilkins (1989) for these and their type localities. *T. b. cynocephala* (Le Conte 1831:432) is the Florida subspecies, and the type locality is "Georgia, probably in neighborhood of Le Conte Plantation, Liberty Co." (Benson 1944).

DESCRIPTION: This is a relatively small bat with long, narrow wings. Selected external measurements (in mm) of *T. b. cynocephala* (for 16 males and 15 females, respectively) in the Florida Museum of Natural History (FMNH) are: total length 96.4, 94.1 and forearm 41.4, 41.9. Average mass is 11.5 g (for 11 males) and 12.1 g (for 2 nonreproductive females). Other females in the FMNH collection were pregnant when collected or lack data.

The distal portion of the tail extends beyond the uropatagium, as in all molossid bats. The calcar is not keeled, and long, bristly hairs extend from the feet. The ears are large and rounded, are not joined at the midline, have a series of small papillae on their leading edges, and extend to or only slightly beyond the muzzle when laid forward. The upper lip has vertical grooves or wrinkles.

The pelage is short and velvety, and the dorsal hairs in both sexes are uniformly dark brown or gray from base to tip. Hairs on the venter have white tips. White flecks occur in the pelage of some individuals. Cave-inhabiting members of this species can be bleached pale brown or reddish by ammonia fumes. *Eumops glaucinus* is the only other molossid in Florida. It is much larger than *T. b. cynocephala* and has ears that are joined at the midline.

Brazilian free-tailed bat, *Tadarida brasiliensis cynocephala*. (Photo by Stephen R. Humphrey)

T. brasiliensis in Florida occupies buildings. Individuals can number from several thousand to 20,000 in any one location. Roosts of these bats have a characteristically pungent, musky odor that may be apparent within several meters of a roost.

POPULATION SIZE AND TREND: These bats have not been studied extensively in Florida. Therefore, size and overall trend of the state's population are not well known. They appear to have a patchy distribution and are abundant where they occur. Their abundance appears to be limited by the availability of roosts.

Florida *T. brasiliensis* are well represented in collections and in the past were believed to be the most common species in the state. Jennings (1958), in his ecological survey of bats in Florida, stated that "the number of roosts located during the study make it obvious that almost every town with abandoned buildings may shelter a *Tadarida* colony."

T. b. cynocephala is no longer this common. When groups are found, particularly in buildings, their presence is unique enough to garner considerable media attention (e.g., *Stuart News*, 13 November 1987; *Florida Times Union*, 12 September 1989), usually in the context of attempts to destroy colonies.

Elsewhere in its range, *T. brasiliensis* forms the world's largest and densest mammal aggregations. Colonies in Texas and Arizona have been

Tadarida brasiliensis cynocephala

known to number 20 million individuals (Barbour and Davis 1969; McCracken 1986). Since 1963, however, populations of these bats have declined drastically—by as much as 99.9 percent. The principal reasons for this include the effects of pesticide exposure (e.g., Clark et al. 1975; Geluso et al. 1976; McCracken 1986) and human-caused disturbance

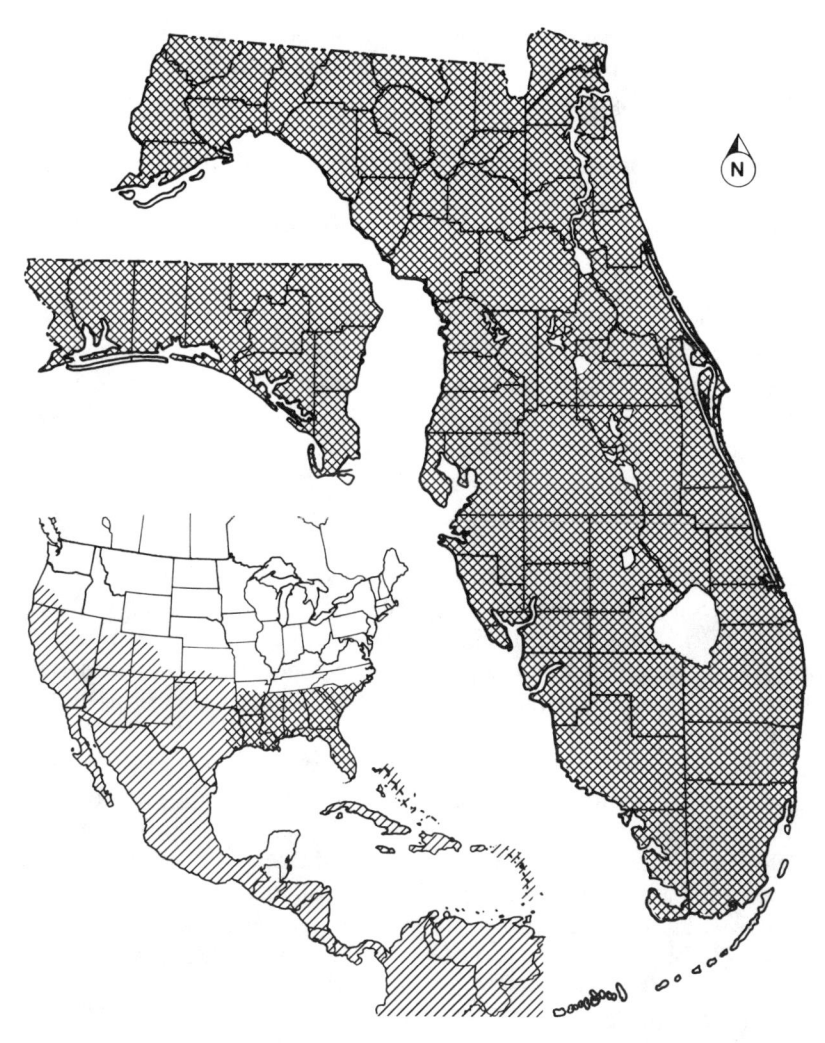

Distribution map of the Brazilian free-tailed bat, *Tadarida brasiliensis cynocephala*. Hatching represents the species; crosshatching represents the subspecies.

and vandalism. These bats are particularly susceptible to the latter because they roost in large numbers in relatively few specific locations. It is likely that these same factors are responsible for the decline of *T. brasiliensis* in Florida.

DISTRIBUTION AND HISTORY OF DISTRIBUTION: *T. brasiliensis* is one of the western hemisphere's most widely distributed bats (Wilkins 1989). In addition to its presence in large portions of the United States, it occurs in most of Central America, the Greater Antilles, and 11 islands in the Lesser Antilles (Baker and Genoways 1978). It also covers a large area in South America. See Wilkins (1989) for specific localities and citations.

In Florida, the species has been collected in most of the mainland but does not appear to occur in the Keys (Jennings 1958). Pleistocene *T. brasiliensis* in Florida have been found in Indian River, Marion, Sarasota, and Dade counties (Webb 1974; Morgan 1985; Morgan and Ridgway 1987). Other North American fossil *T. brasiliensis* are known from Mammoth Cave, Kentucky (Jegla and Hall 1962), Santa Cruz County, Arizona (Skinner 1942), Eddy County, New Mexico (Harris 1977), and Culberson County, Texas (Logan 1983). West Indian fossils have been found in Puerto Rico (Choate and Birney 1968), Cuba (Koopman and Ruibal 1955), and Antigua (Steadman et al. 1984).

GEOGRAPHICAL STATUS: Throughout its range, this bat occurs in a variety of geographical regions, in warm areas between about 42° N and 45° S. In the southwestern United States, it occurs in caves in several life zones, including the Lower and Upper Sonoran (Barbour and Davis 1969), and reaches maximum elevations of 2700 m. In California and the southeastern United States it occurs most often and in the largest numbers in man-made structures, which more often than not occur in cities. Small groups can be found in trees (Lowery 1974). Members of the southwestern population are highly migratory and regularly may move more than 1000 km from summer to winter ranges. Wilkins (1989) reviewed the movement patterns of several populations of this bat. In contrast, the California and southeastern populations appear to be nonmigratory, or move only short distances between summer and winter roosts (Spenrath and LaVal 1974; Lowery 1974). Their overall summer and winter ranges appear to be the same; in Florida, H.B. Sherman's large collection of *T. brasiliensis* at the FMNH indicates that members of both sexes were found in the same roosts during all months of the year. Whether this is the case for all colonies in Florida is not known. Wilkins (1989) describes the faunal regions in which these bats are found in South America as including the Patagonian subregion, eastern Brazilian highlands and coast, eastern slope of the northern Andes, and the Pacific coast of Peru and northern Chile.

HABITAT REQUIREMENTS AND HABITAT TREND: Precise habitat requirements for *T. b. cynocephala* are unknown. The following summarizes the little that is known about actual roosting sites. This is based on old accounts describing the southeastern population (e.g., Sherman 1937; Jennings 1958; Barbour and Davis 1969; Lowery 1974), and on anecdotal observations of extant colonies.

These bats occur in large numbers—at least several hundred, often several thousand, and up to 50,000 (Jennings 1958). A wide variety of roosts, constructed of wood, concrete, brick, or steel, appear to be used. These are usually situated in clear, uncluttered surroundings, allow bats to drop 6–8 m upon exiting, and are near a source of fresh water. Other bats commonly found roosting with *T. b. cynocephala* include the evening bat, *Nycticeius humeralis*, and southeastern brown bat, *Myotis austroriparius*. *T. b. cynocephala* are occasionally found roosting in trees. In Louisiana, *T. b. cynocephala* numbering up to 20,000 individuals used the same buildings as roosts for up to 35 years before these were destroyed.

Internally, the structures used by bat colonies allow individuals either to roost in large clusters in a single layer (Barbour and Davis 1969), or they provide long, continuous crevices in which bats can congregate. Preferred crevice dimensions for bats inhabiting a steel bridge in Texas are 40 cm deep, 2–4 cm wide, and at least 3 m long (Tuttle 1989). *T. b. cynocephala* that inhabit concrete structures in Gainesville, Florida, roost in crevices with similar dimensions (J. J. Belwood, C. L. Bloomcamp, W. H. Kern, and F. J. Bonaccorso, personal observation). Concentration of roosting clusters varies with temperature. Aggregations are loose when temperatures are 24° to 33.5°C but are tighter when temperatures drop below this (Pagels 1975). The highest temperatures in which bats are found roosting is about 35°C (Herreid 1967), although some parts of some roosts can exceed 40°C (Henshaw 1960; Licht and Leitner 1967). The latter are avoided by adults.

VULNERABILITY OF SPECIES AND HABITAT: *T. brasiliensis* rely solely on insects for food and consequently are susceptible to pesticide poisoning. Several heavy metals (mercury, lead, selenium) have been found incorporated in their tissues (Reidinger 1972; Clark 1981), especially in cities or agricultural areas. These bats have higher pesticide residues than their counterparts in caves or in areas away from human influence. Mercury has also been detected in bat guano deposits in an Arizona cave. These substances probably were ingested when the animals fed (Petit and Altenbach 1973).

Residues of DDT, DDE, DDD, dieldrin, endrin, toxaphene, and Aroclor 1254 and 1250 also have been found in *T. brasiliensis* from California, Arizona, New Mexico, and Texas (Reidinger 1976; Geluso et al. 1976, 1981; White and Krynitsky 1986). DDE and other contaminants are lipid-soluble and are stored mainly in the animals' fatty tissues. The concentrations of these substances increase when body fat is depleted during the increased energy demands of migration, lactation, or temporary food shortages. Under such circumstances, DDE is mobilized and concentrates in the brain, which is particularly sensitive to pesticides (Geluso et al. 1976).

Young-of-the-year are particularly susceptible to pesticide poisoning (Clark et al. 1975). In addition to environmental exposure, they can ingest these chemicals from their mothers' milk. Pesticide poisoning is probably the major cause of the drastic reduction observed in some populations of *T. brasiliensis* in the southwestern United States (Geluso et al. 1976; McCracken 1986). It is reasonable to assume that pesticides have the potential to affect bats in Florida in a manner similar to that found elsewhere.

Because *T. brasiliensis* in Florida occur in human habitations in large numbers, they are particularly vulnerable to eviction, roost destruction, vandalism, harassment, or large-scale colony destruction. In recent years, several such episodes have occurred (e.g., *Stuart News*, 13 November 1987). In this way, large segments of the population are lost in a short period of time.

CAUSES OF THREAT: Unknown.

RESPONSES TO HABITAT MODIFICATION: Largely unknown. Females have been known to carry their young from roosts following disturbances (Davis 1970). Where surviving *T. b. cynocephala* retreat after their roosts are destroyed is not known.

DEMOGRAPHIC CHARACTERISTICS: Most bats, including *T. brasiliensis*, have low reproductive rates. In Florida, females and males become sexually mature at about 9 months and 2 years of age, respectively (Sherman 1937). Spermatogenesis begins in September and peaks in February and March. Ovulation occurs in a brief period in March and lasts from 1 to 3 weeks. Mating occurs at about this time. The gestation period is about 11–12 weeks. One young is usually born, although a high incidence of multiple births has been reported in animals from Alabama (DiSalvo et al. 1969). The sex ratio is 1:1 at birth. Parturition in Florida occurs over a 2–3-week period, usually in June (Sherman 1937).

Newborns are about two-thirds the length of adults and have hairless bodies. Development of young in cave colonies occurs in maternity roosts, where babies are placed in communal creches by their mothers. Maternity roosts in the southeastern United States number only about 1000 (Barbour and Davis 1969). Males can be present in these. Contrary to early reports (e.g., Davis et al. 1962), females in maternity roosts do not nurse young randomly (McCracken 1984). Females do not carry their young on foraging flights, and most young are capable of flight at 38 days. Young reach adult length and mass between 3 weeks and 2 months of age. Fast growth of young is contingent on high ambient roost temperatures.

The longevity record for *T. brasiliensis* is eight years (LaVal 1973). Jennings (1958) reports that both sexes experience the same mortality rates.

KEY BEHAVIORS: *T. b. cynocephala* differ from *T. brasiliensis* in the southwestern United States in several important ways. First, they occupy buildings rather than caves. Second, the size of their populations is much smaller. Third, they do not appear to migrate long distances, if at all (Barbour and Davis 1969; Lowery 1974; Wilkins 1989).

These bats subsist wholly on insects. On warm evenings, each bat takes 3–4 g of food. Therefore a colony consisting of 10,000 individuals in Florida consumes 30–40 kg of insects per night, 900–1200 kg per month, and at least 6300–8400 kg per year (assuming a minimum of 7 months per year of summer-like weather).

In Louisiana, the bats emerge 7–59 min after sunset and return several hours later. On cold evenings, when insects are scarce, the bats either do not leave to forage (they enter torpor in their roosts), or they forage for only a short period of time. In Louisiana in winter, these bats store fat to overcome periods of food shortage (Pagels 1975).

Wilkins (1989) summarized the diet of *T. brasiliensis* as consisting of small moths, primarily gelechiids 5–9 mm long. This probably is an oversimplification of the dietary habits of this animal. Recently obtained fecal pellets from Stuart, Florida (collected in February and October 1988), and from Gainesville, Florida (collected in October 1989) reveal that a wide variety of small insects, including large numbers of flies and beetles, are eaten (J. J. Belwood, unpubl. data). Wilkins' (1989) statement that mosquitoes are not fed on by *T. brasiliensis* remains to be substantiated.

These bats, like most molossids, have small pointed wings, are extremely agile, have rapid flight, and usually fly in open areas (Norberg 1981). They are also capable of relatively efficient long-distance flight, which is adaptive given the great distances these highly concentrated animals must travel to feeding grounds. *T. brasiliensis* hawk high-flying insects at considerable altitudes above forest canopies and within large clearings. In essence, they are convergent with swifts in their flying behavior and habits.

The echolocation calls of these bats are adapted for foraging and flight in uncluttered areas. The "cruising" calls of these bats, those that are likely to be picked up by an ultrasound detector used for acoustic censuses, sweep from 50 to 25 kHz (Simmons et al. 1979). These are "basic pursuit strategy calls."

Jennings (1958) believed that *T. b. cynocephala* in Florida probably frequent more than one roost. Seasonally, there is evidence that these bats move from one roost to another. However, the distances moved and the reason for this move are not known. Movements are likely to reflect seasonally changing requirements in optimal roost microenvironments.

Key behaviors and related aspects of biology that are not known for these bats include: the complete range of habitat associations and roosts in which these bats can occur (including delineation of optimal conditions); the average number of roosts used by individuals; information on seasonal movement patterns, nature of foraging sites, and distances traveled to these.

CONSERVATION MEASURES TAKEN: Florida does not allow the use of Rozol (Chlorophacinone) to kill bats (Bat Conservation International, unpubl.). Florida Wildlife Code (Title 39, 39-4.001; effective 1 July 1989) prohibits the poisoning or "wanton or willful waste of wildlife." Permits are required to poison or trap wildlife determined to be "destructive" (Florida Wildlife Code, Title 39, 39-12.009).

Several institutions, including the University of Central Florida and the University of Florida (UF), have recently adopted policies to remove, rather than destroy, resident "nuisance" bats. UF has gone a step further by attempting to relocate bats, including *T. brasiliensis*, that occupy buildings on its campus. A bat house was designed and built for this purpose.

CONSERVATION MEASURES PROPOSED: Better knowledge of the summer and winter roosting habits of *T. brasiliensis* in Florida is needed immediately because these bats roost in large numbers, their older roosts are disappearing rapidly, and they are routinely evicted from the newer buildings in which they settle. Every attempt should be made to preserve known bat roosts. Otherwise, construction of alternate roosts should seriously be considered. These should be built in close proximity to old roosts and should be in place before known roosts are destroyed (Tuttle 1989). Because this has not yet been attempted in Florida, the case study designed to relocate roosts should be documented thoroughly.

If it is necessary to rid bats from roosts, care should be taken to do this at the least critical times. For example, bats should not be evicted from roosts when females are pregnant or lactating.

Studies are needed on the effects of pesticides on bats where applications of these chemicals are known to be high. An attempt also should be made to educate the public about the importance of bats, particularly *T. brasiliensis*, in their role as abundant and important insect consumers. With this in mind, a thorough study of the feeding habits of free-tailed bats in Florida is needed.

Companies and businesses that promote pesticides and other chemical means to kill bats should be discouraged from this practice. These methods have been shown to be ineffective and dangerous to human beings living in houses with bats (Kunz et al. 1977; Tuttle and Kern 1981).

Wildlife laws that pertain specifically to bats do not exist in Florida. It is becoming increasingly clear that such laws are necessary. Bats are poorly known animals that fall victim to tales and superstitions that are based on fear and misunderstanding. Consequently, bats often are persecuted and their roosts (e.g., in caves) vandalized. The propensity of some species to roost in large concentrations in a restricted number of localities poses special problems for the purpose of conservation. The precipitous decline of most of the large colonies of *T. brasiliensis* in Florida, which were once ubiquitous, indicates that existing laws to protect bats are ineffective.

ACKNOWLEDGMENTS: This was prepared during a Smithsonian postdoctoral fellowship. M. D. Tuttle provided information on artificial bat roosts, G. S. Morgan on fossil distributions, P. D. Southall on wildlife laws, and C. L. Bloomcamp on pesticides. Matt Crowder provided fecal material and newspaper clippings from Stuart, Florida, and allowed access to the roost from which the feces were obtained. Mark S. Robson and Jeffery A. Gore read the manuscript. This account is dedicated to Lee Bloomcamp on her retirement as head of Pest Control at the University of Florida, where she has worked tirelessly to preserve bat colonies.

Literature Cited

Baker, R. J., and H. H. Genoways. 1978. Zoogeography of Antillean bats. Pp. 53–97 in F. B. Gill, ed., Zoogeography in the Caribbean. Acad. Nat. Sci. Philadelphia, Spec. Publ. 13:1–128.

Barbour, R. W., and W. H. Davis. 1969. Bats of America. University Press of Kentucky, Lexington. 286 pp.

Benson, S. B. 1944. The type locality of *Tadarida mexicana*. J. Wash. Acad. Sci. 34:159.

Choate, J. R., and E. C. Birney. 1968. Sub-recent Insectivora and Chiroptera from Puerto Rico, with the description of a new bat of the genus *Stenoderma*. J. Mammal. 48:400–412.

Clark, D. R., Jr. 1981. Bats and environmental contaminants: A review. U.S. Dept. Interior, Fish Wildl. Serv., Spec. Sci. Rep. Wildl. 235:1–27.

Clark, D. R., Jr., C. O. Martin, and D. M. Swineford. 1975. Organochloride insecticide residue in the free-tailed bat (*Tadarida brasiliensis*) at Bracken Cave, Texas. J. Mammal. 56:429–443.

Davis, R. 1970. Carrying of young by flying female North American bats. Amer. Midland Nat. 83:186–196.

Davis, R. B., C. F. Herreid, and H. L. Short. 1962. Mexican free-tailed bat in Texas. Ecol. Monogr. 32:311–346.

DiSalvo, A. F., J. Palmer, and L. Ajello. 1969. Multiple pregnancy in *Tadarida brasiliensis cynocephala*. J. Mammal. 50:152.

Geluso, K. N., J. S. Altenbach, and D. E. Wilson. 1976. Bat mortality: Pesticide poisoning and migratory stress. Science 194:184–186.

Geluso, K. N., J. S. Altenbach, and D. E. Wilson. 1981. Organochlorine residues in young Mexican free-tailed bats from several roosts. Amer. Midland Nat. 105:249–257.

Geoffroy Saint-Hilaire, I. 1824. Memoire sur une chauve-souris americaine, formant une nouvelle espece dans le genre Nyctinome. Annales des Sciences Naturelles, Paris 1:337–347.

Harris, A. H. 1977. Wisconsin age environments in the northern Chihuahuan Desert: Evidence from the higher vertebrates. Pp. 23–52 in R. H. Wauer and D. H. Riskind, eds., Transactions of the symposium on the biological resources of the Chihuahuan Desert region, United States and Mexico. National Park Service Trans. Proc. ser. 3, U.S. Govern. Printing Office, Washington, D.C. 658 pp.

Henshaw, R. E. 1960. Responses of free-tailed bats to increases in cave temperature. J. Mammal. 41:396–398.

Herreid, C. F., II. 1967. Temperature regulation, temperature preference and tolerance, and metabolism of young and adult free-tailed bats. Physiol. Zool. 40:1–22.

Jegla, T. C., and J. S. Hall. 1962. A Pleistocene deposit of the free-tailed bat in Mammoth Cave, Kentucky. J. Mammal. 43:477–481.

Jennings, W. L. 1958. The ecological distribution of bats in Florida. Unpubl. Ph.D. diss., University of Florida, Gainesville. 125 pp.
Koopman, K. F., and R. Ruibal. 1955. Cave-fossil vertebrates from Camaguey, Cuba. Breviora 46:1–8.
Kunz, T. H., E. L. P. Anthony, and W. T. Rummage, III. 1977. Mortality of little brown bats following multiple pesticide applications. J. Wildl. Manage. 41:476–483.
LaVal, R. K. 1973. Observations on the biology of *Tadarida brasiliensis cynocephala* in southeastern Louisiana. Amer. Midland Nat. 89:112–120.
Le Conte, J. L. 1831. Appendix, p. 432 *in* G. Cuvier, Animal Kingdom arranged in conformity with its organization with notes and additions by H. M'Murtrie. G., C., and H. Carvill 1:1–448.
Lee, D. S., and C. Marsh. 1978. Range expansion of the Brazilian free-tailed bat into North Carolina. Amer. Midland Nat. 100:240–241.
Licht, P., and P. Leitner 1967. Behavioral responses to high temperatures in three species of California bats. J. Mammal. 48:52–61.
Logan, L. E. 1983. Paleoecological implications of the mammalian fauna of Lower Sloth Cave, Guadalupe Mountains, Texas. Natl. Speleol. Soc. Bull. 45:3–11.
Lowery, G., Jr. 1974. The mammals of Louisiana and its adjacent waters. Louisiana State University Press, Baton Rouge. 565 pp.
McCracken, G. F. 1984. Communal nursing in Mexican free-tailed bat maternity colonies. Science 223:1090–1091.
McCracken, G. F. 1986. Why are we losing our Mexican free-tailed bats? Bats 3(3):2–4.
Morgan, G. S. 1985. Fossil bats (Mammalia: Chiroptera) from the late Pleistocene and Holocene Vero fauna, Indian River County, Florida. Brimleyana 11:97–117.
Morgan, G. S., and R. B. Ridgway. 1987. Late Pliocene (late Blancan) vertebrates from the St. Petersburg Times site, Pinellas County, Florida, with a brief review of Florida Blancan faunas. Papers Fla. Paleontol. 1:1–22.
Norberg, U.M. 1981. Allometry of bat wings and legs and comparison with bird wings. Phil. Trans. Soc. Lond. (B) 292:359–398.
Pagels, J. F. 1975. Temperature regulation, body weight and changes in total body fat of the free-tailed bat, *Tadarida brasiliensis cynocephala* (Le Conte). Comp. Biochem. Physiol. 50:237–246.
Petit, M. G., and J. S. Altenbach. 1973. A chronological record of environmental chemicals from analysis of stratified vertebrate excretion deposited in a sheltered environment. Environ. Res. 6:339–343.
Reidinger, R. F., Jr. 1972. Factors influencing Arizona bat population levels. Unpubl. Ph.D. diss., University of Arizona, Tucson. 172 pp.
Reidinger, R. F., Jr. 1976. Organochlorine residues in adults of six southwestern bat species. J. Wildl. Manage. 40:677–680.
Sherman, H.B. 1937. Breeding habits of the free-tailed bat. J. Mammal. 18:176–187.

Simmons, J. A., M. B. Fenton, and M. J. O'Farrell. 1979. Echolocation and pursuit of prey by bats. Science 203:16–21.

Skinner, M. F. 1942. The fauna of Papago Springs Cave, Arizona, and a study of *Stockoceros*; with three antilocaprines from Nebraska and Arizona. Bull. Amer. Mus. Nat. Hist. 80:143–220.

Spenrath, C. A., and R. K. LaVal. 1974. An ecological study of a resident population of *Tadarida brasiliensis* in eastern Texas. Occas. Papers Mus., Texas Tech Univ. 21:1–14.

Steadman, D. W., G. K. Pregill, and S. L. Olson. 1984. Fossil vertebrates from Antigua, Lesser Antilles: Evidence for late Holocene human-caused extinctions in the West Indies. Proc. Natl. Acad. Sci. USA 81:4448–4451.

Tuttle, M. D. 1989. Extending an invitation to bats. Bats 7(2):5–13.

Tuttle, M. D., and S. J. Kern. 1981. Bats and public health. Milwaukee Pub. Mus. Contrib. Biol. Geol. 48:1–11.

Webb, S. D. 1974. Chronology of Florida Pleistocene mammals. Pp. 5–31 *in* S. D. Webb, ed., Pleistocene mammals of Florida. University Presses of Florida, Gainesville. 270 pp.

White, D. H., and A. J. Krynitsky. 1986. Wildlife in some areas of New Mexico and Texas accumulate elevated DDE residues, 1983. Arch. Environ. Contam. Toxicol. 15:149–157.

Wilkins, K. T. 1989. *Tadarida brasiliensis*. Mammalian Species 331:1–10.

Prepared by: Jacqueline J. Belwood, Bat Conservation International, P.O. Box 162603, Austin, TX 78716.

Contributors

Writers

Jacqueline J. Belwood, Bat Conservation International, P.O. Box 162603, Austin, TX 78716

Jeffery A. Gore, Florida Game and Fresh Water Fish Commission, 6938 Highway 2321, Panama City, FL 32409

Philip A. Frank, Department of Wildlife and Range Sciences, University of Florida, Gainesville, FL 32611

Nicholas R. Holler, U.S. Fish and Wildlife Service, Alabama Cooperative Fish and Wildlife Research Unit, 331 Funchess Hall, Auburn University, Auburn, AL 36849

Stephen R. Humphrey, Florida Museum of Natural History, University of Florida, Gainesville, FL 32611

Francis C. James, Department of Biological Science, Florida State University, Tallahassee, FL 32306

Patrick G. R. Jodice, Florida Game and Fresh Water Fish Commission, Rt. 7 Box 440, Lake City, FL 32055

Cheri A. Jones, Mississippi Museum of Natural History, 111 N. Jefferson Street, Jackson, MS 39202

Clyde Jones, Department of Biological Sciences and The Museum, Texas Tech University, Lubbock, TX 79409

Angela T. Kantola, U.S. Fish and Wildlife Service, Enhancement, P.O. Box 25486, Denver Federal Center, Lakewood, CO 80225

William H. Kern, Jr., Department of Entomology and Nematology, University of Florida, Gainesville, FL 32611

Willard D. Klimstra, Cooperative Wildlife Research Unit, Southern Illinois University, Carbondale IL 62901

James N. Layne, Archbold Biological Station, P.O. Box 2057, Lake Placid, FL 33852

Lynn W. Lefebvre, U.S. Fish and Wildlife Service, 412 NE 16th Avenue, Room 250, Gainesville, FL 32609

Mark E. Ludlow, Florida Department of Natural Resources, District V Administration, Clermont, FL 34711

David S. Maehr, Florida Game and Fresh Water Fish Commission, 566 Commercial Boulevard, Naples, FL 33942

Daniel K. Odell, Sea World of Florida, 7007 Sea World Drive, Orlando, FL 32821

Thomas J. O'Shea, U.S. Fish and Wildlife Service, 412 NE 16th Avenue, Room 250, Gainesville, FL 32601
Mark S. Robson, Florida Game and Fresh Water Fish Commission, 551 North Military Trail, West Palm Beach, FL 33406
I. Jack Stout, Department of Biological Science, University of Central Florida, Orlando, FL 32816
James T. Tilmant, South Florida Research Center, Everglades National Park, P.O. Box 279, Homestead, FL 33090
Elizabeth S. Wing, Florida Museum of Natural History, University of Florida, Gainesville, FL 32611
James L. Wolfe, Division of Biological Science, Emporia State University, Emporia, KS 66801
John B. Wooding, Florida Game and Fresh Water Fish Commission, 4005 South Main Street, Gainesville, FL 32601
Charles A. Woods, Florida Museum of Natural History, University of Florida, Gainesville, FL 32611

Photographers

Roger W. Barbour, Zoology Department, University of Kentucky, Lexington, KY 40506
Douglas M. Burn, U.S. Fish and Wildlife Service, 1011 East Tudor Road, Anchorage, AK 99503
George W. Folkerts, Department of Zoology and Wildlife Science, 331 Funchess Hall, Auburn University, Auburn, AL 36849
Philip A. Frank, Department of Wildlife and Range Sciences, University of Florida, Gainesville, FL 32611
Patrick G. R. Jodice, Florida Game and Fresh Water Fish Commission, Rt. 7 Box 440, Lake City, FL 32055
John E. Heyning, Natural History Museum of Los Angeles County, 900 Exposition Boulevard, Los Angeles, CA 90007
Stephen R. Humphrey, Florida Museum of Natural History, University of Florida, Gainesville, FL 32611
Willard D. Klimstra, Cooperative Wildlife Research Unit, Southern Illinois University, Carbondale, IL 62901
James N. Layne, Archbold Biological Station, P.O. Box 2057, Lake Placid, FL 33852
Paul W. Lefebvre, 1615 SW 58th Avenue, Gainesville, FL 32608
David S. Maehr, Florida Game and Fresh Water Fish Commission, 566 Commercial Boulevard, Naples, FL 33942
Alan S. Maltz, Light Flight Productions, 819 Peacock Plaza Suite 620, Key West, FL 33040
Barry W. Mansell, 2826 Rosselle Street, Jacksonville, FL 32205

Contributors

James G. Mead, National Museum of Natural History, Smithsonian Institution, Washington, DC 20560

Robert M. Rattner, 106-15 Jamaica Avenue, Richmond Hill, NY 11418

A. Ken Womble, Department of Biological Science, Florida State University, Tallahassee, FL 32306

John B. Wooding, Florida Game and Fresh Water Fish Commission, 4005 South Main Street, Gainesville, FL 32601

Charles A. Woods, Florida Museum of Natural History, University of Florida, Gainesville, FL 32611

Terry L. Zinn, 10031 Loquat Street, Miramar, FL 33025

Index

Agriculture, and habitat destruction, 5; by citrus agriculture, 5, 114, 228, 255, 324; by logging, 114, 297; by pineapple cultivation, 114; by sugarcane industry, 280, 281

Alabama Administrative Code, Title 87-GF-7, 106

Alabama beach mouse (*Peromyscus polionotus ammobates*): demographic characteristics of, 82; description of, 102; diet of, 80; habitat for, 79; residential behavior of, 83; responses to habitat modification, 81

Alligator River National Wildlife Refuge (ARNWR), North Carolina: reintroduction of red wolf (*Canis rufus*) into, 30, 33

American alligator (*Alligator mississippiensis*): geographical status of, 303; as predator of southern Florida population of mink (*Mustela vison mink*), 323

American bison. *See* Plains bison

Anastasia Island beach mouse (*Peromyscus polionotus phasma*), 94–101; current status of, 7; description of, 19; responses to habitat modification, 81

Anastasia Island population of cotton mouse (*Peromyscus gossypinus gossypinus*), 41–46; causes of threat to, 115; current status of, 6; reasons for extinction of, 2

Anastasia State Recreation Area, St. Johns County: deterioration of dune vegetation at, 100; population of Anastasia Island beach mouse (*Peromyscus polionotus phasma*) at, 97, 98, 99

Antillean manatee (*Trichechus manatus manatus*): geographical status of, 193; taxonomy of, 190

Apalachicola National Forest: population of Florida black bear (*Ursus americanus floridanus*) at, 272

Archbold Biological Station, Highlands County: population of Florida mouse (*Podomys floridanus*) at, 259; population of genus Podomys at, 256; population of short-tailed shrew (*Blarina carolinensis*) at, 331

Armadillo (*Dasypus novemcinctus*): as competitor of Florida panther (*Felis concolor coryi*) at, 180; as prey of Florida black bear (*Ursus americanus floridanus*), 269; as prey of Florida panther (*Felis concolor coryi*) at, 180

Auburn University, Alabama: breeding colony of Choctawhatchee beach mouse (*Peromyscus polionotus allophrys*) at, 83, 84; breeding colony of Perdido Key beach mouse (*Peromyscus polionotus trissyllepsis*) at, 107, 108

Balaenoptera borealis. *See* Sei whale
Balaenoptera physalus. *See* Fin whale
Bald eagle (*Haliaeetus leucicephalus*): nesting population of, x; as predator of Sherman's fox squirrel (*Sciurus niger shermani*), 239
Barbour, Thomas, xi

Note: Numbers in *italic type* denote illustration

Index

Barn bat. *See* Big brown bat
Barn owl (*Tyto alba*): as predator of northern yellow bat (*Lasiurus intermedius floridanus*), 354; as predator of round-tailed muskrat (*Neofiber alleni*), 279
Bartram, William: and record of Seminole Indian hunting of Florida black bear (*Ursus americanus floridanus*), 271–72; sighting of Plains bison (*Bison bison bison*) by, 48
Bat Cave, Alachua County: population of southeastern brown bat (*Myotis austroriparius*) at, 335, 336, 339
Bat Cave, Mammoth Cave National Park, Kentucky: flooding of bat hibernation sites in, 58
Bats (order Chiroptera): *Lasiurus borealis*, 292, 352; *Lasiurus ega*, 350; *Pipistrellus subflavus*, 292; *Plecotus alleganiensis*, 289; *Plecotus subflavus*, 292; *Plecotus townsendii*, 287; *Tadarida midas*, 221. *See also* Big-eared bat; Brown bat; Free-tailed bat; Gray bat; Indiana bat; Mastiff bat; Southeastern brown bat; Yellow bat
Bay swamps: definition of, xxvii
Beach mouse (*Peromyscus polionotus*): causes of threat to, 246; compared to Florida mouse (*Podomys floridanus*), 259; current status of, 3; demographic characteristics of, 81–82; description of, 87, 251; diet of, 80; distribution maps of, *21, 78, 89, 96, 104*; key behaviors of, 82–83; responses to habitat modification, 246
—subspecies of, 76, 242; *P. p. albifrons*, 76; *P. p. leucocephalus*, 76, 80–82, 87, 88, 91; *P. p. polionotus*, 102; *P. p. sumneri*, 76; Alabama beach mouse (*P. p. ammobates*), 79–83, 102; Anastasia Island beach mouse (*P. p. phasma*), 7, 94–101; Choctawhatchee beach mouse (*P. p. allophrys*), x, 7, 76–86; Gulf coast subspecies (*P. polionotus*), 94; pallid beach mouse (*P. p. decoloratus*), 2, 6, 19–23; Perdido Key beach mouse (*P. p. trissyllepsis*), ix–x, 7, 102–9; St. Andrew beach mouse (*P. p. peninsularis*), 7, 87–93; southeastern beach mouse (*P. p. niveiventris*), 8, 242–49
Bessie bugs (*Odontotaenius disjunctus*), 269
Big brown bat (*Eptesicus fuscus fuscus*), 343–48; current status of, 9
Big Cypress fox squirrel (*Sciurus niger avicennia*), 224–33; current status of, 8; taxonomy of, 234
Big Cypress National Preserve: established and expanded, 185, 232; habitat for Big Cypress fox squirrel (*Sciurus niger avicennia*) at, 229, 231; habitat for Florida mastiff bat (*Eumops glaucinus floridanus*) at, 220; hunting of Big Cypress fox squirrel (*Sciurus niger avicennia*) at, 231; population of Big Cypress fox squirrel (*Sciurus niger avicennia*) at, 225, 227, 232; population of Florida black bear (*Ursus americanus floridanus*) at, 272; population of Florida panther (*Felis concolor coryi*) at, 179, 180; population of southern Florida mink (*Mustela vison mink*) at, 325; population of white-tailed deer (*Odocoileus virginianus*) at, 180
Big Cypress Seminole Indian Reservation: population of Florida black bear (*Ursus americanus floridanus*) at, 272; population of Florida panther (*Felis concolor coryi*) at, 179
Big-eared bat (*Plecotus rafinesquii*): demographic characteristics of, 290–91; description of, 287; distribution map of, *289*; fossil remains of, 289; geographical status of, 290; key behaviors of, 291–92
—subspecies of: *P. r. rafinesquii*, 287,

Big-eared bat (*continued*)
288; southeastern big-eared bat (*P. r. macrotis*), 8, 287–93
Big-eared mouse. *See* Florida mouse
Big Pine Key: habitat loss for Key deer (*Odocoileus virginianus clavium*) at, 206, 208; habitat management for Key deer at, 210, 211; population of Key deer at, 203, 204, 206
Big Pine Key ringneck snake (*Diadophis punctatus acricus*), 303
Big Torch Key: and habitat management for Key deer (*Odocoileus virginianus clavium*) at, 211
Big yellow bat. *See* Northern yellow bat
Bird Rookery Swamp: habitat for southern Florida mink (*Mustela vison mink*) at, 325
Bird Rookery Swamp–Flintpen Strand: land acquisition at, 232
Birds: pileated woodpecker (*Dryocopus pileatus*), 219; red-cockaded woodpecker (*Picoides borealis*), 219; red-headed woodpecker (*Melanerpes erythrocephalus*), 220; red-tailed hawk (*Buteo jamaicensis*), 239; scrub jay (*Aphelocoma c. coerulescens*), 256
Bison (*Bison bison*): distribution map of, 49; fossil remains of (*Bison antiquus*), 50
—subspecies of: Plains bison (*B. b. bison*), 2, 6, 47–53
Bison bison bison. *See* Plains bison
Black bear (*Ursus americanus*): distribution and history of distribution of, 266; distribution map of, 267
—subspecies of: Florida black bear (*U. a. floridanus*), 3, 8, 265–75
Black panthers, 177
Black rat (*Rattus rattus*): as competitor of cotton mouse, 27, 45, 115; as competitor of Key Largo woodrat (*Neotoma floridana smalli*), 124–25; as competitor of Lower Keys population of rice rat (*Oryzomys palustris natator*), 304, 306
Blackwater State Forest, Santa Rosa and Okaloosa counties: habitat management at, 298
Black wolf. *See* Florida red wolf
Blarina brevicauda carolinensis. *See* Short-tailed shrew
Blarina brevicauda peninsulae. *See* Short-tailed shrew
Blarina brevicauda shermani. *See* Sherman's short-tailed shrew
Blarina carolinensis. *See* Short-tailed shrew
Blarina carolinensis shermani. *See* Sherman's short-tailed shrew
Blind cricket (*Typhloceuthophilus floridanus*), 16
Blind staphylinid (*Cubanotyphlus largo*), 127
Blind weevil (*Caecossonus dentipes*), 127
Blue Spring, Volusia County: population of Florida manatee (*Trichechus manatus latirostris*) at, 191; as sanctuary for Florida manatee (*Trichechus manatus latirostris*), 197
Boats: as threat to fin whale (*Balanopters physalus*), 156; as threat to Florida manatee (*Trichechus manatus latirostris*), x, 194, 195, 197; as threat to humpback whale (*Megaptera novaeangliae*), 163; as threat to right whale (*Eubalaena glacialis*), 143, 144
Bobcat (*Felis rufus*): as competitor of Florida panther (*Felis concolor coryi*), 186; confused with Florida panther, 177; description of, 176; disease in, 182; as predator of Florida mouse (*Podomys floridanus*), 258; as predator of Key Largo woodrat (*Neotoma floridana smalli*), 125
Bottlenose dolphins, 144

Index

Bowhead whale, 140
Brazilian free-tailed bat (*Tadarida brasiliensis cynocephala*), 357–368; current status of, 9
Brown bat (*Eptesicus fuscus*): cohabitation with big-eared bat (*Plecotus rafinesquii*), 292; distribution map of, *345*
—subspecies of: *E. f. osceola*, 343, 344; big brown bat (*E. f. fuscus*), 9, 343–48
Brown pelican, delisted, x
Bryde's whale (*Balaenoptera edeni*), 150
Buffalo: as common name for Plains bison (*Bison bison bison*), 47; true, description of, 47
Bulow Plantation Historical Site, Flagler County: habitat for pallid beach mouse (*Peromyscus polionotus decoloratus*) at, 20
Burrowing owl (*Athene cunicularia*), 253

Cachalot. *See* Sperm whale
Calderón, Gabriel Díaz Vara: sighting of Plains bison (*Bison bison bison*) by, 49
California sea lions, feral (*Zalophus californicus*), 36
California vole (*Microtus californicus*), 136
Canaveral National Seashore, Brevard County: habitat of southeastern beach mouse (*Peromyscus polionotus niveiventris*) at, 243, 245, 247; population of southeastern beach mouse (*Peromyscus polionotus niveiventris*) at, 243
Canines (family Canidae). *See* Coyote; Domestic dog; Gray wolf; Red wolf
Canis floridanus. *See* Florida red wolf
Canis niger. *See* Florida red wolf
Canis niger niger. *See* Florida red wolf
Canis rufus floridanus. *See* Florida red wolf
Cape Sable: population of round-tailed muskrat (*Neofiber alleni*) at, 280
Capillaria aerophila, 270
Captive breeding: of Choctawhatchee beach mouse (*Peromyscus polionotus allophrys*), 83, 84; of Florida red wolf (*Canis rufus floridanus*), 30, 32–33; of Florida saltmarsh vole (*Microtus pennsylvanicus dukecampbelli*), 138; of Perdido Key beach mouse (*Peromyscus polionotus trissyllepsis*) at, 107, 108
Caribbean monk seal. *See* West Indian monk seal
CARL. *See* Conservation and Recreation Lands
Carpenter ants (*Campanotus abdominalis floridanus*), 269
Catamount. *See* Florida panther
Cattle (*Bos taurus*): as competitor of Florida panther (*Felis concolor coryi*) at, 179; infection of Plains bison (*Bison bison bison*) by, 52; and sighting of Plains bison, 50
Cave bat. *See* Gray bat
Cecil Webb Wildlife Management Area: Florida mastiff bat (*Eumops glaucinus floridanus*) at, 219, 220
Cedar Key Scrub State Preserve: population of Florida mouse (*Podomys floridanus*) at, 259
Cestode (*Anoplocephaloides neofibrinus*), 282
Chadwick Beach cotton mouse (*Peromyscus gossypinus restrictus*), 24–28; causes of threat to, 2; current status of, 6
Chassahowitzka National Wildlife Refuge, Citrus County: population of Florida black bear (*Ursus americanus floridanus*) at, 272
Chigger (*Euschoengastia splendens*), 282
Chipmunk (*Tamias striatus*): description of, 294
—subspecies of: eastern chipmunk (*T. s. striatus*), 8, 294–99
Choctawhatchee beach mouse (*Peromyscus polionotus allophrys*),

Choctawhatchee (*continued*) 76–86; causes of threat to, 91; conservation measures proposed for, 92; current status of, 7; demographic characteristics of, 22; description of, 88, 102; expanded range of, x; habitat requirements and habitat trend for, 105; key behaviors of, 22, 91, 99, 106, 247; responses to habitat modification, 106; taxonomy of, 102; vulnerability of species and habitat, 91, 105

Citrus agriculture: and habitat of Key Largo cotton mouse (*Peromyscus gossypinus allapaticola*), 114; and "scrub-rubbing" practice, 5; as threat to Big Cypress fox squirrel (*Sciurus niger avicennia*), 228; as threat to Florida mouse (*Podomys floridanus*), 255; as threat to southern Florida population of mink (*Mustela vison mink*), 324

Coach Cave, Kentucky: bat hibernation sites in, 58

Coastal hammocks: definition of, xxiii

Coastal marsh habitat, xxiv

Coastal strand habitat, xviii

Cockroaches, 340

Collier-Seminole State Park, Collier County: population of Big Cypress fox squirrel (*Sciurus niger avivennia*) at, 232; population of Florida black bear (*Ursus americanus floridanus*) at, 272; population of southern Florida mink (*Mustela vison mink*) at, 325

Colossal Cave, Kentucky: bat hibernation sites in, 58

Columbus, Christopher: sighting of West Indian monk seals (*Monachus tropicalis*) by, 36

Commercial fishing. *See* Fishing, commercial

Conservation, current regulation in Florida, 5

Conservation and Recreation Lands (CARL), land acquisition by, x, 127, 128, 198

Convention on International Trade in Endangered Species (CITES, 1989), 144, 151, 157, 164, 172

Cookie-cutter shark (genus *Isistius*), 148

Corbett Wildlife Management Area: habitat for Florida mastiff bat (*Eumops glaucinus floridanus*) at, 220

Corkscrew Swamp Sanctuary. *See* National Audubon Society, Corkscrew Swamp Sanctuary, Collier County

Corn snake, 220

Cotton mouse (*Peromyscus gossypinus*): capture of, 132–34; compared to Florida mouse (*Podomys floridanus*), 259; confused with Florida mouse, 251–52; distribution maps of, 26, 43, 112; habitat for, 25, 44; key behaviors of, 27; mainland type trapped on Anastasia Island in 1989, 44; responses to habitat modification, 256

—subspecies of: *P. g. palmarius*, 24, 41–42, 115–16; Gulf Hammock population of cotton mouse (*P. g. gossypinus*), 116–17; Anastasia Island population of cotton mouse (*P. g. gossypinus*), 2, 6, 41–46; Chadwick Beach cotton mouse (*P. g. restrictus*), 2, 6, 24–28; Cumberland Island population of cotton mouse (*P. g. gossypinus*), 41–42, 44; Key Largo cotton mouse (*P. g. allapaticola*), 7, 110–18

Cotton rat (*Sigmodon hispidus*): capture of, 132–34; as competitor of Lower Keys population of rice rat (*Oryzomys palustris natator*), 306; intestinal helminths in, 282

—subspecies of: insular cotton rat (*S. h. insulicola*), 4, 9; Lower Keys cotton rat (*S. h. exsputus*), 4, 9

Cottontail rabbit, 74

Index

Cougar. *See* Florida panther
Coyote (*Canis latrans*): description of, 30; taxonomy of, 29; vulnerability of, 32
Crested caracara (*Polyborus plancus*), 253
Crocodile Lake National Wildlife Refuge: land acquisition for, 127; threat to Key Largo woodrat (*Neotoma floridana smalli*) at, 125
Cross Florida Barge Canal, proposed: habitat of round-tailed muskrat (*Neofiber alleni*) along, 280
Crowned snake (*Tantilla coronata*), 16–17
Crystal River, Citrus County: population of Florida manatee (*Trichechus manatus latirostris*) at, 191; as sanctuary for Florida manatee, 197
Cumberland Island population of cotton mouse (*Peromyscus gossypinus gossypinus*), 41–42, 44
Cypress swamp habitat, xxv–xxvi

Dasypterus floridanus. See Northern yellow bat
de la Barreda, Rodrigo: sighting of Plains bison (*Bison bison bison*) by, 49
de Luna, Tristan: failure to sight Plains bison (*Bison bison bison*), 47
de Soto, Hernando: failure to sight Plains bison (*Bison bison bison*), 47
Devaux Bank, South Carolina: introduction of Anastasia Island beach mouse (*Peromyscus polionotus phasma*) at, 99
Development of Regional Impact reviews, 259
Devil's Den, Levy County: population of southeastern brown bat (*Myotis austroriparius*) at, 336
Devil's Head and Horns, Suwannee County: population of southeastern brown bat (*Myotis austroriparius*) at, 336

Diamondback rattlesnake (*Crotalus adamanteus*), 125
Disease: in bobcat (*Felis rufus*), 182; in Florida black bear (*Ursus americanus floridanus*), 270; in Florida manatee (*Trichechus manatus latirostris*), 194; in Florida mouse (*Podomys floridanus*), 258; in Florida panther (*Felis concolor coryi*), 182–83, 186; in Florida red wolf (*Canis rufus floridanus*), 32; in Key deer (*Odocoileus virginianus clavium*), 207, 210; in Key Largo woodrat (*Neotoma floridana smalli*), 127; in marsh rice rat (*Oryzomys palustris*), 306; in Plains bison (*Bison bison bison*), 52; in round-tailed muskrat (*Neofiber alleni*), 282–83; in southeastern brown bat (*Myotis austroriparius*), 339, 340; in southern Florida population of mink (*Mustela vison mink*), 323; in swine (*Sus scrofa*), 182; in West Indian monk seal (*Monachus tropicalis*), 39; in yellow bat (*Lasiurus intermedius*), 354
Domestic dog (*Canis familiaris*): confused with Florida panther (*Felis concolor coryi*), 177; as predator of Goff's pocket gopher (*Geomys pinetis goffi*), 15; as predator of Lower Keys marsh rabbit (*Sylvilagus palustris hefneri*), 74; as threat to Key deer (*Odocoileus virginianus clavium*), 206, 207, 210, 211; as threat to southeastern beach mouse (*Peromyscus polionotus niveiventris*), 246, 247
Dry prairie habitat, xix
Duke's saltmarsh vole (*Microtus pennsylvanicus dukecampbelli*). *See* Florida saltmarsh vole

Eastern big-eared bat. *See* Southeastern big-eared bat
Eastern chipmunk (*Tamias striatus*

Eastern chipmunk (*continued*)
striatus), 294–99; current status of, 8
Eastern gray wolf (*Canis lupus lycaon*), delisted, 4, 9
Eastern indigo snake (*Drymarchon corais*): geographical status of, 253; as predator of Key Largo woodrat (*Neotoma floridana smalli*), 125
Eastern lump-nosed bat. *See* Southeastern big-eared bat
Eastern mule-eared bat. *See* Southeastern big-eared bat
Eastern Test Range, 247
Eastern woodrat (*Neotoma floridana*): distribution map of, *121*; fossil remains of, 122; geographical status of, 122; taxonomy of, 119
—subspecies of: *N. f. floridana*, 122; Key Largo woodrat (*N. f. smalli*), 7, 119–30
Eastern yellow bat. *See* Northern yellow bat
Eglin Air Force Base: population of Florida black bear (*Ursus americanus floridanus*) at, 272
Eptesicus fuscus. *See* Brown bat
Eptesicus fuscus fuscus. *See* Big brown bat
Eptesicus fuscus osceola: geographical status of, 344; taxonomy of, 343
Euarctos americanus floridanus. *See* Florida black bear
Eubalaena glacialis. *See* Right whale
Eumops glaucinus. *See* Mastiff bat
Eumops glaucinus floridanus. *See* Florida mastiff bat
Evening bat (*Nycticeius humeralis*): cohabitation with Brazilian free-tailed bat (*Tadarida brasiliensis cynocephala*), 361; cohabitation with southeastern big-eared bat (*Plecotus rafinesquii macrotis*), 292; cohabitation with southeastern brown bat (*Myotis austroriparius*), 338
Everglades Agricultural Area: population of round-tailed muskrat (*Neofiber alleni*) at, 279, 281; preservation of habitat for round-tailed muskrat at, 283
Everglades Drainage District: and habitat of southern Florida population of mink (*Mustela vison mink*), 323
Everglades National Park: population of Big Cypress fox squirrel (*Sciurus niger avivennia*) at, 232; population of Florida black bear (*Ursus americanus floridanus*) at, 272; population of Florida mastiff bat (*Eumops glaucinus floridanus*) at, 220; population of Florida panther (*Felis concolor coryi*) at, 179; population of round-tailed muskrat (*Neofiber alleni*) at, 279, 282; population of southern Florida mink (*Mustela vison mink*) at, 325; preservation of habitat for round-tailed muskrat (*Neofiber alleni*) at, 283, 284; radiotelemetry of Florida panther (*Felis concolor coryi*) at, 185
Everglades population of mink (*Mustela vison mink*). *See* Southern Florida population of mink

Fakahatchee Strand State Preserve Collier County: established, 185; habitat for Florida mastiff bat (*Eumops glaucinus floridanus*) at, 220; land acquisition at, 184, 232; population of Big Cypress fox squirrel (*Sciurus niger avivennia*) at, 232; population of Florida black bear (*Ursus americanus floridanus*) at, 272; population of Florida panther (*Felis concolor coryi*) at, 179, 180; population of round-tailed muskrat (*Neofiber alleni*) at, 279; population of southern Florida mink (*Mustela vison mink*) at, 325
Fear's Cave, Jackson County: population of southeastern brown bat (*Myotis austroriparius*) at, 336
Felis concolor. *See* Puma

Index

Felis concolor coryi. See Florida panther
Fin whale (*Balaenoptera physalus*), 154–59; current status of, 7
—subspecies of: northern hemisphere fin whale (*B. p. physalus*), 154–55; southern hemisphere fin whale (*B. p. quoyi*), 154–55
Fishing, commercial: as threat to fin whale (*Balaenoptera physalus*), 156–57; as threat to Florida manatee (*Trichechus manatus latirostris*), 194, 195; as threat to humpback whale (*Megaptera novaeangliae*), 163; as threat to right whale (*Eubalaena glacialis*), 143, 144; as threat to sei whale (*Balaenoptera borealis*), 150, 151; as threat to sperm whale (*Physeter macrocephalus*), 171. See also Whaling, commercial
Fleas: *Orchopeas howardi*, 127; *Polygenis floridanus*, 254; *Polygenis gwyni*, 282
Florida Administrative Code: Title 39-4.001, 354, 364; Title 39-12.009, 364; Title 39-27.002, 83, 106
Florida bat. See Northern yellow bat
Florida black bear (*Ursus americanus floridanus*), 265–75; as competitor of Florida panther (*Felis concolor coryi*), 186; causes of threat to, 3; confused with Florida panther, 177; current status of, 8
Florida brown snake (*Storeria dekayi victa*), 303
Florida Caverns State Park, Jackson County: as hibernaculum for gray bat (*Myotis grisescens*), 60, 64, 68–69; Indiana bat (*Myotis sodalis*) specimen from, 55, 56; population of southeastern brown bat (*Myotis austroriparius*) at, 336, 341; specimen of southeastern big-eared bat (*Plecotus rafinesquii macrotis*) at, 290; suitability for bat hibernation, 57, 60
Florida Coastal Setback Law, 98

Florida Committee on Rare and Endangered Plants and Animals (*FCREPA*), xiii–xiv
Florida deer mouse. See Florida mouse; White-bellied Florida deermouse
Florida Endangered and Threatened Species Act of 1977, 197
Florida Keys mole skink (*Eumeces egregius egregius*), 303
Florida Local Government Comprehensive Planning and Land Development Regulation Act of 1985, 197
Florida manatee (*Trichechus manatus latirostris*), 190–200; current status of, 7; sanctuaries established for, x
Florida Manatee Recovery Plan, 197
Florida Manatee Sanctuary Act of 1978, 197, 198
Florida mastiff bat (*Eumops glaucinus floridanus*) 216–23; current status of, 8
Florida mink (*Mustela vison*), subspecies of: Atlantic coast subspecies (*M. v. lutensis*), 4, 9, 322; Gulf coast subspecies (*M. v. halilimnetes*), 4, 322; southern mink (*M. v. mink*), 8, 319–27. See also Southern Florida population of mink
Florida mouse (*Podomys floridanus*), 250–64; causes of threat to, 3; current status of, 8; habitat requirements and habitat trend for, 3
Florida panther (*Felis concolor coryi*), 176–89; conservation measures proposed for, 3; current status of, 7; preservation of, 3
Florida Panther Interagency Committee, 185
Florida Panther National Wildlife Refuge: land acquisition at, 184, 185; population of Big Cypress fox squirrel (*Sciurus niger avivennia*) at, 232; population of Florida black bear (*Ursus*

Florida panther (*continued*)
americanus floridanus) at, 272; population of Florida panther (*Felis concolor coryi*) at, 179; population of southern Florida mink (*Mustela vison mink*) at, 325
Florida Panther Trust Fund and Advisory Council, 185
Florida red wolf (*Canis rufus floridanus*), 29–34; causes of threat to, 2
Florida ribbon snake (*Thamnophis sauritus sackeni*), 303
Florida saltmarsh vole (*Microtus pennsylvanicus dukecampbelli*), 131–39; current status of, 7
Florida water rat. *See* Round-tailed muskrat
Florida weasel (*Mustela frenata peninsulae*), 315–18; current status of, 8; description of, 310; taxonomy of, 310, 313
Florida yellow bat. *See* Northern yellow bat
Fly, parasitic: *Bailia boardmani*, 340; *Trichobius major*, 340
Flying squirrel: as competitor of Florida mastiff bat (*Eumops glaucinus floridanus*), 220
Folly Island, South Carolina: introduction of Anastasia Island beach mouse (*Peromyscus polionotus phasma*) at, 99
Fontaneda, d'Escalante: sighting of Plains bison (*Bison bison bison*) by, 48
Fort Matanzas National Monument: population of Anastasia Island beach mouse (*Peromyscus polionotus phasma*) at, 97, 98, 99
Ft. Pierce Inlet State Recreation Area, St. Lucie County: population of southeastern beach mouse (*Peromyscus polionotus niveiventris*) at, 243, 245
Fox: as predator of Florida mouse (*Podomys floridanus*), 258
Fox squirrel (*Sciurus niger*): description of, 224; distribution maps of, 226, 236
—subspecies of: *S. n. niger*, 234; Big Cypress fox squirrel (*S. n. avicennia*), 8, 224–33; midwestern fox squirrel (*S. n. rufiventer*), 238, 239; Sherman's fox squirrel (*S. n. shermani*), 3, 8, 234–41
Free-tailed bat (*Tadarida brasiliensis*): cohabitation with big brown bat (*Eptesicus fuscus fuscus*), 347; cohabitation with southeastern brown bat (*Myotis austroriparius*), 338; demographic characteristics of, 362–63; description of, 216; distribution and history of distribution of, 360; distribution map of, 359; effect of pesticides on, 222; fossil remains of, 360; key behaviors of, 363–64; population size and trend for, 217, 358–60; vulnerability of species and habitat, 361–62
—subspecies of: Brazilian free-tailed bat (*T. b. cynocephala*), 9, 357–68
Freshwater marsh and wet prairie habitats, xxiv–xxv

Geomys pinetis goffi. *See* Goff's pocket gopher
Geomys pinetis pinetis, 11
Gerome's Cave, Jackson County: gray bat (*Myotis grisescens*) maternity colony in, 68, 69; population of southeastern brown bat (*Myotis austroriparius*) at, 336, 338, 341
Girard's Cave, Jackson County: population of southeastern brown bat (*Myotis austroriparius*) at, 336, 338, 341
Global warming: as threat to Florida saltmarsh vole (*Microtus pennsylvanicus dukecampbelli*), 137; as threat to Lower Keys population of rice rat (*Oryzomys palustris natator*), 305
Goff's pocket gopher (*Geomys pinetis goffi*), 11–18; causes of threat to, 2; current status of, 6

Index

Golden mouse (*Ochrotomys nuttalli*), 256
Gopher mouse. See Florida mouse
Gophers. See Southeastern pocket gopher
Gopher tortoise (*Gopherus polyphemus*): geographical status of, 254; reclamation of habitat for, 260; use of burrows by Florida mouse (*Podomys floridanus*), 250, 258; use of burrows by southeastern weasel (*Mustela frenata olivacea*), 313
Grant's Cave, Alachua County: population of southeastern brown bat (*Myotis austroriparius*) at, 335, 336, 341
Gray bat (*Myotis grisescens*), 63–70; conservation measures taken for, 341; current status of, 7; hibernaculum in Florida for, 60; population size and trend for, 336
Gray myotis. See Gray bat
Gray squirrel (*Sciurus carolinensis*), 227, 228
Grayton Beach State Recreation Area, Walton County: habitat of Choctawhatchee beach mouse (*Peromyscus polionotus allophrys*) at, 80, 81; translocation of Choctawhatchee beach mouse (*Peromyscus polionotus allophrys*) to, 77, 79, 80, 83, 84
Gray wolf (*Canis lupus*): description of, 29–30; taxonomy of, 4, 29
—subspecies of: eastern gray wolf (*C. l. lycaon*), 4, 9
Greater yellow bat. See Northern yellow bat
Great White Heron National Wildlife Refuge: population of marsh rice rat (*Oryzomys palustris*) at, 306
Green Key Largo woodrat dung beetle (*Onthophagus orpheus orpheus*), 127
Guana River State Park, St. John County: failure to sight Anastasia Island beach mouse (*Peromyscus polionotus phasma*) at, 97; introduction of Anastasia Island beach mouse (*Peromyscus polionotus phasma*) at, 100
Guano bat. See Brazilian free-tailed bat
Gulf Islands National Seashore: habitat at, 105; population of Perdido Key beach mouse (*Peromyscus polionotus trissyllepsis*) at, 103, 104–5, 107
Gulf State Park, Alabama: population of Perdido Key beach mouse (*Peromyscus polionotus trissyllepsis*) at, 103, 105, 106

Habitat Conservation Plan (HCP): and protection of Key Largo woodrat (*Neotoma floridana smalli*), 127–28
Habitats: coastal marshes, xxiv; coastal strand, xviii; cypress swamps, xxv–xxvi; dry prairies, xix; freshwater marshes and wet prairies, xxiv–xxv; hardwood hammocks, xxii–xxiii; hardwood swamps, xxvi–xxvii; longleaf pine—xerophytic oak woodlands (sandhill communities), xxi–xxii; mangrove swamps, xxvii–xxviii; mixed hardwood—pine, xxii; pine flatwoods, xix–xx; sand pine scrub, xx–xxi; scrub cypress, xxv; tropical hammocks, xxiii–xxiv
Hardwood hammock habitat, xxii–xxiii
Hardwood swamp habitat, xxvi–xxvii
Hawaiian monk seal (*Monachus schauinslandi*), 35, 38
Hesperomys floridanus. See Florida mouse
Hesperomys macropus. See Florida mouse
Hesperomys niveiventris. See Southeastern beach mouse
Highlands Hammock State Park, Highlands County: population of Florida mouse (*Podomys floridanus*) at, 259
Hillsborough River State Park,

Hillsborough (*continued*)
  Hillsborough County: population of round-tailed muskrat (*Neofiber alleni*) at, 279
Histoplasma capsulatum, 339
Hoary bat (*Lasiurus cinereus*), delisted, 4, 9, 352
Hogs. *See* Swine
Hog Sink, Alachua County: population of southeastern brown bat (*Myotis austroriparius*) at, 335, 336
Homosassa shrew (*Sorex longirostris eionis*), delisted, 4, 9
Honey bee (*Apis mellifera*), 268, 269, 271, 272
Hooded seal (*Cystophora cristata*), 36
Hookworm (*Ancylostoma pluridentatus*), 182
House bat. *See* Big brown bat
House cat (*Felis catus*): as predator of beach mouse (*Peromyscus polionotus*), 22, 80, 81, 84, 98–100, 106, 246, 247; as predator of cotton mouse (*Peromyscus gossypinus*), 27, 115; as predator of Florida mouse (*Podomys floridanus*), 258; as predator of Key Largo woodrat (*Neotoma floridana smalli*), 124, 125; as predator of Lower Keys marsh rabbit (*Sylvilagus palustris hefneri*), 74; as threat to Sherman's short-tailed shrew (*Blarina carolinensis shermani*), 331, 332
House mouse (*Mus musculus*): as competitor of beach mouse (*Peromyscus polionotus*), 21–22, 80, 81, 84, 90, 98–100, 106, 246, 247; at Perdido Key State Recreation Area, 105, 108
Howell's bat. *See* Gray bat
Humpback whale (*Megaptera novaeangliae*), 160–67; current status of, 7
Hunting: of Big Cypress fox squirrel (*Sciurus niger avicennia*), 228, 230, 231; as cause of extermination of some species, 2; of Florida black bear (*Ursus americanus floridanus*), 269, 270, 271–72; of Florida manatee (*Trichechus manatus latirostris*), 191, 194, 195; of Florida panther (*Felis concolor coryi*) at, 181, 184, 185; of Florida red wolf (*Canis rufus floridanus*), 32; of Key deer (*Odocoileus virginianus clavium*), 202, 206, 207, 209, 211; of Lower Keys marsh rabbit (*Sylvilagus palustris hefneri*), 74; by Native Americans, 156, 161, 163, 271–72; of Plains bison (*Bison bison bison*), 51; of Sherman's fox squirrel (*Sciurus niger shermani*), 238, 239, 240; of West Indian monk seal (*Monachus tropicalis*), 38. *See also* Fishing, commercial; Whaling
Hurricanes. *See* Tropical storms and hurricanes

Indiana bat (*Myotis sodalis*), 54–62; current status of, 7
Indiana Bat Recovery Plan, 60
Insular cotton rat (*Sigmodon hispidus insulicola*), delisted, 4, 9
Insularization of a population: Florida panther (*Felis concolor coryi*) as example of, 183
International Convention for the Regulation of Whaling (*1946*), 144, 151, 157, 164, 172
Island Field salt marsh, Levy County: preservation of habitat in, 137; as type locality for Florida saltmarsh vole (*Microtus pennsylvanicus dukecampbelli*), 131, 132, 134, 135
IUCN *Mammal Red Data Book*. *See Mammal Red Data Book* (International Union for Conservation of Nature and Natural Resources)

John Pennekamp Coral Reef State Park: habitat for Key Largo woodrat (*Neotoma floridana smalli*) at, 124
Jonathan Dickinson State Park:

Index

population of Florida mouse (*Podomys floridanus*) at, 259
Judge's Cave, Jackson County: maternity colony of gray bat (*Myotis grisescens*) at, 64, 68, 69; population of southeastern brown bat (*Myotis austroriparius*) at, 336, 338, 341

Keen's bat (*Myotis septentrionalis*), delisted, 4, 9
Key deer (*Odocoileus virginianus clavium*), 201–15; current status of, 7; population size and trend for, 3
Key Largo cotton mouse (*Peromyscus gossypinus allapaticola*), 110–18; current status of, 7; demographic characteristics of, 27, 45; habitat requirements and habitat trend for, 123, 124; key behaviors of, 45
Key Largo woodrat (*Neotoma floridana smalli*), 119–30; conservation measures proposed for, 117; conservation measures taken for, 117; current status of, 7; vulnerability of species and habitat, 115
Key silverside (*Menidia conchorum*), 303
Key Vaca raccoon (*Procyon lotor auspicatus*), delisted, 4, 9

Lake Arbuckle Wildlife Management Area: population of Florida mouse (*Podomys floridanus*) at, 259
Lake Manatee State Recreation Area, Manatee County: population of Florida mouse (*Podomys floridanus*) at, 259
Lasiurus intermedius. See Yellow bat
Lasiurus intermedius floridanus. See Northern yellow bat
Le Challeux, Nicolas: sighting of Plains bison (*Bison bison bison*) by, 48
Lignumvitae Key, Monroe County: introduction of Key Largo cotton mouse (*Peromyscus gossypinus allapaticola*) at, 113; introduction of Key Largo woodrat (*Neotoma floridana smalli*) at, 122, 126, 127, 128
Lignumvitae Key State Botanical Site. See Lignumvitae Key, Monroe County
Little brown bat (*Myotis lucifugus*), 335. See also Southeastern brown bat (*Myotis austroriparius*)
Live oak–cabbage palm hammocks, xxiii
Logging: and habitat of Key Largo cotton mouse (*Peromyscus gossypinus allapaticola*), 114; as threat to eastern chipmunk (*Tamias striatus striatus*), 297
Longleaf pine flatwoods, xix–xx
Longleaf pine–turkey oak woodlands, xxi
Longleaf pine–xerophytic oak woodland habitat (sandhill communities), xxi–xxii, 234, 237
Long-tailed weasel (*Mustela frenata*): demographic characteristics of, 311; distribution maps of, *312, 317*; in tunnels of gopher tortoise (*Gopherus polyphemus*), 313; in tunnels of southeastern pocket gopher (*Geomys pinetis*), 16, 313
—subspecies of: Florida weasel (*M. f. peninsulae*), 8, 315–18; southeastern weasel (*M. f. olivacea*), 8, 310–14
Lower Keys cotton rat (*Sigmodon hispidus exsputus*), delisted, 4, 9
Lower Keys marsh rabbit (*Sylvilagus palustris hefneri*), 71–75; current status of, 7; geographical status of, 393
Lower Keys population of rice rat (*Oryzomys palustris natator*), 300–309; current status of, 8
Loxahatchee National Wildlife Refuge, Palm Beach County: population of round-tailed muskrat (*Neofiber alleni*) at, 279

MacArthur Beach State Park:

MacArthur (*continued*)
population of southeastern beach mouse (*Peromyscus polionotus niveiventris*) at, 245
Macracanthorhyncus ingens, 270
Mammal Red Data Book (International Union for Conservation of Nature and Natural Resources), 39
Mammoth Cave National Park, Kentucky: bat hibernation sites in, 58; fossil remains of free-tailed bat (*Tadarida brasiliensis*) at, 360
Manalapan Beach, Palm Beach County: population of southeastern beach mouse (*Peromyscus polionotus niveiventris*) at, 245
Manasota Key: habitat for Chadwick Beach cotton mouse (*Peromyscus gossypinus restrictus*) at, 25, 27
Manatee Springs State Park, Levy County: big-eared bat (*Plecotus rafinesquii*) from, 292
Mangrove fox squirrel. *See* Big Cypress fox squirrel
Mangrove swamp habitat, xxvii–xxviii
Maritime hammocks, xxiii
Marsh rabbit (*Sylvilagus palustris*): demographic characteristics of, 74; distribution map of, 73; key behaviors of, 74–75; as prey of Florida red wolf (*Canis rufus floridanus*), 32
—subspecies of: Lower Keys marsh rabbit (*S. p. hefneri*), 7, 71–75; Upper Keys marsh rabbit (*S. p. paludicola*), 71, 73–74
Marsh rice rat (*Oryzomys palustris*): demographic characteristics of, 137, 305; description of, 300–301; disease in, 306; distribution and history of distribution of, 301–3; distribution map of, *302*; fossil remains of, 303; geographical status of, 303; habitat requirements and habitat trend for, 304; population size and trend for, 132–34; swimming ability of, 306
—subspecies of: *O. p. palustris*, 301; Lower Keys population (*O. p. natator*), 8, 300–309; Pine Island rice rat (*O. p. planirostris*), 4, 9
Mastiff bat (*Eumops glaucinus*): description of, 349, 357; distribution map of, *218*; habitat requirements and habitat trend for, 219–20; key behaviors of, 221; population size and trend for, 216–17; vulnerability of species and habitat, 220
—subspecies of: *E. g. glaucinus*, 216, 218, 219, 221; Florida mastiff bat (*E. g. floridanus*), 8, 216–23
Meadow mouse. *See* Meadow vole
Meadow vole (*Microtus pennsylvanicus*): demographic characteristics of, 137; description of, 131; distribution and history of distribution of, 134–35; distribution map of, *133*; fossil remains of, 134; taxonomy of, 131
—subspecies of: Georgia population (*M. p. pennsylvanicus*), 131, 135; Grand Manan Island population (*M. p. copelandi*), 131; Gulf Coast prairie vole (*M. p. ludovicianus*), 135, 137; Magdalen Island population (*M. p. magdalenensis*), 131; Maryland population (*M. p. nigrans*), 131, 135; Florida saltmarsh vole (*M. p. dukecampbelli*), 7, 131–39
Mediterranean monk seal (*Monachus monachus*), 35, 38
Megaptera novaeangliae. *See* Humpback whale
Melbourne Regional Airport: habitat of Goff's pocket gopher (*Geomys pinetis goffi*) at, 14
Merritt Island National Wildlife Refuge, Brevard County: habitat for southeastern beach mouse (*Peromyscus polionotus niveiventris*) at, 246, 247; population of Florida mouse (*Podomys floridanus*) at, 259;

Index

population of round-tailed muskrat (*Neofiber alleni*) at, 279, 280
Mexican free-tailed bat. *See* Brazilian free-tailed bat
Micco beach mouse. *See* Southeastern beach mouse
Microtus pennsylvanicus. *See* Meadow vole
Microtus pennsylvanicus dukecampbelli. *See* Florida saltmarsh vole
Midwestern fox squirrel (*Sciurus niger rufiventer*): home range size of, 239; litter size of, 238
Miller Cave, Jackson County: population of southeastern brown bat (*Myotis austroriparius*) at, 336
Mink. *See* Florida mink
Minute Key Largo woodrat dung beetle (*Ataenius brevicollis*), 127
Mississippi myotis. *See* Southeastern brown bat
Mites: *Androlaelaps farenholzi*, 282; *Ichoronyssus quadridentatus*, 340; *Laelaps evansi*, 282; *Listrophorus caudatus*, 282; *Listrophorus layni*, 282; *Olabidocarpus amaericanus*, 354; *Prolistrophorus birkenholzi*, 282–83; *Pteracarus chalinolobus*, 354; *Radfordia* sp., 282; *Spinturnix* sp., 340; *Steatonyssus radovskyi*, 354; blood-sucking mites (family Macronyssidae), 283; hair mites (family Listrophotidae), 283
Mixed hardwood–pine habitat, xxii
Mole (*Scalopus aquaticus*), 331
Mole skink (*Eumeces egregius*), 17
—subspecies of: Florida Keys mole skink (*E. e. egregius*), 303
Monachus tropicalis. *See* West Indian monk seal
Monk seals (genus *Monachus*). *See* Hawaiian monk seal; Mediterranean monk seal; West Indian monk seal
Monroe County Comprehensive Land Use Plan, 211
Mosquito ditches: and habitat management for Key deer (*Odocoileus virginianus clavium*), 208; as threat to Key deer, 206, 207, 211
Mountain lion. *See* Florida panther
Mud Cave, Jackson County: population of southeastern brown bat (*Myotis austroriparius*) at, 336, 339
Muskrat (*Ondatra zibethicus*): description of, 276. *See also* Round-tailed muskrat
—subspecies of: muskrat (*O. z. macrodon*), 135–36
Muskrat (*Ondatra zibethicus macrodon*), 135–36
Mustela frenata. *See* Long-tailed weasel
Mustela frenata olivacea. *See* Southeastern weasel
Mustela peninsulae olivacea. *See* Southeastern weasel
Mustela frenata peninsulae. *See* Florida weasel
Mustela vison evergladensis. *See* Southern Florida population of mink
Mustela vison mink. *See* Southern Florida population of mink
Myotis spp.: hibernaculum in Old Indian Cave for, 68–69; population size of, 64
Myotis austroriparius. *See* Southeastern brown bat
Myotis grisescens. *See* Gray bat
Myotis sodalis. *See* Indiana bat

National Audubon Society, Corkscrew Swamp Sanctuary: population of Big Cypress fox squirrel (*Sciurus niger avivennia*) at, 225, 227, 229, 230, 231
National Audubon Society, Corkscrew Swamp Sanctuary, Collier County: established, 185
National Key Deer Wildlife Refuge: established, 210; population of Key deer (*Odocoileus virginianus clavium*) at, 204; population of marsh rice rat (*Oryzomys palustris*) at, 306

Nature Conservancy: and preservation of Judge's Cave gray bat (*Mytis grisescens*) maternity colony, 68; at Saddleblanket Lake Scrub Preserve, 259; at Tiger Creek Preserve, 259
Nematode (*Carolinensis kinsellai*), 282. *See also* Hookworm
Neofiber alleni. See Roundtailed muskrat
Neotoma floridana. See Eastern woodrat
Neotoma floridana smalli. See Key Largo woodrat
No Name Key: and habitat management for Key deer (*Odocoileus virginianus clavium*) at, 211
North Atlantic right whale. *See* Right whale
Northern yellow bat (*Lasiurus intermedius floridanus*), 349–56; current status of, 9
North Key Largo State Preserve: and protection of Key Largo woodrat (*Neotoma floridana smalli*), 127
North Pacific right whale, 143

Ocala National Forest: habitat for Florida black bear (*Ursus americanus floridanus*) at, 268; population of Florida black bear (*Ursus americanus floridanus*) at, 270, 271, 272; population of Florida mouse (*Podomys floridanus*) at, 259
Odocoileus virginianus clavium. See Key deer
Oil development, offshore: as threat to humpback whale (*Megaptera novaeangliae*), 163; as threat to sperm whale (*Physeter macrocephalus*), 171–72
Oil exploration, inland: as threat to Florida panther (*Felis concolor coryi*), 181
Okefenokee Swamp: preservation of habitat for round-tailed muskrat (*Neofiber alleni*) at, 283

Oldfield mouse. *See* Beach mouse
Old Indian Cave, Jackson County. *See* Florida Caverns State Park, Jackson County
Opossum (*Diselphis virginiana*), 340
Ordway-Swisher Preserve, Putnam County: population of Florida mouse (*Podomys floridanus*) at, 259; population of genus Podomys at, 256; population of round-tailed muskrat (*Neofiber alleni*) at, 283; population of Sherman's fox squirrel (*Sciurus niger shermani*) at, 238
Oryzomys palustris. See Marsh rice rat
Oryzomys palustris natator. See Lower Keys population of rice rat
Osceola National Forest: habitat for Florida black bear (*Ursus americanus floridanus*) at, 268; habitat of round-tailed muskrat (*Neofiber alleni*) at, 280; population of Florida black bear (*Ursus americanus floridanus*) at, 271, 272
Owls: as predator of Anastasia Island beach mouse (*Peromyscus polionotus phasma*), 99; as predator of Florida mouse (*Podomys floridanus*), 258; as predator of Lower Keys population of rice rat (*Oryzomys palustris natator*), 306; as predator of southeastern brown bat (*Myotis austroriparius*), 340. *See also* Barn owl; Burrowing owl

Pallid beach mouse (*Peromyscus polionotus decoloratus*), 19–23; causes of threat to, 2; current status of, 6; description of, 94; geographical status of, 245
Pardo, Juan: failure to sight Plains bison (*Bison bison bison*), 47
Payne's Prairie State Preserve, Alachua County: Plains bison (*Bison bison bison*) herd at, 52; population of round-tailed muskrat (*Neofiber alleni*) at, 277, 283; preservation of habitat for round-

Index

tailed muskrat (*Neofiber alleni*) at, 283, 284
Peña, Diego: sighting of Plains bison (*Bison bison bison*) by, 49
Pepper Park, St. Lucie County: population of southeastern beach mouse (*Peromyscus polionotus niveiventris*) at, 243, 244
Perdido Key beach mouse (*Peromyscus polionotus trissyllepsis*), 102–9; current status of, 7; demographic characteristics of, 82
Perdido Key State Recreation Area: habitat for Perdido Key beach mouse (*Peromyscus polionotus trissyllepsis*) at, 105, 108
Peromyscus anastasae. See Anastasia Island population of cotton mouse
Peromyscus gossypinus anastasae. See Anastasia Island population of cotton mouse
Peromyscus gossypinus allapaticola. See Key Largo cotton mouse
Peromyscus gossypinus gossypinus. See Anastasia Island population of cotton mouse
Peromyscus gossypinus restrictus. See Chadwick Beach cotton mouse
Peromyscus insulanus. See Anastasia Island population of cotton mouse
Peromyscus maniculatus, 77, 102
Peromyscus phasma. See Anastasia Island beach mouse
Peromyscus polionotus. See Beach mouse
Peromyscus polionotus allophrys. See Choctawhatchee beach mouse
Peromyscus polionotus decoloratus. See Pallid beach mouse
Peromyscus polionotus niveiventris. See Southeastern beach mouse
Peromyscus polionotus peninsularis. See St. Andrew beach mouse
Peromyscus polionotus phasma. See Anastasia Island beach mouse
Peromyscus polionotus trissyllepsis. See Perdido Key beach mouse
Phoca tropicalis. See West Indian monk seal
Physeter catodon. See Sperm whale
Physeter macrocephalus. See Sperm whale
Pigs. *See* Swine
Pileated woodpecker (*Dryocopus pileatus*), 219
Pineapple cultivation: and habitat of Key Largo cotton mouse (*Peromyscus gossypinus allapaticola*), 114
Pine Barrens treefrog, delisted, x
Pine flatwood habitat, xix–xx
Pine Island rice rat (*Oryzomys palustris planirostris*), delisted, 4, 9
Pine snake (*Pituophis melanoleucas mugitus*), 16
Plains bison (*Bison bison bison*), 47–53; causes of threat to, 2; current status of, 6, 265
Playboy Foundation, 71
Plecotus rafinesquii. See Big-eared bat
Plecotus rafinesquii macrotis. See Southeastern big-eared bat
Poaching. *See* Hunting
Pocket gophers. *See* Southeastern pocket gopher
Podomys floridanus. See Florida mouse
Poisoning: of big brown bat (*Eptesicus fuscus fuscus*), 346; of Florida mastiff bat (*Eumops glaucinus floridanus*), 220, 222; of Florida red wolf (*Canis rufus floridanus*), 32; of free-tailed bat (*Tadarida brasiliensis*), 359, 361–62, 365; of Goff's pocket gopher (*Geomys pinetis goffi*), 15; of gray bat (*Myotis grisescens*), 67, 69; of Key Largo cotton mouse (*Peromyscus gossypinus allapaticola*), 115; of Key Largo woodrat (*Neotoma floridana smalli*), 125; of southeastern big-eared bat (*Plecotus rafinesquii macrotis*) at, 290; of southern Florida population of mink (*Mustela vison mink*), 324
Pollution: regulation of, 5; as threat to fin whale (*Balaenoptera physalus*), 157; as threat to Florida manatee (*Trichechus manatus latirostris*), 194; as threat to gray bat (*Myotis*

Pollution (*continued*)
grisescens), 67, 69; as threat to
humpback whale (*Megaptera
novaeangliae*), 163; as threat to
right whale (*Eubalaena glacialis*),
143; as threat to sei whale
(*Balaenoptera borealis*), 150, 151; as
threat to southern Florida
population of mink (*Mustela vison
mink*), 324; as threat to sperm
whale (*Physeter macrocephalus*), 171
Ponce de León, Juan: sighting of
West Indian monk seals (*Monachus
tropicalis*) by, 36
Ponce Park, Volusia County: habitat
for pallid beach mouse (*Peromyscus
polionotus decoloratus*) at, 19, 20
Pond pine flatwoods, xix–xx
Prairie vole (*Microtus ochrogaster*),
135; Gulf Coast (*Microtus
pennsylvanicus ludovicianus*, 135,
137
Preservation, of wildlife versus
habitat, 6
Puma (*Felis concolor*): demographic
characteristics of, 182; distribution
and history of distribution of,
177–78, 179; distribution map of,
178; fossil remains of, 179
—subspecies of: Florida panther (*Felis
concolor coryi*), 3, 7, 176–89
Putorius peninsulae. See Florida weasel

Rabbits (family Leporidae): as prey of
Florida red wolf (*Canis rufus
floridanus*), 32. *See also* Cottontail
rabbit; Marsh rabbit
Rabies: in Florida panthers (*Felis
concolor coryi*), 183
Raccoon (*Procyon lotor*): as predator
of Florida mouse (*Podomys
floridanus*), 258; as predator of Key
Largo woodrat (*Neotoma floridana
smalli*), 125; as predator of Lower
Keys population of rice rat
(*Oryzomys palustris natator*), 304,
305; as prey of Florida panther
(*Felis concolor coryi*), 180; as prey of
Florida red wolf (*Canis rufus
floridanus*), 32
—subspecies of: Key Vaca raccoon
(*Procyon lotor auspicatus*), 4, 9
Radiotelemetry: of Big Cypress fox
squirrel (*Sciurus niger avivennia*),
231; of Florida black bear (*Ursus
americanus floridanus*), 271; of
Florida manatee (*Trichechus
manatus latirostris*), 197; of Florida
panther (*Felis concolor coryi*), 179,
180, 183, 184, 185, 186; of silver
rice rat (*Oryzomys argentatus*), 304,
306
Rafinesque's big-eared bat. *See*
Southeastern big-eared bat
Rat snakes: *Elaphe guttata*, 340;
Elaphe obsoleta, 340
Rats. *See* Black rat; Cotton rat;
Eastern woodrat; Marsh rice rat;
Silver rice rat
Red-cockaded woodpecker (*Picoides
borealis*), 219
Red fox (*Vulpes vulpes*), 106
Red-headed woodpecker (*Melanerpes
erythrocephalus*), 220
Red rat snake (*Elaphe guttata
guttata*), 303
Red-tailed hawk (*Buteo jamaicensis*),
239
Red tides: as threat to Florida
manatee (*Trichechus manatus
latirostris*), 194
Red wolf (*Canis rufus*): demographic
characteristics of, 32; distribution
map of, *31*; habitat for, 30–31; key
behaviors of, 32; vulnerability of
species and habitat, 32
—subspecies of: Florida red wolf (*C.
r. floridanus*), 2, 6, 29–34
Rice rats. *See* Marsh rice rat; Silver
rice rat
Right whale (*Eubalaena glacialis*),
140–47; current status of, 7. *See
also* North Pacific right whale;
Southern Ocean right whale
Road kills: of Florida black bear
(*Ursus americanus floridanus*), 269,
270, 271, 272–73; of Florida
panther (*Felis concolor coryi*), 179,

Index

181, 182, 183, 184; of Key deer (*Odocoileus virginianus clavium*), 206, 207, 209, 210; of Lower Keys marsh rabbit (*Sylvilagus palustris hefneri*), 74; of round-tailed muskrat (*Neofiber alleni*), 283; of southern Florida population of mink (*Mustela vison mink*), 319, 322, 324

Robert's Cave, Gilchrist County: population of southeastern brown bat (*Myotis austroriparius*) at, 336

Round-tailed muskrat (*Neofiber alleni*), 276–86; current status of, 8; distribution and history of distribution of, 136; preservation of, 3–4

—subspecies of: *N. a. alleni*, 276; *N. a. apalachicolae*, 276; *N. a. exoristus*, 276; *N. a. nigrescens*, 276; *N. a. struix*, 276

Saddleblanket Lake Scrub Preserve of the Nature Conservancy: population of Florida mouse (*Podomys floridanus*) at, 259

St. Andrew beach mouse (*Peromyscus polionotus peninsularis*), 87–93; current status of, 7

St. Andrews State Recreation Area, Bay County: as critical habitat for Choctawhatchee beach mouse (*Peromyscus polionotus allophrys*), 80, 84

St. Joseph Peninsula State Park, Gulf County: population of St. Andrew beach mouse (*Peromyscus polionotus peninsularis*) at, 90, 91

St. Lucie Inlet State Park, Martin County: population of southeastern beach mouse (*Peromyscus polionotus niveiventris*) at, 245

St. Marks National Wildlife Refuge: population of Florida black bear (*Ursus americanus floridanus*) at, 272

Saltmarsh meadow vole. *See* Florida saltmarsh vole

Sand pine scrub habitat: causes of threat to, xxi; definition of, xx–xxi

San Felasco State Preserve, Alachua County: population of Florida mouse (*Podomys floridanus*) at, 259

Save Our Coasts, x

Save Our Everglades, x, 185

Save Our Rivers, x, 260

Save the Manatee Club, 197

Sciurus niger. *See* Fox squirrel (*Sciurus niger*)

Sciurus niger avicennia. *See* Big Cypress fox squirrel

Sciurus niger shermani. *See* Sherman's fox squirrel

Scrubby flatwood: definition of, xx

Scrub cypress habitat: definition of, xxv; preservation of, xxv

Scrub jay (*Aphelocoma c. coerulescens*), 256

Scrub rubbing, 5

Sea cow. *See* Florida manatee

Sebastian Inlet State Recreation Area, Indian River County: population of southeastern beach mouse (*Peromyscus polionotus niveiventris*) at, 243–44, 245, 247

Sei whale (*Balaenoptera borealis*), 148–53; current status of, 7

—subspecies of: northern hemisphere morph (*B. b. borealis*), 148, 151; southern hemisphere morph (*B. b. schleglii*), 148, 150–51

Seminole Indians: hunting of Florida black bear (*Ursus americanus floridanus*) by, 271–72

Sherman's fox squirrel (*Sciurus niger shermani*), 234–41; causes of threat to, 3; current status of, 8; description of, 224; distribution and history of distribution of, 227; habitat requirements and habitat trend for, 3; taxonomy of, 224

Sherman's short-tailed shrew (*Blarina carolinensis shermani*), 328–34; current status of, 8; failure to determine status of, 4

Short-tailed shrew (*Blarina carolinensis*): distribution map of,

Short-tailed shrew (*continued*) 330; taxonomy of, 328
—subspecies of: *B. brevicauda*, 328–29; *B. peninsulae*, 329; *B. c. carolinensis*, 328–29; *B. c. peninsulae*, 328–29, 331; Sherman's short-tailed shrew (*B. c. shermani*), 4, 8, 328–34
Shrews (family Soricidae). *See* Homosassa shrew; Short-tailed shrew; Southeastern shrew
Sigmodon hispidus. *See* Cotton rat
Silver rice rat (*Oryzomys argentatus*), 300, 306
Slash pine flatwoods, xix–xx
Smithsonian Institution, Marine Mammal Event Program, 171
Snakes: *Elaphe guttata*, 340; *Elaphe obsoleta*, 340; Big Pine Key ringneck snake (*Diadophis punctatus acricus*), 303; corn snake, 220; crowned snake (*Tantilla coronata*), 16–17; diamondback rattlesnake (*Crotalus adamanteus*), 125; eastern indigo snake (*Drymarchon corais*), 125, 253; Florida brown snake (*Storeria dekayi victa*), 303; Florida ribbon snake (*Thamnophis sauritus sackeni*), 303; pine snake (*Pituophis melanoleucas mugitus*), 16; as predator of Florida mouse (*Podomys floridanus*), 258; as predator of Lower Keys population of rice rat (*Oryzomys palustris natator*), 305; red rat snake (*Elaphe guttata guttata*), 303
Snapping turtle (*Chelydra serpentina*), 323
Snead's Cave, Jackson County: population of southeastern brown bat (*Myotis austroriparius*) at, 336, 338, 341
Southeast Deer Disease Study Group, 210
Southeastern beach mouse (*Peromyscus polionotus niveiventris*), 242–49; current status of, 8; demographic characteristics of, 82; description of, 19, 94; population size and trend for, 79
Southeastern big-eared bat (*Plecotus rafinesquii macrotis*), 287–93; current status of, 8
Southeastern brown bat (*Myotis austroriparius*), 335–42; cohabitation with Brazilian free-tailed bat (*Tadarida brasiliensis cynocephala*), 361; current status of, 8; hibernaculum in Old Indian Cave for, 69; key behaviors of, 292
Southeastern myotis. *See* Southeastern brown bat
Southeastern pocket gopher (*Geomys pinetis*): distribution map of, *13*; geographical status of, 254; photograph of, *12*; as primary prey of the pine snake (*Pituophis melanoleucas mugitus*), 16; taxonomy of, 11; tunnel systems of, 15–16
—subspecies of: *G. p. pinetis*, 11; Goff's pocket gopher (*G. p. goffi*), 2, 6, 11–18
Southeastern shrew (*Sorex longirostris longirostris*), delisted, 4, 9
Southeastern U.S. Marine Mammal Stranding Network, 171
Southeastern weasel (*Mustela frenata olivacea*), 310–14; current status of, 8; demographic characteristics of, 317; habitat requirements and habitat trend for, 316; key behaviors of, 317; population size and trend for, 315; taxonomy of, 315; vulnerability of species and habitat, 316
Southern Florida population of mink (*Mustela vison mink*), 319–27; current status of, 8
Southern mink (*Mustela vison mink*): delisted, 9; distribution map of, *321*
—subspecies of: southern Florida population of mink (*Mustela vison mink*), 8, 319–27
Southern Ocean right whale (*Eubalaena australis*): demographic

Index

characteristics of, 144; description of, 140–41; geographical status of, 143; population size and trend for, 141; taxonomy of, 140
Southern slash pine forest: definition of, xix–xx
South Fork State Park: population of Florida mouse (*Podomys floridanus*) at, 259
Spanish moss (*Tillandsia usneoides*): as habitat of northern yellow bat (*Lasiurus intermedius floridanus*), 351, 352, 353, 354
Sperm whale (*Physeter macrocephalus*), 168–75; current status of, 7
Striped mud turtle (*Kinosternon bauri*), 303
Strongyloides spp., 270
Sugarcane industry: impact on round-tailed muskrat (*Neofiber alleni*), 280, 281
Sunday Sunk, Marion County: population of southeastern brown bat (*Myotis austroriparius*) at, 336, 338
Sweetgum Cave, Citrus County: population of southeastern brown bat (*Myotis austroriparius*) at, 335
Swine (*Sus scrofa*): as competitor of Florida panther (*Felis concolor coryi*) at, 179–80; disease in, 182; as prey of Florida black bear (*Ursus americanus floridanus*), 269; as prey of Florida panther (*Felis concolor coryi*), 180, 181, 186; as prey of Florida red wolf (*Canis rufus floridanus*), 32

Tadarida brasiliensis. *See* Free-tailed bat
Tadarida brasiliensis cynocephala. *See* Brazilian free-tailed bat
Tamias striatus striatus. *See* Eastern chipmunk
Tiger Creek Preserve of the Nature Conservancy: population of Florida mouse (*Podomys floridanus*) at, 259
Titi swamps, xxvii
Trapping. *See* Hunting
Tree snails (*Liguus fasciatus*), 126–27
Trematode (*Quinqueserialis floridensis*), 282
Trichechus manatus. *See* West Indian manatee
Trichechus manatus latirostris. *See* Florida manatee
Tropical hammock habitat, xxiii–xxiv
Tropical storms and hurricanes: and restoration of beach mouse (*Peromyscus polionotus*) habitat, 81; as threat to beach mouse (*Peromyscus polionotus*) habitat, 80, 89, 105–6; as threat to Florida saltmarsh vole (*Microtus pennsylvanicus dukecampbelli*), 135–36
Turtles: snapping turtle (*Chelydra serpentina*), 323; striped mud turtle (*Kinosternon bauri*), 303
Turtle Trail Public Beach Access, St. Lucie County: population of southeastern beach mouse (*Peromyscus polionotus niveiventris*) at, 243, 244
Tyndall Air Force Base: population of St. Andrew beach mouse (*Peromyscus polionotus peninsularis*) at, 90, 91

U.S. Endangered Species Act of 1973, xv, 83, 106, 144, 151, 157, 164, 172, 197, 210, 246
U.S. Marine Mammal Protection Act of 1972, 144, 151, 157, 163, 164, 172, 197
University of Central Florida: bat removal at, 364
University of Florida (UF): bat removal at, 364
Upland plant communities: causes of threat to, 5; coastal strand, xviii; dry prairies, xix; hardwood hammocks, xxii–xxiii; longleaf pine—xerophytic oak woodlands (sandhill communities), xxi–xxii; mixed hardwood—pine, xxii; pine flatwoods, xix–xx; sand pine

Upland plant (*continued*)
   scrub, xx–xxi; tropical hammocks, xxiii–xxiv
Upper Keys marsh rabbit (*Sylvilagus palustris paludicola*): description of, 71; distribution of, 73–74; taxonomy of, 71
Ursus americanus. See Black bear
Ursus americanus floridanus. See Florida black bear

Voles (genus *Microtus*). See California vole; Meadow vole; Prairie vole

Wagner's mastiff bat. See Florida mastiff bat
Walking sticks (*Anisomorpha buprestoides*), 269
Wekiwa Springs State Park, Seminole County: population of Florida mouse (*Podomys floridanus*) at, 259
Welaka Preserve: population of Florida mouse (*Podomys floridanus*) at, 259
West Indian manatee (*Trichechus manatus*): conservation measures taken for, 197; description of, 190; distribution and history of distribution of, 192–93; distribution map of, *192*; population size and trend for, 190; taxonomy of, 190
—subspecies of: Antillean manatee (*T. m. manatus*), 190, 193; Florida manatee (*T. m. latirostris*), 7, 190–200
West Indian monk seal (*Monachus tropicalis*), 35–40; current status of, 6; reasons for extinction of, 2
West Indian seal. See West Indian monk seal
Wetland plant communities: coastal marshes, xxiv; cypress swamps, xxv–xxvi; freshwater marshes and wet prairies, xxiv–xxv; hardwood swamps, xxvi–xxvii; mangrove swamps, xxvii–xxviii; preservation of, 5; scrub cypress, xxv
Whales. See Fin whale; Humpback whale; Right whale; Sei whale; Sperm whale

Whale-watching: as threat to fin whale (*Balaenoptera physalus*), 157; as threat to humpback whale (*Megaptera novaeangliae*), 163
Whaling: commercial, 150, 151, 155, 161, 163, 171; of fin whales (*Balaenoptera physalus*), 143; by Native Americans, 156, 161, 163; and population count of sperm whale (*Physeter macrocephalus*), 171; of right whale (*Eubalaena glacialis*), 143; of sperm whale (*Physeter macrocephalus*), 171. See also Hunting
White-bellied Florida deermouse (*Peromyscus niveiventris*), 242
White-crowned pigeon (*Columba leucocephala*), 114–15, 128
White-tailed deer (*Odocoileus virginianus*): description of, 201; dispersal pattern of, 209; distribution map of, *203*; as prey of Florida black bear (*Ursus americanus floridanus*), 269; as prey of Florida panther (*Felis concolor coryi*) at, 179, 180, 181, 186; as prey of Florida red wolf (*Canis rufus floridanus*), 32
—subspecies of: *O. v. osceola*, 201; *O. v. seminolus*, 201; Key deer (*O. v. clavium*), 3, 7, 201–15
Wingate Creek State Park: population of Florida mouse (*Podomys floridanus*) at, 259
Wolves. See Gray wolf; Red wolf
Wyandotte Cave, Indiana: bat hibernation sites in, 58

Yellow bat (*Lasiurus intermedius*): description of, 349–50; disease in, 354; distribution and history of distribution of, 352; fossil remains of, 352; population size and trend for, 350–51; taxonomy of, 349
—subspecies of: *L. i. insularis*, 349, 352; *L. i. intermedius*, 349; northern yellow bat (*L. i. floridanus*), 9, 349–56
Yellow jackets (*Vespula* spp.), 269